DNA Viruses

The Practical Approach Series

SERIES EDITOR

B. D. HAMES
Department of Biochemistry and Molecular Biology
University of Leeds, Leeds LS2 9JT, UK

See also the Practical Approach web site at **http://www.oup.co.uk/PAS**
★ **indicates new and forthcoming titles**

Affinity Chromatography
Affinity Separations
Anaerobic Microbiology
Animal Cell Culture (2nd edition)
Animal Virus Pathogenesis
Antibodies I and II
Antibody Engineering
★ Antisense Technology
Applied Microbial Physiology
Basic Cell Culture
Behavioural Neuroscience
Bioenergetics
Biological Data Analysis
Biomechanics—Materials
Biomechanics—Structures and Systems
Biosensors
Carbohydrate Analysis (2nd edition)
Cell-Cell Interactions
The Cell Cycle
Cell Growth and Apoptosis
★ Cell Separation

Cellular Calcium
Cellular Interactions in Development
Cellular Neurobiology
★ Chromatin
★ Chromosome Structural Analysis
Clinical Immunology Complement
★ Crystallization of Nucleic Acids and Proteins (2nd edition)
Cytokines (2nd edition)
The Cytoskeleton
Diagnostic Molecular Pathology I and II
DNA and Protein Sequence Analysis
DNA Cloning 1: Core Techniques (2nd edition)
DNA Cloning 2: Expression Systems (2nd edition)
DNA Cloning 3: Complex Genomes (2nd edition)
DNA Cloning 4: Mammalian Systems (2nd edition)

- ★ Drosophila (2nd edition)
- Electron Microscopy in Biology
- Electron Microscopy in Molecular Biology
- Electrophysiology
- Enzyme Assays
- Epithelial Cell Culture
- Essential Developmental Biology
- Essential Molecular Biology I and I
- ★ Eukaryotic DNA Replication
- Experimental Neuroanatomy
- Extracellular Matrix
- Flow Cytometry (2nd edition)
- Free Radicals
- Gas Chromatography
- Gel Electrophoresis of Nucleic Acids (2nd edition)
- ★ Gel Electrophoresis of Proteins (3rd edition)
- Gene Probes 1 and 2
- Gene Targeting
- Gene Transcription
- ★ Genome Mapping
- Glycobiology
- ★ Growth Factors and Receptors
- Haemopoiesis
- ★ High Resolution Chromotography
- Histocompatibility Testing
- HIV Volumes 1 and 2
- ★ HPLC of Macromolecules (2nd edition)
- Human Cytogenetics I and II (2nd edition)
- Human Genetic Disease Analysis
- ★ Immobilized Biomolecules in Analysis
- Immunochemistry 1
- Immunochemistry 2
- Immunocytochemistry
- ★ *In Situ* Hybridization (2nd edition)
- Iodinated Density Gradient Media
- Ion Channels
- ★ Light Microscopy (2nd edition)
- Lipid Modification of Proteins
- Lipoprotein Analysis
- Liposomes
- Mammalian Cell Biotechnology
- Medical Parasitology
- Medical Virology
- MHC Volumes 1 and 2
- ★ Molecular Genetic Analysis of Populations (2nd edition)
- Molecular Genetics of Yeast
- Molecular Imaging in Neuroscience
- Molecular Neurobiology
- Molecular Plant Pathology I and II
- Molecular Virology
- Monitoring Neuronal Activity
- Mutagenicity Testing
- ★ Mutation Detection
- Neural Cell Culture
- Neural Transplantation
- Neurochemistry (2nd edition)
- Neuronal Cell Lines

NMR of Biological Macromolecules
Non-isotopic Methods in Molecular Biology
Nucleic Acid Hybridisation
Oligonucleotides and Analogues
Oligonucleotide Synthesis
PCR 1
PCR 2
★ PCR 3: PCR In Situ Hybridization
Peptide Antigens
Photosynthesis: Energy Transduction
Plant Cell Biology
Plant Cell Culture (2nd edition)
Plant Molecular Biology
Plasmids (2nd edition)
Platelets
Postimplantation Mammalian Embryos
★ Post-Translational Modification Preparative Centrifugation

Protein Blotting
★ Protein Expression
Protein Engineering
Protein Function (2nd edition)
Protein Phosphorylation
Protein Purification Applications
Protein Purification Methods
Protein Sequencing
Protein Structure (2nd edition)
Protein Structure Prediction
Protein Targeting
Proteolytic Enzymes
Pulsed Field Gel Electrophoresis
RNA Processing I and II
★ RNA–Protein Interactions
Signalling by Inositides
Subcellular Fractionation
Signal Transduction
★ Transcription Factors (2nd edition)
Tumour Immunobiology

DNA Viruses
A Practical Approach

Edited by
ALAN J. CANN
*Department of Microbiology and Immunology,
University of Leicester.*

OXFORD
UNIVERSITY PRESS

Great Clarendon Street, Oxford OX2 6DP
Oxford University Press is a department of the University of Oxford
and furthers the University's aim of excellence in research, scholarship,
and education by publishing worldwide in

Oxford New York
Athens Auckland Bangkok Bogotá Buenos Aires Calcutta
Cape Town Chennai Dar es Salaam Delhi Florence Hong Kong Istanbul
Karachi Kuala Lumpur Madrid Melbourne Mexico City Mumbai
Nairobi Paris São Paulo Singapore Taipei Tokyo Toronto Warsaw
and associated companies in Berlin Ibadan

Oxford is a registered trade mark of Oxford University Press

Published in the United States
by Oxford University Press Inc., New York

© Oxford University Press, 1999

All rights reserved. No part of this publication may be reproduced, stored in a retrieval system, or transmitted, in any form or by any means, without the prior permission in writing of Oxford University Press. Within the UK, exceptions are allowed in respect of any fair dealing for the purpose of research or private study, or criticism or review, as permitted under the Copyright, Designs and Patents Act, 1988, or in the case of reprographic reproduction in accordance with the terms of licenses issued by the Copyright Licensing Agency. Enquiries concerning reproduction outside those terms and in other countries should be sent to the Rights Department, Oxford University Press, at the address above.

This book is sold subject to the condition that it shall not, by way of trade or otherwise, be lent, re-sold, hired out, or otherwise circulated without the publisher's prior consent in any form of binding or cover other than that in which it is published and without a similar condition including this condition being imposed on the subsequent purchaser

Users of books in the Practical Approach Series are advised that prudent laboratory safety procedures should be followed at all times. Oxford University Press makes no representation, express or implied, in respect of the accuracy of the material set forth in books in this series and cannot accept any legal responsibility or liability for any errors or omissions that may be made.

A catalogue record for this book is available from the British Library

Library of Congress Cataloging in Publication Data
DNA Viruses: a practical approach / edited by Alan J. Cann.
(The practical approach series ; 214)
Includes bibliographical references and index.
1. DNA viruses Laboratory manuals.   I. Cann, Alan.   II. Series.
[DNLM: 1. DNA Viruses–physiology.   2. DNA Virus Infections–
virology.   3. DNA Viruses–isolation & purification. QW 165 D629
2000]
QR394.5.D63 2000   616'.0194–dc21   99–32190

ISBN 0 19 963718 0 (Pbk)
0 19 963719 9 (Hbk)

Typeset by Footnote Graphics,
Warminster, Wilts
Printed in Great Britain by Information Press, Ltd,
Eynsham, Oxon.

Preface

The properties of viruses are distinct from those of living organisms, which makes the study of virology different from other areas of biology. Many specialized techniques have been developed to study viruses, and yet these have almost invariably found their way into mainstream biology. It is not possible to define in a single volume the techniques of virology, but this book and its companions set out to illustrate the major experimental methods currently employed by virologists. In particular, it is hoped that by grouping methods used by those who study DNA viruses and those who study RNA viruses–the same groupings which can frequently be observed at scientific meetings–the maximum utility has been gained.

Inevitably there will be those who feel that this or that should have been included or left out. It is not possible to include everything within the format of this series! Nevertheless, this volume is wide ranging in scope, from the fundamentals of virus culture to novel techniques such as surface plasmon resonance spectrometry and real-time PCR analysis of drug resistance mutations in clinical isolates. Chapter 1 provides an overview of the extraction, purification, and characterization of virus DNA, but also covers the fundamentals of DNA-virus culture. Chapters 2 and 3 describe approaches to the molecular investigation and mutagenesis of DNA-virus genomes. Chapter 4 considers DNA-virus replication, and Chapters 5 and 6 the linked topic of transcription control in DNA viruses. Chapter 7 to 9 consider aspects of the pathogenesis of DNA-virus infections. The final chapter describes the current technology being applied to the development of DNA-virus vectors for gene delivery.

My thanks go to David Hames (series editor) for his guidance in shaping this and the accompanying volumes, and to the editorial staff at Oxford University Press. Most importantly, thanks must go to the contributors who were prepared to share their combined expertise with the wider research community.

Leicester University   A. J. Cann
July 1999

Contents

List of Contributors xvii
Abbreviations xix

1. Extraction, purification, and characterization of virus DNA   1

Ian W. Halliburton

1. Introduction	1
2. Tissue culture	2
Cell lines	2
3. Virus culture	4
Preparation of virus working stocks	4
Titration of virus stocks	5
Virus purification	6
4. Manipulation of virus DNA	6
Extraction of virus DNA	6
Restriction enzyme digestion of virus DNA	9
5. Mapping DNA virus genes	9
Marker rescue	10
Transcript mapping	11
Mapping by use of intertypic recombinants	12
Sequencing	12
References	12

2. Investigation of DNA virus genome structure   15

Michael A. Skinner and Stephen M. Laidlaw

1. Introduction	15
Scope	16
2. Genome mapping	16
Southern blotting	17
Pulsed-field agarose gel electrophoresis	19
3. Correlation of genotype with phenotype by marker rescue	21
Design of marker rescue	21
4. Nucleotide sequencing	22
Primer walking	23

Library construction	23
Minipreparation of sequence grade DNA	27
Sequence determination	29
Data collection	30

5. Sequence assembly — 31
- The Staden package — 31

6. Sequence analysis — 39
- Prediction of open reading frames — 39
- Homology searches — 41
- Sequence alignment — 42
- Phylogenetic analysis — 43
- Mutational analysis — 44

7. Future directions — 44

Acknowledgements — 44

References — 44

3. Mutagenesis of DNA virus genomes — 47

Keith N. Leppard

1. Introduction — 47
- The diversity of DNA viruses — 47
- Alternative mutagenesis strategies — 47
- Suiting mutagenesis strategy to the virus — 48

2. Preparation and titration of virus stocks — 51
- Cell culture techniques — 51
- Generating and titrating virus stocks — 52

3. Chemical mutagenesis of viral DNA *in vitro* — 56

4. Chemical mutagenesis of virus particles *in vitro* — 56

5. Mutagenesis through growth of virus in the presence of nucleoside analogues — 59

6. Site-directed mutagenesis *in vitro* — 60

7. Reintroducing mutagenized sequences into virus — 64
- Complementing cell lines — 68
- Polyomaviruses — 70
- Parvoviruses — 71
- Adenoviruses — 72
- Herpes viruses and pox viruses — 74

8. Selection of mutant phenotypes in randomly mutagenized stocks — 75

9. Mapping mutations in isolates obtained by random mutagenesis — 76

Contents

10. Characterization of mutant viruses	77
Single step growth curve	77
DNA replication assay	77
Late protein expression	79
Virus assembly	80
References	80

4. Interactions between viral and cellular proteins during DNA virus replication   83

Catherine H. Botting and Ronald T. Hay

1. Introduction	83
2. Identification and purification of proteins involved in replication	84
Obtaining replication-active extracts	84
Assaying for activity	86
Purification of proteins required for adenovirus DNA replication from HeLa cells	92
3. Overexpression of replication components	97
Overexpression and purification of pTP and pol	97
Expression and purification of the cellular factors	99
4. Mapping interactions	101
Protein–DNA interactions	101
Protein–protein interactions	104
5. Investigating the dynamics of the replication process	109
Immobilized replication assay	109
Glycerol gradient centrifugation	109
References	110

5. Analysis of transcriptional control in DNA virus infections   113

S. K. Thomas and D. S. Latchman

1. Introduction	113
2. Analysis of viral gene expression during infection	114
Preparation of RNA samples	115
Analysis of transcripts by Northern blotting	118
Analysis of transcripts by reverse transcription-polymerase chain reaction (RT-PCR)	122
Analysis of transcripts by *in situ* hybridization (ISH)	124

Preface

3. Analysis of cloned promoter sequences using reporter
gene constructs — 128
    Methods of transfection — 129
    Analysis of promoter activity — 130
    Preparation of indicator viruses for the analysis of promoter activity
during infection — 134
    Using transgenic animals to analyse the tissue-specific expression of
viral promoters — 140

4. Analysis of transcriptional control by mutagenesis — 141
    Types of mutation — 141
    Linker scanning mutagenesis — 142
    Oligonucleotide insertion — 142
    Site-directed mutagenesis — 144

5. Identification of cellular transcription factors involved in
the control of viral transcription — 148
    The DNA mobility-shift assay — 148
    South-western blotting — 153
    Methods for isolating cloned transcription factors — 154

Acknowledgements — 154

References — 154

6. Identification and analysis of *trans*-acting proteins involved in the regulation of DNA virus gene expression — 157

Adrian Whitehouse and David M. Meredith

1. Introduction — 157

2. Identification of transactivating proteins — 158
    Transfection of mammalian cells — 158

3. Analysis of the mechanism of transactivation — 159
    Characterization of RNA — 160

4. Identification of *cis*-acting elements — 164
    Mobility shift assays — 164
    Purification of transcription factors — 169
    Conventional purification procedures — 173

Acknowledgements — 174

References — 174

Contents

7. Interaction of DNA virus proteins with host cytokines — 177

Alshad S. Lalani, Piers Nash, Bruce T. Seet, Janine Robichaud and Grant McFadden

1. Introduction	177
2. Identification of novel virus soluble cytokine-binding proteins	178
Generation of secreted virus proteins from infected cells	178
Use of chemical cross-linking to detect viral cytokine-binding proteins	179
Ligand blot overlays for detection of viral cytokine-binding proteins	180
Immunoprecipitation of viral cytokine-binding proteins	182
Use of plasmon resonance for analysis of interactions of cytokines with viral proteins	184
3. Synthesis and purification of cytokine-binding proteins	186
Vaccinia virus expression system	187
Synthesis of cytokine binding proteins by Baculovirus expression systems	189
Fc fusion protein production	189
Purification of secreted viral cytokine-binding proteins by fast protein liquid chromatography (FPLC)	191
Purification by affinity chromatography	196
4. Analysis of cytokine-binding partners	197
Solid phase binding	197
Scintillation proximity assay	199
The use of surface plasmon resonance (SPR) for detailed kinetic studies	201
Inhibition of cell-surface binding	202
Inhibition of cytokine-induced cytolysis	204
Growth inhibition/proliferation assay	205
Measuring the effect of viral chemokine-binding proteins on chemokine-induced calcium flux	206
References	207

8. Analysis of DNA virus proteins involved in neoplastic transformation — 209

Kersten T. Hall, Maria E. Blair Zajdel and G. Eric Blair

1. Introduction	209
Human adenovirus oncoproteins	209
Simian virus 40 and polyomavirus oncoproteins	211
Human papillomavirus oncoproteins	214

Contents

2. Cell systems utilized for study of DNA tumour viruses — 216
    DNA transfection of virus oncogenes into mammalian cells — 217
    Selection of virus-transformed cells — 217
    Cloning of transformed cells, and assay of cell growth — 219

3. Characterization of virus-transformed cells — 221
    Analysis of virus oncoproteins by radiolabelling of transformed cells and immunoprecipitation — 224
    Sub-cellular distribution of virus oncoproteins — 230

4. Biological activity of virus oncoproteins — 235
    Studies of interactions between virus oncoproteins and cellular proteins — 238
    Use of the yeast two-hybrid screen to identify interactions between virus oncoproteins and cellular proteins — 240

Acknowledgements — 244

References — 244

9. Chemotherapy of DNA virus infections — 247

Patricia A. Cane and Deenan Pillay

1. Introduction — 247

2. Antivirals effective against herpesviruses — 247
    Aciclovir — 247
    Penciclovir — 249
    Foscarnet — 249
    Cidofovir — 250
    Testing susceptibility of HSV to antiviral drugs — 250
    Resistance assays for therapeutic antivirals used for HSV — 252

3. Resistance to antivirals of human cytomegalovirus (HCMV) — 254
    Phenotypic CMV drug susceptibility assay — 256
    Genotypic assays for detection of ganciclovir resistance-associated mutations in CMV — 257

4. Antivirals active against hepatitis B virus — 260
    Genotypic assay for resistance associated mutations in HBV using nucleotide sequencing. — 261

5. Real time PCR and fluorimetry for detection of mutations — 263

Acknowledgements — 265

References — 265

Contents

10. Herpes simplex virus and adenovirus vectors — 267
Cinzia Scarpini, Jane Arthur, Stacey Efstathiou, Yvonne McGrath and Gavin Wilkinson

1. Introduction — 267

2. Herpes simplex virus — 267
    Biological properties — 267
    Gene expression during lytic infection — 268
    The latent state — 268
    Basic techniques of virus handling — 269
    Construction of recombinant virus genomes — 273
    The use of wild-type and replication-defective viruses as vectors to deliver genes to the peripheral and central nervous system — 280
    In vitro culture of neurones to study the biology of HSV — 284

3. Adenovirus — 287
    Biological properties — 287
    Pattern of gene expression during lytic adenovirus infection — 288
    Adenovirus vectors — 288
    The helper cell line — 292
    Basic adenovirus handling techniques — 292
    Quantification of adenovirus stocks — 295
    Construction of replication-deficient adenovirus recombinants — 296
    Characterization of virus — 299
    Infection of cells with adenovirus vectors — 301
    Enhanced infection — 301

References — 302

Appendix — 307

Index — 313

Contributors

JANE ARTHUR
Division of Virology, Department of Pathology, University of Cambridge, Tennis Court Road, Cambridge CB2 1QP, UK.

G. ERIC BLAIR
School of Biochemistry and Molecular Biology, University of Leeds, Leeds LS2 9JT, UK.

MARIA E. BLAIR ZAJDEL
Biomedical Research Centre, School of Science and Mathematics, Sheffield Hallam University, Sheffield S1 1WB

CATHERINE H. BOTTING
School of Biomedical Sciences, University of St Andrews, Biomolecular Sciences Building, North Haugh, St Andrews, Fife KY16 9ST, UK.

PATRICIA A. CANE
Division of Immunity and Infection, University of Birmingham Medical School, Edgbaston, Birmingham B15 2TA, UK.

STACEY EFSTATHIOU
Division of Virology, Department of Pathology, University of Cambridge, Tennis Court Road, Cambridge CB2 1QP, UK.

KERSTEN T. HALL
Molecular Medicine Unit, Research School of Medicine, St James's University Hospital, Leeds LS9 7TF, UK.

IAN W. HALLIBURTON
Department of Microbiology, University of Leeds, Leeds LS2 9JT, UK.

RONALD T. HAY
School of Biomedical Sciences, University of St Andrews, Biomolecular Sciences Building, North Haugh, St Andrews, Fife KY16 9ST, UK.

STEPHEN M. LAIDLAW
Institute for Animal Health, Compton Laboratory, Newbury, Berkshire RG20 7NN, UK.

ALSHAD S. LALANI
Department of Biochemistry, University of Alberta, Edmonton, Canada, and Department of Microbiology and Immunology, and The J. P. Robarts Research Institute, London, Ontario, Canada.

D. S. LATCHMAN
Department of Molecular Pathology, University College London Medical School, 46 Cleveland Street, London W1P 6DB, UK.

Contributors

KEITH N. LEPPARD
Department of Biological Sciences, University of Warwick, Coventry CV4 7AL, UK.

GRANT MCFADDEN
Department of Microbiology and Immunology, The University of Western Ontario and The J. P. Robarts Research Institute, London, Ontario, Canada.

YVONNE MCGRATH
Department of Medicine, University of Wales College of Medicine, Tenovus Building, Cardiff CF4 4XX, UK.

DAVID M. MEREDITH
Molecular Medicine Unit, University of Leeds, St James's University Hospital, Leeds LS9 7TF, UK.

PIERS NASH
Department of Biochemistry, University of Alberta, Edmonton, Canada, and Department of Microbiology and Immunology and The J. P. Robarts Research Institute, London, Ontario, Canada.

DEENAN PILLAY
PHLS Antiviral Susceptibility Reference Unit, Division of Immunity and Infection, University of Birmingham Medical School, Birmingham B15 2TT, UK.

JANINE ROBICHAUD
Department of Microbiology and Immunology, The University of Western Ontario and The J. P. Robarts Research Institute, London, Ontario, Canada.

CINZIA SCARPINI
Division of Virology, Department of Pathology, University of Cambridge, Tennis Court Road, Cambridge CB2 1QP, UK.

BRUCE T. SEET
Department of Microbiology and Immunology, The University of Western Ontario and The J. P. Robarts Research Institute, London, Ontario, Canada.

MICHAEL A. SKINNER
Institute for Animal Health, Compton Laboratory, Newbury, Berkshire RG20 7NN, UK.

S. K. THOMAS
Department of Molecular Pathology, University College London Medical School, 46 Cleveland Street, London W1P 6DB, UK.

ADRIAN WHITEHOUSE
Molecular Medicine Unit, University of Leeds, St James's University Hospital, Leeds LS9 7TF, UK.

GAVIN WILKINSON
Department of Medicine, University of Wales College of Medicine, Tenovus Building, Cardiff CF4 4XX, UK.

Abbreviations

Ad	adenovirus
AP	alkaline phosphatase
AraC	cytosine arabinoside
BSA	bovine serum albumin
CAT	chloramphenicol acetyl transferase
CMC	carboxymethyl cellulose
CPE	cytopathic effect
CR	conserved region
CSF	colony stimulating factor
DBD	DNA-binding domain
DBP	DNA-binding protein
DEAE	diethylaminoethyl
DIG	digoxigenin
DMEM	Dulbecco's modified Eagle's medium
DMP	dimethylpimelimidate
DMS	dimethyl sulfide
DMSO	dimethyl sulfoxide
DRG	dorsal root ganglion
DTT	dithiothreitol
E. coli	*Escherichia coli*
EDTA	ethylenediamine tetraacetic acid
ELISA	enzyme-linked immunosorbent assay
FCS	fetal calf serum
FGF	fibroblast growth factor
FPLC	fast protein liquid chromatography
GAPDH	glyceraldehyde-6-phosphate dehydrogenase
GFP	green fluorescent protein
GST	glutathione S transferase
HBS	HEPES-buffered saline
HCF	host cell factor
HCMV	human cytomegalovirus
HPLC	high pressure liquid chromatography
HPV	human papilloma virus
HSV	herpes simplex virus
IAA	isoamyl alcohol
IE	immediate early
IEC	ion-exchange chromatography
IFN	interferon
Ig	immunoglobulin
IL	interleukin

Abbreviations

IPTG	isopropyl-β-D-thiogalactopyranoside
ISH	*in situ* hybridization
ITR	inverted terminal repeat
LAT	latency-associated transcript
LMP	low melting point
MIEP	immediate early promoter
m.o.i.	multiplicity of infection
NCS	newborn calf serum
NFI	nuclear factor I
NFIII	nuclear factor III
NP-40	Nonidet P40
ONPG	*o*-nitrophenylgalactoside
PAGE	polyacrylamide gel electrophoresis
PAS	protein-A Sepharose
PBS	phosphate-buffered saline
PCR	polymerase chain reaction
PFC	plaque-forming cells
PFU	plaque-forming unit
PI 3-K	phosphatidylinositol 3-kinase
PLC-γ-1	phospholipase C-gamma-1
PMSF	phenylmethylsulfonyl fluoride
PP2A	protein phosphatase 2A
pRb	retinoblastoma susceptibility protein
pTP	preterminal protein
PVDF	polyvinylidine difluoride
RT	reverse transcriptase
SA7	simian adenovirus 7
SAC	*Staphylococcus aureus* (Cowan 1) strain
SDM	site-directed mutagenesis
SDS	sodium dodecyl sulfate
SEC	size-exclusion chromatography
SH2	src homology 2
Shc	src homology and collagen
SMBS	sodium metabisulfite
SOS	son of sevenless
SPA	scintillation proximity assay
SPR	surface plasmon resonance
ss	single-stranded
SV40	simian virus 40
TCID	tissue-culture infectious dose
TLC	thin-plate liquid chromatography
TNF	tumour necrosis factor
TP	terminal protein
ts	temperature sensitive

1

Extraction, purification, and characterization of virus DNA

IAN W. HALLIBURTON

1. Introduction

Since viruses will only replicate within living cells, special methods have to be employed for growth *in vitro*. Four main systems are used for growth of viruses in the laboratory:

Laboratory animals: Before other techniques were available, viruses were mainly isolated by inoculation of laboratory animals. The obvious choice of experimental animal is the natural host, but for human infections this is obviously not generally possible, and animals such as mice, rabbits, ferrets, and monkeys have been commonly used. The cost of upkeep of animals is, however, high, and there tends to be a large variation in response between individual animals even if inbred, so large numbers are therefore needed. To a growing extent, ethical considerations are also important, but animals are still required for the isolation of some viruses which will not grow in tissue culture, and animals still have to be used to study virus–host interactions or symptoms of a virus infection in its natural host.

Fertile hens' eggs: Some viruses can be grown in the cells of the chick embryo on the chorio-allantoic membrane, in the allantoic cavity, or in the amniotic cavity; for example, herpes simplex virus can be grown on the chorio-allantoic membrane.

Organ culture: Not often used, but viruses have been grown, for example, in pieces of brain, gut, or trachea. If the organ is from the natural host, and particularly if it is the organ in which the virus multiplies in the animal, then one could argue that it is close to a natural infection. However, organ culture involves a mixed population of several cell types and is obviously far from natural, since the culture is isolated from regulatory substances such as hormones. Large-scale work is also very difficult.

Tissue culture: Cells obtained from man or animals are grown in artificial growth medium in glass or plastic containers, usually at 37 °C. Most, but not all, viruses can be propagated in cultures of suitable cells.

The system of choice by far for virus isolation is tissue culture. This chapter describes methods for the propagation of herpesviruses, and extraction and characterization of DNA from them. By substituting alternative culture techniques appropriate to other viruses, these methods may be applied to a range of DNA-containing viruses.

2. Tissue culture

Three main types of tissue culture are used in virology:

- primary cultures
- semi-continuous cell lines
- continuous cell lines

Primary cultures are prepared from tissue fragments such as monkey kidney or human amnion by dispersing cells with trypsin. These are particularly useful, because they can support a wide range of animal viruses. There is little cell division in such cultures, the cells simply settling on the vessel surface, and spreading out to form a monolayer. After two or three weeks, the cells deteriorate and die. With other cell lines, which are usually fibroblastic in origin, such as human embryo lung cells, there is rapid growth, and the cells can be subcultured up to about 50 times in culture. These are known as semi-continuous cell strains. With cell lines derived from malignant or cancerous tissue (e.g. HeLa, Hep-2, or Vero cells), there is rapid growth and the cells can be subcultured indefinitely. Such cells are usually but not always epithelial cells, and the chromosomes are heteroploid, with more than the diploid number of chromosomes but not an exact multiple of it. They are referred to as established or continuous cell lines.

The usefulness of any particular cell line cannot be predicted; any new cell line has to be tested extensively before it can be recommended for widespread use. The sensitivity of many cell lines for cultivation of viruses does not, in general, depend on the organ of origin. Respiratory viruses, for example, do not necessarily grow best in lines from respiratory tissues such as lung. Diploid cell lines which are very sensitive for isolation of respiratory viruses have been derived from skin, intestine, and tonsil.

2.1 Cell lines

Herpes simplex virus (HSV) or equine herpesvirus type 1 (EHV-1) strains are propagated in baby hamster kidney or rabbit kidney type 13 cells (RK-13), supplied, for example, by Flow Laboratories.

2.1.1 Growth media

BHK or RK-13 cells can be grown in Dulbecco's modified Eagle's medium (DMEM), supplemented with 10% newborn calf serum and 2 mM glutamine, and buffered with 0.1% sodium bicarbonate. Penicillin and streptomycin are added to final concentrations of 200 units ml^{-1} (approximately 0.12 mg ml^{-1}) and 0.1 mg ml^{-1} respectively. For growth of BHK cells, 10% (v/v) tryptose–phosphate broth is usually added to the growth medium, but it is not essential for growth, the cells simply growing better in its presence. COS 7 cells can be grown in DMEM containing 10% fetal calf serum, with glutamine, sodium bicarbonate, and antibiotics added to the same concentration as above.

2.1.2 Growth of cells

For production of large stocks of cells or virus, BHK or RK-13 cells are cultured as monolayers in 2-litre roller bottles seeded with approximately 4.5×10^7 cells in 200 ml of DMEM, incubated at 37°C, and rotated at 0.1 r.p.m. until fully confluent. The culture medium is then decanted, and monolayers are rinsed twice with phosphate-buffered saline (PBS) containing 0.05% trypsin and 0.02% EDTA (*Protocol 1*). Cells are harvested in approximately 15 ml DMEM per roller bottle, and their concentration is determined with the aid of a haemocytometer. Cells can also be grown in plastic dishes (Falcon), incubated at 37°C and gassed at 4% CO_2. Stocks of RK-13 or BHK cells are kept in liquid nitrogen.

Protocol 1. Growth and trypsinization of cells

Equipment and reagents

- Dulbecco's modified Eagle's medium (DMEM) supplemented with 10% newborn bovine serum and 2 mM glutamine, buffered with 0.1% (w/v) sodium bicarbonate, and containing penicillin and streptomycin at 200 units ml^{-1} (approximately 0.12 mg ml^{-1} and 0.10 mg ml^{-1} respectively) (Sigma, Life Technologies, ICN Flow, etc.)
- Roller machine at 37°C
- Haemocytometer
- Phosphate-buffered saline (PBS) containing 0.05% (w/v) trypsin and 0.02% (w/v) EDTA

Method

1. Seed 2-litre roller bottles with 4.5×10^7 RK13 cells in 200 ml DMEM.
2. Incubate the bottles at 37°C with rotation at 0.1 r.p.m.
3. When a confluent monolayer of cells has formed (3 or 4 days), decant the medium from the bottle, and add about 20 ml PBS, containing trypsin-EDTA, to the cell monolayer.
4. Rotate the bottle slowly and gently by hand for about 20 s.
5. Decant PBS-trypsin-EDTA into a waste container, and repeat the wash with another 20 ml PBS-trypsin-EDTA.
6. Leave the roller bottles on the bench until the cells have detached from the glass.
7. Add about 20 ml DMEM to the roller bottle.

Protocol 1. *Continued*

8. Pipette the medium up and down, washing the sides of the roller bottle with it.
9. Transfer medium containing the cells to a fresh container, and count the cells in a haemocytometer.

3. Virus culture

3.1 Preparation of virus working stocks

Herpes simplex virus (HSV) or equine herpesvirus type 1 (EHV-1) strains are propagated in baby hamster kidney (BHK) or rabbit kidney type 13 (RK-13) cells, as described in *Protocol 2*.

Protocol 2. Preparation of virus working stocks

Equipment and reagents
- Roller machine at 37°C
- Refrigerated centrifuge
- Ultrasonic water bath
- Dulbecco's modified Eagle's medium (DMEM) (*Protocol 1*)

Method
1. Seed 2-litre roller bottles with 4.5×10^7 RK13 or BHK cells in 200 ml DMEM.
2. Incubate bottles at 37°C with rotation at 0.1 r.p.m.
3. When a confluent monolayer of cells has formed (3 or 4 days), decant the medium from the bottle, and add virus inoculum in 20 ml DMEM at a multiplicity of infection (m.o.i.) of 0.02 plaque-forming units (PFU) per cell (see *Protocol 3*).
4. Allow adsorption to proceed for 1 h at 37°C.
5. Add 180 ml DMEM, and incubate with rotation for 24–48 h, until 100% cytopathic effect (CPE) is observed.
6. Gently scrape any remaining infected cells into the culture medium, and transfer to 250 ml polycarbonate centrifuge pots.
7. Centrifuge at 1000 g for 5 min at 4°C.
8. Carefully remove the supernatant, and retain it for extraction of DNA or for virus purification.
9. Resuspend the cell pellet in 5 ml DMEM, and transfer to a sterile 50 ml flat glass bottle.

1: Extraction, purification, and characterization of virus DNA

10. Sonicate in an ice-cold ultrasonic water bath until the suspension has lost its granular appearance (approximately 5 min).
11. Transfer disrupted cells to a 20 ml universal bottle, and centrifuge at 1500 g for 5 min to pellet cell debris.
12. Dispense the supernatant into sterile plastic vials in 1 ml aliquots, and store at $-70\,°C$.

3.2 Titration of virus stocks

Virus stocks are titrated using the plaque assay of Russell (1), as described in *Protocol 3*.

Protocol 3. Titration of virus stocks

Equipment and reagents

- Shaking platform at 37°C
- Dulbecco's modified Eagle's medium (DMEM) (*Protocol 1*)
- Formaldehyde-saline solution (10% formaldehyde, 0.85% NaCl)
- 0.4% w/v high-viscosity carboxymethyl cellulose (Sigma) in DMEM *or* molten 4% agarose in PBS
- 0.1% w/v Gentian violet

Method

1. Prepare tenfold serial dilutions (to 10^{-8}) of the virus stock in DMEM.
2. Add 2 ml diluted virus to 7×10^6 RK-13 cells in separate 20 ml universal bottles.
3. Mix by inversion, and incubate on a shaking platform at 37°C for 45 min.
4. Add 8 ml of DMEM containing 0.4% high viscosity carboxymethyl cellulose to each tube. Mix thoroughly by inversion. Alternatively, agarose can be used as an overlay medium by mixing 1 volume of 4% agarose in PBS (45–50°C) with five volumes of DMEM (42–44°C) and dispensing immediately.
5. Divide each suspension equally between two 60 mm Petri dishes, and incubate at 37°C in 4% CO_2.
6. After 3 days, fix the cell sheets for 15 min in formaldehyde-saline solution.
7. Stain the fixed cell sheet for 15 min in 0.1% Gentian violet, and count the plaques.

3.3 Virus purification

Protocol 1. Virus purification

Equipment and reagents
- Ultracentrifuge with appropriate rotors
- Ultrasonic water bath
- Gradient former (e.g. BRL)
- 10% w/v sucrose in PBS
- 40% w/v sucrose in PBS

Method
1. Extracellular virus is recovered from infected cell supernatants (*Protocol 2*) by centrifugation at 15 000 g for 2 h at 4°C.
2. Resuspend virus pellets in 1 ml DMEM per roller bottle, and briefly sonicate in an ultrasonic water bath to disrupt any clumps.
3. Layer the suspension onto 35 ml 10–40% sucrose gradients in 38.5 ml ultracentrifuge tubes.
4. Fill tubes to the top with PBS, then centrifuge at 40 000 g for 45 min at 8°C, using a swinging-bucket rotor.
5. Remove the upper portion of the gradient to within 2 cm of the virus band, approximately halfway down the tube.
6. Harvest the virus in a volume of less than 10 ml of sucrose.
7. Transfer to a fresh ultracentrifuge tube, top up with PBS, and pellet by centrifugation at 110 000 g for 45 min at 8°C.
7. Resuspend virus pellet in 100 µl PBS per roller bottle, and store at −70°C.

4. Manipulation of virus DNA

4.1 Extraction of virus DNA

Infected cells are used to prepare DNA either from extracellular virus (*Protocol 5*), or from intracellular virus (*Protocol 6*).

Protocol 5. Extraction of DNA from extracellular virus

Equipment and reagents
- Ultracentrifuge with appropriate rotors
- Sterile (autoclaved) 30 ml Corex™ tubes (Sorvall)
- TE buffer (0.01 M Tris, 0.001 M EDTA, pH 8.0)
- Ribonuclease A (e.g. Sigma), 10 mg ml^{-1} in TE buffer
- Sodium dodecyl sulfate (SDS), 10% w/v
- Sodium acetate, 3 M, adjusted to pH 5.2 with glacial acetic acid.

1: Extraction, purification, and characterization of virus DNA

- Phenol: melt 100 g ultrapure phenol (e.g. BRL, molecular biology grade) at 37 °C with 11 ml distilled water and 0.1 g 8-hydroxyquinoline (e.g. Sigma), dispense into 5 ml aliquots, and store at −20 °C. Before use, thaw, and equilibrate with an equal volume of 10 × TE.
- Proteinase K (e.g. Boehringer), 20 mg ml^{-1} in TE; self digest for 30 min at 37 °C, aliquot, and store at −20 °C.
- Water-saturated ether
- Pasteur pipette heated in a Bunsen burner flame to seal the tip and form it into a hook.

Method

1. Infect RK-13 or BHK cell monolayers in 2-litre roller bottles at an m.o.i. 0.1 PFU per cell, as described in *Protocol 2*, then harvest when approximately 90% CPE is observed.
2. Transfer culture supernatants to 250 ml polycarbonate centrifuge pots, and centrifuge at 15 000 g for 2 h at 4 °C.
3. Remove all traces of medium, then resuspend the virus pellets in 2 ml TE per roller bottle, and transfer to sterile 30 ml Corex tubes.
4. Add 40 μl RNase A (10 mg ml^{-1}) per tube, and incubate for 30 min at 37 °C (cover tubes with plastic film).
5. Add 400 μl SDS (10%) and 100 μl proteinase K (20 mg ml^{-1}), and incubate for a further 30 min at 37 °C.
6. Add 4 ml TE-saturated phenol to each tube, mix carefully by inversion, and separate the phases by centrifugation at 1500 g for 10 min in a bench-top centrifuge.
8. Remove the (lower) phenol layer, taking care not to take any of the (upper) aqueous phase, and repeat phenol extraction as necessary until no debris is seen between phases after centrifugation.
9. Transfer the upper aqueous layer to a fresh (sterile) Corex tube, and extract four times with an equal volume of water-saturated ether to remove all traces of phenol. Remove any remaining ether by gently blowing nitrogen gas over the surface of the sample. **CAUTION**: Ether is highly flammable.
10. Precipitate DNA by slow addition of 0.4 ml 3 M sodium acetate and 5 ml ice-cold ethanol. On swirling the tube, virus DNA precipitates at the interphase. Transfer DNA to a sterile Eppendorf tube using a sealed Pasteur pipette.
11. Wash pellets with 70% ethanol, air-dry, and resuspend in 200 μl TE. Check a 1 μl aliquot on a 0.8% agarose gel for integrity.
12. Store DNA at −20 °C.

Protocol 6. Extraction of DNA from intracellular virus

Equipment and reagents
- Ultracentrifuge with appropriate rotors
- Sterile (autoclaved) 30 ml Corex™ tubes (Sorvall)
- LCM buffer: 0.5% NP-40, 30 mM Tris-HCl pH 7.5, 3.06 mM $CaCl_2$, 5 mM $MgCl_2$, 125 mM KCl, 0.5 mM EDTA, 6 mM 2-mercaptoethanol
- TNE buffer: 50 mM Tris-HCl pH 7.5, 100 mM NaCl, 10 mM EDTA
- 1,1,2-trichloro-1,2,2-trifluoroethane (Freon) (e.g. Aldrich)
- Glycerol step gradient consisting of 3 ml each of 5% and 45% glycerol in LCM buffer in a 38.5 ml centrifuge tube
- Ultrasonic water bath
- Phenol:chloroform:isoamyl alcohol, 25:24:1, v/v/v
- Chloroform:isoamyl alcohol, 24:1, v/v
- Ethanol
- 70% v/v ethanol

Method

1. Infect RK-13 or BHK cell monolayers in 2-litre roller bottles at an m.o.i. of 0.1 PFU per cell, and harvest virus as described in *Protocol 2*.
2. Resuspend the cell pellet in 10 ml LCM buffer per roller bottle, and transfer to sterile 30 ml Corex tubes.
3. Extract the lipid envelope by carefully adding 1 ml Freon, and mixing gently by inversion.
4. Centrifuge at 1500 g for 5 min. Transfer the upper phase to a fresh tube, and re-extract with 1 ml Freon.
5. Layer the aqueous phase onto a glycerol 'step' in a cellulose tube. Pellet the nucleocapsids through the glycerol by centrifugation at 72 000 g for 2 h at 4°C, using a swinging-bucket rotor.
6. Pour the glycerol off, and drain the pellet by inverting the tube on a paper towel for 2–3 min.
7. Resuspend the pellet in 1 ml TNE per roller bottle by mild sonication in an ultrasonic water bath.
8. Add SDS to a final concentration of 0.5%, and extract the samples with an equal volume of TE-saturated phenol.
9. Continue phenol extractions until no interphase is present between the phases. Transfer the upper aqueous layer to a fresh tube.
10. Extract the aqueous layer twice with phenol:chloroform:isoamyl alcohol, and once with chloroform:isoamyl alcohol. Transfer the aqueous phase to a clean tube.
11. Precipitate DNA by adding 2 volumes of ethanol, mix gently by inversion, and store at −20°C overnight.
12. Recover DNA by centrifugation at 10 000 g for 30 min at 4°C.
13. Wash pellets with 70% ethanol, air-dry, and resuspend in 200 µl TE.

1: Extraction, purification, and characterization of virus DNA

4.2 Restriction enzyme digestion of virus DNA

To perform restriction enzyme digestions, virus DNA is purified using the methods given in *Protocols 5 and 6*. Small scale digestions are first set up, containing about 1 μg DNA in a total reaction volume of 20 μl. Larger scale digestions, containing approximately 10 μg DNA, are set up to provide fragments for cloning or for Southern blotting. All restriction digestions are carried out under conditions recommended by the enzyme supplier (e.g. Gibco-BRL or Boehringer).

Protocol 7. Restriction enzyme digestion of virus DNA

Equipment and reagents

- Waterbath
- Sterile (autoclaved) plastic microfuge tubes
- Sterile (autoclaved) plastic pipette tips
- Micropipettes (e.g. Gilson)
- Chloroform:isoamyl alcohol, 24:1, v/v
- Phenol:chloroform:isoamyl alcohol, 25:24:1, v/v/v
- Ethanol
- 70% v/v ethanol

Method

1. Incubate samples in 1x manufacturer's reaction buffer, 0.1 mg ml^{-1} gelatin, and approximately 5 units of enzyme per μg DNA, for 3–6 h at 37 °C. Spermidine can be included at a concentration of 4 mM to help overcome enzyme inhibition where necessary.

2. After incubation, check an aliquot on an agarose gel. If digestion is incomplete, add more enzyme, and continue incubation overnight.

3. Extract samples once with phenol:chloroform:isoamyl alcohol, twice with chloroform:isoamyl alcohol, and precipitate with ethanol (*Protocol 5*).

4. After centrifugation, wash DNA pellets with 70% ethanol, and resuspend in 50 μl TE.

---

5. Mapping DNA virus genes

Extracted DNA is needed for mapping DNA virus genes by a number of methods, such as marker rescue (2–4), transcript mapping by hybridising viral RNA to restriction endonuclease fragments using the method of Southern (5), transcript mapping by RNA displacement loop (R-loop) mapping (6, 7), mapping by use of intertypic recombinants (8–12), use of plasmids such as λgt11, or sequencing. For all methods, knowledge of the restriction enzyme maps of the virus DNA is an essential prerequisite.

5.1 Marker rescue

To map a ts mutation (1, 2), or any other phenotypically identifiable marker such as drug resistance or plaque morphology marker (3), by marker rescue, individual ts⁺ restriction enzyme fragments of the virus DNA are first purified, and each fragment is cotransfected separately with purified ts mutant DNA. The progeny are titrated at non-permissive and at permissive temperatures, and any cotransfection giving a higher yield than the transfection yield with ts DNA alone indicates rescue of the marker and mapping of the function involved. Use of several different restriction enzymes with cleavage sites within the gene in question can narrow the map position. The method is described in *Protocol 8* for physical mapping of a ts mutation.

Protocol 8. Marker rescue

Equipment and reagents
- Hepes-buffered saline (HBS): 8.0 g l^{-1} NaCl, 0.37 g l^{-1} KCl, 0.125 g l^{-1} Na$_2$HPO$_4$.2H$_2$O, 1.0 g l^{-1} dextrose, 5.0 g l^{-1} N-2-hydroxyethylpiperazine-N'-2-ethane-sulfonic acid (Hepes) pH 7.05
- 2M CaCl$_2$
- 25% v/v dimethylsulfoxide (DMSO) in HBS

Method

1. Mix DNA extracted from a ts mutant at a concentration of 0.5 to 1.0 µg ml^{-1} with 10 µg salmon sperm or calf thymus carrier DNA and wild-type DNA fragments (resulting from digestion of 1 µg intact virus DNA) per ml of HBS. Stir at room temperature for 2–4 h.
2. Add CaCl$_2$ to a final concentration of 130 mM.
3. Leave at room temperature for 10–20 min for a precipitate to form.
4. Add 0.4 ml of the fine suspension to a suitable slightly subconfluent cell monolayer for assay of the virus in a 60 mm diameter Petri dish.
5. Incubate at 38°C (non-permissive temperature) for 30 min.
6. Add 4.6 ml growth medium.
7. After 4 h, drain the monolayers, and wash once with growth medium.
8. Add 1 ml 25% DMSO for 3 min.
9. Wash monolayers once with growth medium.
10. Add 5 ml growth medium containing overlay medium (*Protocol 3*, Step 4).
11. Incubate dishes at 38°C for 2 days.
12. Harvest cells in their own growth medium, and disrupt by ultrasonic treatment or by three rounds of freezing and thawing.
13. Titrate virus yield (*Protocol 3*) at the non-permissive and at the permissive temperatures.

1: Extraction, purification, and characterization of virus DNA

5.2 Transcript mapping

The transcript to be mapped is purified by some combination of size separation on sucrose density or methylmercury gradients, size fraction by electrophoresis on agarose gels containing 10mM methylmercury hydroxide, and separation on a poly-dT column. The DNA of the virus is digested with a restriction enzyme, and the fragment(s) hybridizing to the mRNA are identified by Southern blotting (2). The method for this approach to mapping a purified transcript is described in *Protocol 9.* Alternatively, a partial cDNA copy of the mRNA can be prepared, and hybridized to a restriction digest with a different enzyme to identify the direction of transcription. Use of the RNA in an *in vitro* translation system, and separation of the products on SDS-PAGE, also allows at least a determination of the M_r of the product of the mapped transcript. However, by running tracks with infected cell extracts and with purified virus side by side, it should be possible to identify the product.

Protocol 9. Transcript mapping

Equipment and reagents

- Standard saline citrate (SSC) (20×): 175.3 g NaCl and 98 g trisodium citrate dissolved in 800 ml double-distilled water, adjusted to pH 7.0 using concentrated HCl, and made up to 1 litre
- Hybridization buffer: 65% formamide, 0.4 M NaCl, 0.1 M Hepes, 5 mM EDTA
- Nitrocellulose paper (BA85, Schleicher & Schuell)

Method

1. Digest the virus DNA with a restriction endonuclease, separate the fragments on an agarose gel, and transfer the DNA to nitrocellulose paper in 10 × SSC by the method of Southern (5) (see Chapter 2, *Protocol 1*).

2. Size-fractionate ^{32}P-labelled RNA by electrophoresis on 1.2% agarose gels containing 10 mM methylmercury hydroxide.

3. Soak the gels in 50 mM 2-mercaptoethanol for 20 min, and slice at 2 or 3 mm intervals.

4. Dissolve each RNA gel slice in 1.5 ml hybridization buffer containing 65% formamide, by heating at 75°C for 5 min.

5. Seal the dissolved RNA gel samples and strips from the nitrocellulose DNA blots in plastic bags, and incubate at 57°C for 48 h.

6. Remove unhybridized RNA by three 1 h rinses in 65% formamide-2 × SSC at 45°C.

7. Wrap the nitrocellulose strips in Saran Wrap™ and expose to X-ray film for the required exposure time. Develop the autoradiogram.

5.3 Mapping by use of intertypic recombinants

This approach has been of considerable use in the mapping of genes of adenovirus and herpes simplex virus, for which intertypic recombinants can be isolated (8–11). The restriction endonuclease maps of the intertypic recombinants are first obtained. The phenotype of the protein, function, or marker to be mapped is then identified in each recombinant, and the results correlated with the restriction maps (8–12). To describe this entire technique in detail would require an entire chapter. However, the principle of this approach is described in *Protocol 10*. Readers interested in pursuing the method further are referred to references 8–12.

Protocol 10. Mapping by use of intertypic recombinants

Equipment and reagents

- Ultracentrifuge with appropriate rotors
- Ultrasonic water bath
- Gradient former (e.g. BRL)
- 10% w/v sucrose in PBS
- 40% w/v sucrose in PBS
- Waterbath
- Sterile (autoclaved) plastic microfuge tubes
- Sterile (autoclaved) plastic pipette tips
- Micropipettes (e.g. Gilson)
- Dulbecco's Modified Eagle's medium (DMEM) (*Protocol 1*)
- 0.4% w/v high-viscosity carboxymethyl cellulose in DMEM (*Protocol 3*)
- Phenol:chloroform:isoamyl alcohol, 25:24:1 v/v
- Chloroform:isoamyl alcohol, 24:1 v/v
- Ethanol
- 70% v/v ethanol

Method

1. Isolate intertypic recombinants, and determine the restriction endonuclease maps of their DNA, identifying the fragments of each parental virus and the position of the crossovers (8–12).
2. Identify in the recombinants the phenotype of the protein, function, or marker to be mapped.
3. Correlate the results of Step 1 with those from Step 2.

5.4 Sequencing

DNA sequencing is the ultimate way of mapping animal virus genes, but is outside the remit of this chapter (see Chapter 2).

References

1. Russell, W. C. (1962). *Nature*, **195**, 1028.
2. Bookout, J. B., Schaffer, P. A., Purifoy, D. J. M., and Biswal, N. (1978). *Virology*, **89**, 528.
3. Knipe, D. M., Ruyechan, W. T., and Roizman, B. (1979). *J. Virol.*, **29**, 698.

1: Extraction, purification, and characterization of virus DNA

4. Stow, N. D., Subak-Sharpe, J. H., and Wilkie, N. M. (1978). *J. Virol.*, **28**, 182.
5. Southern, E. M. (1975). *J. Mol. Biol.*, **98**, 503.
6. May, E., Maizel, J. V., and Salzman, N. P. (1977). *Proc. Natl. Acad. Sci. USA*, **74**, 496.
7. Meyer, J., Neuwald, P. D., Lai, S. P., Maizel, Jr., J. V., and Westphal, H. (1977). *J. Virol.*, **21**, 1010.
8. Williams, J., Grodzicker, T., Sharp, P. and Sambrook, J. (1975). *Cell*, **4**, 113.
9. Morse, L. S., Pereira, L., Roizman, B., and Schaffer, P. A. (1978). *J. Virol.*, **26**, 389.
10. Marsden, H. S., Stow, N. D., Preston, V. G., Timbury, M. C., and Wilkie, N. M. (1978). *J. Virol.*, **28**, 624.
11. Halliburton, I. W. (1980). *J. Gen. Virol.*, **48**, 1.
12. Halliburton, I. W., Morse, L. S., Roizman, B., and Quinn, K. E. (1980). *J. Gen. Virol.*, **49**, 235.

2

Investigation of DNA virus genome structure

MICHAEL A. SKINNER and STEPHEN M. LAIDLAW

1. Introduction

The nucleotide sequence of the DNA genome of bacteriophage φX174 (5386 bp) was determined in 1977 (1), and that of bacteriophage λ (48 502 bp) followed in 1982 (2). In these early years of sequencing, virus genomes provided the targets to push the development of nucleotide sequencing technology forward. This situation continued until the completion of the sequences of the genomes of the large DNA viruses such as the herpesviruses and the poxviruses. Notable landmarks have been the first complete sequence of a herpesvirus, varicella zoster virus (124 884 bp) in 1986 (3), and of a poxvirus, vaccinia virus (191636 bp) in 1990 (4). The intervening years have seen the determination of the complete sequences of other members of these families of large viruses, representing examples of different genera and with different hosts. Examples of these for the herpesviruses are human and equine herpes viruses I (5, 6) and II (7, 8), cytomegalovirus (9), HSV 7 (10), channel catfish virus (11); and for the poxviruses, smallpox (12), and molluscum contagiosum virus (13). The sequences of other large DNA virus genomes, such as African swine fever (14), and a baculovirus (15), have also been determined.

These projects represent various approaches to the problem; clearly there are alternative strategies that can be employed successfully. Moreover, it is no longer the viruses that provide the targets for the technological push, but the larger genomes of bacteria (16–19) and yeast (20), now superseded by the chromosomes and indeed entire genomes of eukaryotes. The tools currently available to the virus genome sequencer are therefore considerably more powerful than those used to elucidate the first sequences. We can therefore look forward to the routine determination of the sequences of a wide range of large DNA viruses with different host ranges and phenotypes, bringing new opportunities for the study of virus evolution and virus–host interactions.

Table 1. Programs, packages and online services

Title	Attribution	Web or e-mail address
BCM Search Launcher		http://www.hgsc.bcm.tmc.edu/SearchLauncher/
BLIXEM	Erik Sonnhammer, Bethesda, MD, USA (esr@ncbi.nlm.nih.gov)	ftp://ncbi.nlm.nih.gov/pub/esr/MSPcrunch+Blixem
CLUSTALW	Thompson, J. D., Heidelberg, Germany	http://www.embl-heidelberg.de/~aasland/Clustal.html
EBI		http://www2.ebi.ac.uk/services.html
GCG	Genetics Computer Group, Inc.	http://www.gcg.com
GENEMARK		http://genemark.biology.gatech.edu/GeneMark
GHOSTSCRIPT		http://www.cs.wisc.edu/~ghost/index.html
NCBI		http://www.ncbi.nlm.nih.gov/
PAUP	D. L. Swofford, Washington D.C., USA (Sinauer Associates Inc.)	orders@sinauer.com
PHYLIP	J. Felsenstein, Washington, USA	http://evolution.genetics.washington.edu/phylip.html
STADEN		http://www.mrc-lmb.cam.ac.uk/pubseq/index.html
TREEVIEW	Page, R.D.M., Glasgow, UK	http://taxonomy.zoology.gla.ac.uk/rod/treeview.html

1.1 Scope

In this chapter we will describe general approaches to genome analysis, but the bulk of the chapter will concern approaches to genome sequence determination, assembly, and analysis. We will consider the advantages and disadvantages of the various approaches, some of which may be more appropriate to particular situations. Information on sources of software, including web addresses, is provided in *Table 1*.

2. Genome mapping

Although no longer an essential prerequisite for genome sequencing, genome mapping is still useful for the comparison of strains, and for checking the homogeneity of the isolate to be used for genome sequence determination. Ideally such an isolate would, where possible, be biologically cloned to avoid problems during sequence determination and assembly. Essentially, genome mapping involves the determination of the relative order of overlapping restriction enzyme fragments generated by several different restriction enzymes. Suitable enzymes, chosen by trial and error, will ideally produce fragments in the size range between 1 and 25 kb, with minimal overlap between the different fragments. If the GC content of the genomic DNA is

2: Investigation of DNA virus genome structure

known, then restriction enzyme choice can be tailored accordingly. Individual fragments are then isolated, either physically (from agarose gels) or genetically (by subcloning in plasmid or cosmid vectors).

Overlapping fragments can then be identified by labelling the isolated fragments (released from gels or from subclones), and probing them against a library (or libraries) of subclones, or against a Southern blot of the genomic DNA digested with the different restriction enzymes. Southern blotting (*Protocol 1*) is also still the method of choice for monitoring for gross deletions or insertions.

The major problem associated with the physical approach is the overlap of different fragments in a gel, because of their similar size or the limitations in the resolution capacity of the gel systems. The use of pulsed-field electrophoresis can help with the resolution of larger fragments (*Section 2.2*) but smaller fragments may be best resolved by recourse to the subcloning approach. Subcloning is also useful in resolving the primary structure of the internal and terminal repeat regions found in the genomes of herpesviruses and poxviruses. The major limitation of the subcloning approach is the stability of the inserts, which can be influenced by size and sequence content of the insert as well as by host genotype. Plasmid vectors are suitable for routine subcloning of restriction fragments up to about 10 kb. Cosmids offer the ability to clone fragments up to 40 kb, which are difficult to resolve and isolate pure by gel electrophoresis, and can thus be very useful in establishing a map of overlapping fragments (21).

2.1 Southern blotting

With large DNA genomes, restriction enzyme analysis is difficult because of the large size range of digestion products, which makes it difficult to view small fragments reliably and obtain good size estimates for large fragments. This is because conventional methods of observing the DNA (e.g. ethidium bromide staining) are effectively dependent on the mass of DNA, not its molarity. Thus small fragments can be difficult to see without overloading the larger fragments. Southern blotting, particularly with radioactive probes, offers more flexibility in observing large and small fragments by using various different exposure times.

Protocol 1. Southern blotting of genomic DNA

Equipment and reagents

- Agarose (e.g. Type V, Sigma)
- Tris-acetate-EDTA buffer (TAE) 1×: 0.04 M Tris-acetate, 0.001 M EDTA
- Ethidium bromide (1 mg ml^{-1})
- Denaturation solution: 0.5 M NaOH, 1.5 M NaCl
- Neutralization solution: 2 M NaCl, 1 M Tris-HCl pH 7.5
- Filter membranes (Hybond-N, Amersham)
- SSC, 20×: 3 M NaCl, 0.3 M tri-sodium citrate
- Geneclean II or III (Bio 101)

Protocol 1. *Continued*

- Denhardt's solution, 1 ×: 0.2% SDS, 100 μg herring sperm DNA per ml
- Prime-It™ random primer labelling kit (Stratagene)
- Hybridization bottles and oven (Hybaid)
- Sephadex-G50 column (Pharmacia) or Nuctrap column (Stratagene)
- Hybridization solution: 2×SSC, 1× Denhardt's solution
- 3MM paper (Whatman)

A. *Preparation of blot*

1. Electrophorese 1 μg of restriction enzyme-digested genomic DNA on a 0.75% agarose submarine gel in TAE at 20 V for 16 h.

2. Stain with ethidium bromide (0.5 μg ml^{-1}) for 30 min, and photograph, with a ruler alongside, on a transilluminator.

3. Denature the gels for 30 min in denaturation solution, then neutralize for 30 min in neutralization solution.[a]

4. Soak the gel in 20 × SSC, then lay it on a wick of three sheets of 3MM paper soaked in 20 × SSC, avoiding bubbles. Lay the filter, soaked in 20 × SSC, on the gel, then apply three layers of 3MM (cut to the size of the gel and soaked in 20 × SSC). Finally apply a wad of paper towels, cut to size, followed by a 500 g weight. Allow the DNA to transfer overnight.[b]

5. Dismantle the apparatus and rinse the filter in 2 × SSC, then fix DNA by exposure to UV light (365 nm) for 5 min on a transilluminator, or by baking in an 80°C oven for 2 h.

B. *Preparation of probes*

1. Digest 2–3 μg of genomic DNA, and electrophorese as above, normally using 0.5% agarose gels in 1 × TAE.

2. Cut specific bands from the gel under long-wave UV illumination, and then purify using Geneclean™, following the manufacturer's protocol.

3. Label 25 ng of purified fragments with 0.05 mCi (1.85 MBq) α-^{32}P-dCTP (3000 Ci mmol^{-1}, 110 TBq mmol^{-1}) using Prime-It™ (follow the manufacturer's instructions), and separate from unincorporated nucleotides on a Sephadex-G50 or Nuctrap column.

C. *Hybridization of probe to blot*[c]

1. Pre-hybridize the filters at 65°C in hybridization solution for 1 h, and then add all of the prepared probe (see Step B3). Hybridization is allowed to proceed at 65°C overnight.[d]

2. Wash the filters three times for 20 min each in 2 × SSC, 0.1% SDS, at 42°C, and then expose to X-ray film (preflashed and with intensifier screens if necessary).

^a For fragments >10 kbp it is recommended that a depurination step is carried out in 0.125 M HCl for 15 min prior to denaturation, to improve transfer.
^b It is possible to transfer DNA from one gel to two filters by sandwiching the gel between two filters and omitting the use of a wick.
^c Blots can be re-used by heating them to 100°C in 0.1% SDS, 0.1× SSC, for 2 min, then leaving them to cool to room temperature.
^d Hybridization is most conveniently performed in specialist glass bottles in a hybridization oven, as supplied by Hybaid.

2.2 Pulsed-field agarose gel electrophoresis

To facilitate the generation of a genome map, better separation of larger restriction enzyme fragments ($\geq$ 20 kb) for Southern blotting can be achieved by pulsed-field agarose gel electrophoresis (PFGE). For analysis of *Pst*I, *Bam*HI and *Hin*dIII fragments of fowlpox virus between 20 and 50 kb, Mockett *et al* (22) performed PFGE, using a CHEF DR II apparatus (Bio-Rad), through 1% Seakem ME agarose (FMC) in 0.5 × TBE buffer, at 10°C, 180 V, for 22 h using 1 to 5 s pulses. Southern blotting can then be performed as described in *Protocol 1*. The improvement in resolution is illustrated in *Figure 1*.

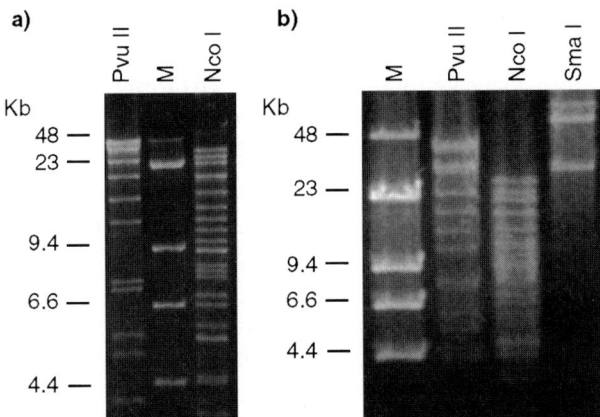

Figure 1. Separation of restriction enzyme digestion products of fowlpox virus genomic DNA by conventional (a) or pulsed field (b) agarose electrophoresis. Products were viewed by UV illumination of ethidium bromide-stained gels. Pulsed field electrophoresis was as described in section 2.2., using 1 to 5 s pulses. Note the increased resolution of the large *Pvu*II fragments by pulsed field electrophoresis and the separation of the *Sma*I fragments above the 48 kbp marker. Markers (M) used in all cases are λ DNA uncut and digested with *Hin*dIII, giving sizes shown.

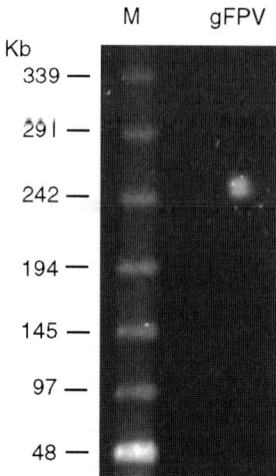

Figure 2. Pulsed-field gel electrophoresis of fowlpox virus genomic DNA (gFPV) following lysis of purified virions embedded in agarose (see *Protocol 2*). Marker (M) sizes are as shown.

2.2.1 Pulsed-field agarose gel electrophoresis of genomic DNA

Pulsed-field agarose gel electrophoresis can also be useful for obtaining a good estimate of the size of the viral genome. Where it proves impossible to extract intact genomic DNA, due to its large size, it may be possible to release it from purified virus by lysis *in situ* in an agarose plug, using a modification of the method of Schwartz and Cantor (23). Such an approach (*Protocol 2*) was used to release fowlpox virus genomic DNA (256 kb) for pulsed-field agarose gel electrophoresis (22) to allow a direct estimate of the genome size, for comparison with the map derived by restriction enzyme mapping (*Figure 2*).

Protocol 2. Pulsed-field agarose gel electrophoresis of poxvirus genomic DNA

Equipment and reagents

- Pulsed-field gel electrophoresis apparatus (e.g. CHEF DR II apparatus, Bio-Rad)
- Resuspension buffer: 10 mM Tris-HCl pH 7.5, 125 mM EDTA, 1% sarkosyl, 10% β-mercaptoethanol
- Tris-borate-EDTA (TBE) buffer: 0.09 M Tris-borate pH 8.3, 0.002 M EDTA
- Low melting-point agarose (BRL): 1% in 10 mM Tris-HCl pH 7.5, 125 mM EDTA, 1% sarkosyl, held at 50°C
- Digestion buffer: 10 mM Tris-HCl pH 7.5, 125 mM EDTA, 1% sarkosyl, 5% 2-mercaptoethanol, proteinase K, 1 mg ml^{-1}

Method

1. Pellet gradient-purified virus (2 × 10^8 PFU) by centrifugation at 160 000 *g* in a swing-out rotor, and resuspend in 20 μl resuspension buffer.

2: Investigation of DNA virus genome structure

2. Prewarm the virus suspension to 37 °C, then add 60 µl molten low melting-point agarose.
3. Suspend the virus plug in 0.5 ml of digestion buffer, and incubate overnight at 37 °C.
4. Pellet the virus in agarose by centrifugation at 6500 r.p.m. for 5 min in a microfuge, and then wash repeatedly in 0.5 ml 50 mM EDTA, pH 8.0, for 15 min at room temperature.
5. Drain the pellet, cut it up, and place it into two wells of a 1% agarose gel.
6. Seal the pellet into the wells using molten agarose, then electrophorese the samples in 0.5 × TBE buffer, at 10 °C, 160 V, for 36 h, using 20–50 s pulses.[a]
7. Stain the gel in ethidium bromide (0.5 µg ml^{-1}) for 30 min prior to viewing on a transilluminator.

[a] For pulsed-field electrophoresis of genomic DNA, λ DNA concatamers (New England Biolabs) are useful as size markers.

3. Correlation of genotype with phenotype by marker rescue

In the investigation of genome structure, there is clearly a need to correlate phenotype with genotype. Marker rescue is a powerful genetic tool for the identification and localization of genes and mutations.

3.1 Design of marker rescue

Marker rescue can be used in a transient complementation form for the identification of *trans*-acting factors (i.e. gene products), or in a permanent recombination form for the identification of both *trans*- and *cis*-acting functions (e.g. regulator sequences, intragenic mutations). Essentially it is marker rescue (in its permanent recombination form) that is used to introduce foreign genes into recombinant poxviruses, herpesviruses, and baculoviruses by recombination. Functions that can be identified can be essential (if conditional mutants are available, e.g. temperature-sensitive mutants) or non-essential (if they can be readily selected or screened for).

Caution should be exercised when using transient complementation for marker rescue with poxviruses. The poxvirus replication cycle is such that early gene expression takes place within the virion core. It is therefore not possible to obtain good temporal expression of poxvirus genes under the control of early promoters in a transient system (some expression may be observed due to late reactivation of some early promoters).

When marker rescue is used to locate viral genes, whole (though not necessarily intact) genome should be used as a positive control donor (with 'recipient' genome providing a useful 'negative' control). The specific gene can then be localized by progressively narrowing down the source of the donor DNA, starting with pools of clones from a library or of size-selected restriction enzyme fragments, then progressing to individual clones or fragments.

3.1.1 Transfection

Transfection protocols for the introduction of DNA into the infected (or to-be-infected) cell will depend on the cell type being used for infection. A wide range of options is available, including electroporation, lipofection, and calcium phosphate/DEAE transfection. For fowlpox virus in chick embryo fibroblasts, we find Lipofectin (Life Technologies) to be easiest and most reliable, if not the most efficient, though this may be a function of the inherent heterogeneity between preparations of the fibroblasts (see Chapter 3).

3.1.2 Selection

Selection protocols will depend on the nature of the marker being rescued, including temperature selection (for rescue of temperature-sensitive mutations), drug selection, or production of chromophores. Whatever the selection, it is important to include relevant controls, such as for the generation of spontaneous revertants of mutants. If the rescued virus is to be used in subsequent experiments, it is also essential to purify it (by plaque purification or limiting dilution) and to try to ensure its purity by molecular analysis. If PCR is used for the latter, it is important to screen for both the absence and the presence of the parental genotype.

4. Nucleotide sequencing

The most favoured method for sequence determination is the dideoxynucleotide or chain-termination method, developed by Sanger (24), applied to sequencing randomly cloned segments of DNA, as used for the determination of the complete bacteriophage λ sequence (2). In the purest form of this strategy, short random fragments are sequenced and assembled into groups, or 'contigs' (25), of overlapping sequences. Early in the project, the number of contigs increases rapidly as overlaps between sequence reads are rare (in the sequence database, a contig can be just one reading). As the project progresses, new readings normally either enter existing contigs or overlap two existing contigs, allowing them to be joined (so the number of contigs decreases). The challenge then is to join all the contigs until there remains only one, representing the complete target sequence. In practice, other contigs also remain, normally with just single readings, representing contaminating DNA sequences (host DNA, bacterial genomic DNA, or vector DNA, depending on the approaches used). Closing the final gaps between contigs can be costly and time consuming unless alternative approaches are introduced to complete the project.

4.1 Primer walking

An alternative approach to random, or 'shotgun', sequencing is that of primer walking, as applied to the determination of the genome of the poxvirus molluscum contagiosum (13). In this approach, randomly cloned segments of DNA are sequenced using the chain-terminator method, initially with primers specific to vector sequences flanking the cloned insert. New, virus-specific primers are then synthesized, based on the results of the previous round of sequencing, and are used to extend the sequence on the cloned templates.

In practice, most sequencing projects will use a blend of both approaches: the random approach being used early to build up a large sequence database (consisting of several contigs), and primer walking being used to close the gaps between contigs. Cloning of templates to close gaps can also be achieved by screening library clones with probes from template clones flanking the gaps, but is more conveniently performed by PCR.

Traditionally, the process of sequencing large virus DNA genomes started with obtaining a digest map of the genome for several restriction enzymes. Appropriate enzymes were then chosen to permit cloning of the genome as a library of large fragments (up to approximately 40 kbp) in plasmid or cosmid vectors. Each cloned fragment would then be sub-cloned, randomly or non-randomly, into bacteriophage or plasmid vectors, for sequencing by the chain-terminator method. As a strategy for whole genome sequencing, this approach is wasteful in that the additional effort of gap closing has to be applied repeatedly to each individual clone, rather than once to the genome sequence as a whole. It was, however, a requirement enforced by the limitations of available computing power. Now computing power does not impose such constraints for sequences the size of viral genomes, so they can be handled as individual projects. Indeed, the nucleotide sequence (1.8 Mbp) of the genome of the bacterium *Haemophilus influenzae* was determined as a single random sequencing project (26). The mapping and cloning approach may still be useful, however, for projects where the aim is to sequence only part of the genome.

4.2 Library construction
4.2.1 Random cloning
Random cloning of genomic DNA fragments, generated by sonication or by other physical shearing methods, offers generation of representative libraries with considerable flexibility in the size of fragment cloned. This can be an important consideration if the random library is to be used after the sequencing project, perhaps for the generation of deletion or insertion mutants by reverse

PCR and recombination. It also has the advantage that the number of libraries that need to be generated can be minimized. The main potential disadvantage is that clones in which two non-contiguous fragments of genome have been inadvertently joined are not readily identified prior to sequence assembly. Care has to be taken, therefore, not to 'force' the ligation reaction with high insert-to-vector ratios. Size selection of fragments for cloning with a minimum size more than twice the average read length will minimize the consequences of such events (unless primer walking is to be performed to complete the clones).

4.2.2 Enzyme digestion

Cloning of restriction enzyme fragments has the advantage that fragments inadvertently and incorrectly joined during the cloning procedure can readily be identified by screening for recognition sites of the relevant restriction enzyme. Any such sequences identified can then be split at the site and re-entered. The main disadvantage of restriction enzyme fragment cloning is that the library is rarely fully representative, mainly due to size constraints, but also due to sequence content. This necessitates the generation of several libraries using different restriction enzymes. The method is also relatively inflexible. Although enzymes with tetranucleotide recognition sites have been used, in general they generate fragments too small (on average 256 bp) for automated sequencing. Partial digestion products can be used (as for the generation of genomic libraries in bacteriophage λ), but then the advantage of being able to identify clones with more than one fragment, by screening for restriction enzyme sites, is lost.

4.2.3 Sonication and end repair of genomic DNA

Cleavage of target DNA by sonication (27), using a cup horn sonicator or a sonicating waterbath, is probably the best way of generating a random set of overlapping templates for cloning (see *Protocol 3*). DNA sheared by sonication has to be 'repaired' to provide blunt ends for efficient cloning. A mixture of Klenow polymerase and T4 polymerase is generally considered to be most effective.

Protocol 3. Sonication and end repair of genomic DNA

Equipment and reagents
- Sonicator (e.g. W-375, Heat Systems Ultrasonics, Inc.)
- Klenow polymerase (5 units μl^{-1}), reaction buffer, and enzyme grade BSA (NEB)
- Deoxynucleotides, dNTPs, 2.5 mM (Pharmacia)
- T4 DNA polymerase (3 units μl^{-1}) (NEB).

A. *Sonication of genomic DNA*
1. Place 50 μg of genomic DNA (in a volume of 50 μl) into a sterile microfuge tube.

2: Investigation of DNA virus genome structure

2. Sonicate the DNA using either a cup horn sonicator or a sonicating waterbath, for approximately 15–30 s, depending on the strength of sonication.[a]

3. Check the size of the fragments following sonication by running 2 μl on a 2% agarose gel, alongside size markers.

B. End repair of sonicated DNA

1. To 25 μl of sonicated DNA, add 5 μl of Klenow reaction buffer, 1 μl dNTPs, 0.6 μl 100 × BSA, 13.4 μl of deionized water, 2.0 μl Klenow polymerase, and 3.0 μl T4 DNA polymerase.

2. Incubate at 16°C overnight.[b]

3. Heat inactivate at 75°C for 15 min.

[a] The power of the sonicator can be calibrated using a sample of genomic DNA at the same concentration, or by using a commercial source of DNA, e.g. herring sperm DNA (Boehringer).
[b] Shorter incubation times (15–30 min) can be used at 37°C, but do not prolong this as exhaustion of dNTPs in the reaction will result in rapid exonucleolytic damage to the template DNA by the T4 DNA polymerase.

4.2.4 Size fractionation

Following end repair, the genomic DNA fragments can be size-selected by agarose-gel electrophoresis, simultaneously purifying the fragments from excess dNTPs in the end-repair reaction. DNA can be size-selected by subjecting it to agarose-gel electrophoresis, then cutting out blocks of agarose to span the desired size range(s). The DNA has then to be removed from the agarose by methods including electroelution or glass milk purification. An easier, more flexible, and more efficient method is described in *Protocol 4*.

Protocol 4. Size fractionation of DNA

Equipment and reagents

- Tris-borate-EDTA buffer (TBE) (*Protocol 2*)
- Phenol:chloroform:isoamyl alcohol; 25:24:1
- Submarine mini-gel electrophoresis kit (e.g. Anachem H1 Set)
- Glycerol loading buffer: 50% v/v glycerol, 125 mM EDTA, 0.1% SDS, Bromophenol blue, Xylene cyanol

Method

1. Prepare a 1.5% agarose gel in TBE, with 1 cm wide, thick wells.

2. Add 5 μl of glycerol loading buffer to the end-repaired DNA (*Protocol 3*), and load all of the sample in one lane in the centre of the gel.

3. Run the gel at 30 V until the bromophenol blue has run approximately 1.5 cm.

Protocol 4. *Continued*

4. Carefully pour off all the buffer, and dry the surface of the gel.
5. Using a scalpel, cut a well (called the collecting well) into the gel, slightly wider than the loading well, immediately above the bromophenol blue band and in line with the loading well.
6. Pour TBE buffer into the reservoirs at either end of the gel, ensuring the buffer does not rise above the upper surface of the gel. Pipette TBE into all of the loading wells, and 50 µl of TBE into the collecting well.
7. Replace the lid on the electrophoresis kit, and run the gel for 40 s at 30 mA.
8. Remove the 50 µl of buffer in the collecting well to a microfuge tube, and replace it with 50 µl of fresh TBE buffer.
9. Replace the lid of the electrophoresis kit, and run the gel for a further 40 s at 30 mA.
10. Repeat this procedure until ten 50 µl aliquots have been collected in one microfuge tube. Label this tube fraction 1.
11. Repeat until four fractions have been collected.
12. Add 500 µl (an equal volume) of phenol/chloroform to each fraction, vortex for 1 min, and centrifuge for 5 min in a microfuge. Remove the top aqueous layer into a microfuge tube, and precipitate with ethanol by adding one-tenth volume of 3 M sodium acetate, pH 5.2, and 2.5 volumes of absolute alcohol. Precipitate the DNA by incubating in a dry-ice bath for 10 min, followed by 10 min centrifugation in a microfuge. Wash the pellet with 70% alcohol.
13. Dry the pellet, and resuspend each fraction in 50 µl TE.
14. Run 5–10 µl of each fraction on a 1% TBE-agarose gel to check the size range of each of the fractions.

4.2.5 Cloning

After size selection, the end-repaired DNA can be cloned into a suitable sequencing vector. Until relatively recently, M13 vectors were preferred, facilitating the production of single-stranded template DNA for manual sequencing with Klenow polymerase or with Sequenase™. With the advent of cycle sequencing methods for sequencing double-stranded DNA templates, plasmid vectors are more commonly used, as the time required for the preparation of template DNA is considerably shorter.

In general it is preferable to clone into dephosphorylated vectors to prevent self-ligation, particularly if blunt-ended fragments are being cloned (as with repaired, sonicated DNA). Where the vector has a system for the selection of inserts, non-dephosphorylated vectors could be used. Suitable vector DNAs may be purchased commercially (e.g. pUC18/Sma, dephosphorylated, Pharmacia), or prepared in the laboratory. Especially in the latter case, extensive controls should be used to ensure that the vector has appropriate termini (using ligation with a control insert such as *Hae*III-digested ϕX174, available commercially). To ensure a high efficiency of ligation we have found that using high concentration ligase (available from NEB) at 3 Weiss units per microlitre of reaction volume gives the best results.

4.2.6 Screening of transformants

Initial screening of transformants can be carried out using methods such as the 'blue/white selection' found in the pUC plasmids (28). Particularly in the case of targeted cloning of restriction fragments, care should be taken as functional LacZ fusions can be produced even with long, in-frame extensions, so positive clones can be missed. Colony hybridization using a total genomic probe is an alternative method for selection of clones, either where the frequency of clones with inserts is low, or where the library is contaminated by, for example, cellular sequences.

Clones can then be picked manually into 2-millilitre broth cultures, either selectively or randomly, for overnight incubation (in the presence of the appropriate antibiotic). Those with access to a suitable robot may prefer to use it to pick the colonies into microtitre plates. Following overnight growth of the culture, 2 μl can be screened by standard PCR, using the M13 forward and reverse primers. This procedure will allow an approximate insert size to be determined for each transformant, allowing the rejection of transformants with insert sizes below 1 kb, before DNA is isolated from the culture.

4.3 Minipreparation of sequence grade DNA

Plasmid DNA prepared using the alkaline lysis method is suitable for DNA sequencing.

Protocol 5. Alkaline lysis DNA miniprep

Equipment and reagents

- GTE buffer: 50 mM glucose, 25 mM Tris pH 8.0, 10 mM EDTA
- NaOH/SDS: 0.2 M NaOH, 1% SDS
- TE buffer: 10 mM Tris, 1 mM EDTA
- Cold ethanol (−20°C)
- Potassium acetate: 60 ml 5 M potassium acetate, adjusted to pH 4.8 with glacial acetic acid, and made up to 100 ml
- Cold 70% ethanol (−20°C)

Method

1. Grow 1.5–2 ml overnight cultures in broth plus appropriate antibiotic.
2. Transfer the cells to microfuge tubes, and pellet by spinning for 1 min. Discard the medium.

Protocol 5. *Continued*

3. Resuspend the cells in 75 μl cold GTE buffer.
4. Add 180 μl NaOH/SDS, and vortex mix.
5. Add 135 μl cold potassium acetate and vortex.
6. Spin down in microfuge for 10 min at 4°C.
7. Transfer the supernatant to fresh tubes containing 0.4 ml TE-saturated phenol, and extract.
8. Remove the aqueous phase to fresh microfuge tubes, and extract with 0.4 ml chloroform.
9. Fill the tubes with cold ethanol, and precipitate at −20°C or on dry ice for 10 min.
10. Spin down the DNA in a microfuge for 10 min.
11. Wash the pellets with cold 70% ethanol. Respin if necessary.
12. Remove the ethanol and dry the pellets briefly under vacuum.
13. Redissolve the DNA in 25 μl TE containing 50 mg ml^{-1} RNAase A. Use 1–5 μl for restriction digests.

Due to the large numbers of transformants that are processed together it is easier, although costlier, to use one of the kits currently on the market. Most of these kits use a resin to bind the DNA, which can result in DNA being eluted in high-concentration salt. As the sequencing reaction is sensitive to salt concentration, it is advisable to purify the DNA by precipitating it with polyethylene glycol (PEG) before sequencing. If commercial minipreps are not used, then PEG precipitation should always be carried out.

Protocol 6. Polyethylene glycol precipitation of DNA

Equipment and reagents
- PEG solution: 30% polyethylene glycol 8000, Sigma, 1.6 M NaCl
- TE buffer (*Protocol 5*)

Method

1. Add 0.4 ml of PEG solution for every 1 ml of DNA solution.
2. Incubate at 4°C for 1–16 h (an overnight incubation is better).
3. Spin in a microfuge for 20 min at 12 000 r.p.m. (usually 10 000 g).
4. Remove all the PEG solution by aspiration, and resuspend the pellet in deionized water or TE buffer.

5. Precipitate with ethanol by adding one-tenth volume of 3 M sodium acetate, pH 5.2, and 2.5 volumes of absolute alcohol. Precipitate the DNA by incubating in a dry ice bath for 10 min followed by 10 min centrifugation in a microfuge. Wash the pellet with 70% alcohol.
6. Resuspend in deionized water or TE.

4.4 Sequence determination
4.4.1 Random versus targeted sequencing

Random sequencing requires just one or two primers (which can therefore be non-conjugated deoxynucleotide primers, or dye-conjugated deoxynucleotide primers), and is amenable to high automated throughput with minimal operator analysis required until later in the project. Targeted sequencing (by primer walking) requires continuous design and synthesis of custom oligonucleotides. For most budgets, dye-conjugated primers would not therefore be an option. Programs are available, however, to 'suggest' suitable primer sequences to extend existing reads. In general though, it must be appreciated that the primers used in walking will frequently be less successful than the universal vector primers, at least unless the conditions are optimized for each primer.

The random sequence method can seem wasteful, in that it generates considerable redundancy of sequence data, but that very redundancy both improves the reliability of the sequence and reduces the amount of time that needs to be spent comparing readings manually to resolve ambiguities. The amount of redundancy required to sequence a genome to near-completion is illustrated in *Table 2*. Five- to sixfold coverage allows more than 99% of the

Table 2. Shotgun sequence statistics

Coverage, m	Number of reads,[a] n	Unsequenced,[b] P_0 (%)	Total gap size,[c] G (b)	Average gap size,[d] g (b)	Average number of gaps,[e] N
1	683	36.8	94177	375	251
2	1365	13.58	34646	188	185
3	2048	5	12745	125	102
4	2731	1.8	4689	94	50
5	3413	0.7	1725	75	23
6	4096	0.25	635	63	10
7	4779	0.1	233	54	4
8	5461	0.03	86	47	2
9	6144	0.01	32	42	1

Following the approach of Lander and Waterman (29), as described by Fleischmann *et al.* (26).
[a] Calculated as n = mL l^{-1}, where L (genome length) = 256 kbp and l (average read length) = 375 bp
[b] Calculated as $P_0 = e^{-m}$
[c] Calculated as $G = Le^{-m}$
[d] Calculated as g = L/n
[e] Calculated as N = G/g

genome to be sequenced with some redundancy on both strands. Strategies for reducing excess redundancy are discussed in Section 5.1.

In practice, the appropriate balance between random and targeted sequencing will depend on the reasons for the sequence determination. If it is to provide a scan of genes within the genome, with sequence accuracy not being a major prerequisite, then the balance can err towards the targeted approach. If higher levels of accuracy are required however, the balance should be skewed towards the random approach with five- to sixfold redundancy.

4.4.2 Manual versus automated sequencing

For overall efficiency, automated sequencing using fluorescent primers is now the method of choice, but this does not mean that manual methods cannot or should not be used where access to automated sequencers is not available. Many sequencing projects were completed using manual sequencing and sequence assembly programs, such as the original STADEN program package (30). Manual data may still be entered into the newer programs, including the latest STADEN program packages (31), taking advantage of most of the features except base accuracy estimation and fluorescent sequence trace display. Methods for data entry will require consideration, as the once ubiquitous digitizers and entry software are now a rare commodity.

4.4.3 Fluorescent sequencing

Of the sequencing approaches amenable to automation, fluorescent sequencing is the most widely available. Dye primer sequencing is capable of providing more reproducible, higher quality sequence data, resulting in longer effective reads. The main advantage of dye terminator sequencing is that the dye is coupled to the dideoxynucleotide terminator. Each dye is coupled to a different terminator, with different emission spectra, so that all four termination reactions can be performed in the same tube and loaded on the same gel lane. Although primers can also be labelled with different fluorochromes, the resulting termination reactions have to be performed separately, and then combined for electrophoresis. We currently sequence by cycle sequencing using Thermo Sequenase dye terminator kits (v2.0) from Amersham, using the manufacturer's protocol.

4.5 Data collection

The ABI 373 sequencer produces two files for each sequence: one a binary ABI sequence source file, which can be submitted straight to an analysis program, and the other a plain text file containing just the resulting machine-interpreted, or 'called', sequence. The sample names can be entered when the sequence run is started, and are thus embedded within the ABI sequence source file. This is a useful feature, which allows each reading to carry a unique identifier. Unfortunately, in our experience with the ABI 373 sequencer, tracking can be poor, often resulting in misnamed readings. After electrophoresis is complete, therefore, the tracking of samples is checked and the sequence source files are named manually. Subsequently it is this manually entered name, rather than the embedded name, that is read by PREGAP (Section 5.1). It is our understanding that more reliable tracking programs are available for newer ABI sequencers. The embedded sequence name can also cause problems with GAP4 (Section 5.1), which will not accept entries with identical names.

Where curiosity dictates, sequences from the raw text files, excluding vector sequences and up to the point where the frequency of ambiguous bases (N)s begins to increase, can be submitted for database searches against the non-redundant protein databases, using BLASTX, or against the non-redundant nucleotide databases, using BLASTN, although this is not necessary.

5. Sequence assembly

Various programs are available for sequence assembly (ABI, GCG, etc.). We shall describe use of the STADEN programs (31). The predecessors of these programs have been used extensively for manual shotgun sequencing of viral and other genomes (30). With a long pedigree, they have been developed over many years in close collaboration with those practising high-throughput shotgun sequencing. The STADEN programs are designed for automated entry of sequences and minimal user intervention. They will mark vector sequences for exclusion, screen for clones containing vector sequences, and mask regions of low reliability at the ends of reads. They also keep track of the reliability of sequence data throughout the project. There are many safeguards built in, based on years of practical experience, to minimize the risk of inadvertently corrupting the sequence data. The databases will also prove useful as a way of keeping track of the location of individual genomic clones used as sequence templates, clones which may well prove to be a valuable experimental resource.

5.1 The Staden package

When STADEN programs are described, we refer to STADEN 97. A new package (STADEN 98) was released in May 1998. These programs run on the UNIX operating system, but can be accessed from PCs or Macintoshes™ using X-window emulators. Extensive online help files are available (see *Table 1*) and can be downloaded for faster local access.

5.1.1 Pre-existing sequences

It may well be the case that pre-existing sequence data exist for parts of the genome, probably gathered by manual sequencing. As long as the sequence is reliable there is no reason why it should not be entered as raw sequence. Such a sequence could be entered as individual, edited gel readings from an old database (using the GELSOUT function of earlier versions of STADEN), or as a preassembled sequence. There is a default limit of 4096 bp for entry into STADEN, but larger sequences can be readily split using the BREAKUP program of GCG9. The command line switch '-SEG=4000 -OVE=50' creates segments starting every 4000 bp, with 50 bp overlaps for convenient entry into STADEN.

5.1.2 Automated sequence entry

Sequence entry involves three main steps:

1. Transfer of the ABI source sequence files to the UNIX host for the STADEN programs, and generation of a lookup table carrying information on the sequence readings (Section 5.1.3).
2. Processing the ABI source sequence files through PREGAP to generate '.scf' and '.exp' files (Section 5.1.3). After this process, the ABI source files can be deleted as the smaller '.scf' files carry all the necessary information.
3. Batch entry of the sequence files into the STADEN database using the 'normal shotgun assembly' mode of GAP4.

5.1.3 Database of readings

PREGAP has the facility to input information regarding a batch of sequences from a lookup table, such as that illustrated in *Table 3*. Such a table can easily be automated in a spreadsheet program to minimize operator input.

(i) *PREGAP* (PREGAP 4 under STADEN 98)
In converting the data from the ABI sequence source file (also from Pharmacia A.L.F. and LiCor sequencer machines and from plain text files) to '.scf' and '.exp' files for entry into the database by GAP4, PREGAP performs a series of tasks, controlled by system and user level configuration 'pregaprc'

Table 3. A lookup table for use by PREGAP

#ID	TN	PR	SP	SI
1007F	1007	1	−41	500..2000
1007R	1007	2	28	500..2000
1027F	1027	1	−41	500..2000
1027R	1027	2	28	500..2000

Sequence read names are copied into the #ID column and the spreadsheet generates the template name (TN), a code (PR) describing the primer (1 for Forward, 2 for Reverse; other options are available), the distance from the primer to the vector cloning site (SP) and the size range of inserts selected for the clones (SI). If multiple sequence or cloning vectors are used, it is also useful to enter their names.

2: Investigation of DNA virus genome structure

files as well as by prompted responses. Additional data on cloning vectors and primers is supplied from the lookup table, or by the user interactively. These tasks include:

1. Estimating and recording the accuracy of base calls (obviously not available for plain text entry from manual sequencing).
2. Masking of vector sequences and unreliable 3' sequences (importantly, although these processes are described as 'clipping', sequence is not removed, only labelled).
3. Screening against clones containing vector-derived inserts or specified repeat sequences (such reads are reported as 'fails' and would not routinely go forward for entry into the database).
4. Generating list files of passed sequences (for entry into the database by GAP4) and of failed sequences.

(ii) *GAP4*

Before entering data, a database has to be created ('new' under the file menu in GAP4). Database size is not a problem, as it will be set and increased automatically as required. At subsequent rounds it will be advisable to make a copy of the database before each new round of entries. GAP4 takes the list file (or 'file of file names'), generated by PREGAP, to input the '.exp' files of passed readings into the database. It is advisable to collect a log of the assembly process for each round, using 'redirect' in GAP4 (but do not forget to 'close' the log file afterwards).

Assembly is initiated from the assembly menu, using 'normal assembly'. We start with the default parameters for 'initial match' to find overlaps, and 'maximum pads' and 'maximum percent mismatch' for reading entry. We routinely select 'show all alignments' and 'reject failures' from the shotgun palette. Entry of failures into the database may, however, be a safer and more manageable option. This is because failure at entry may not be due to problems with the new entry but with earlier entries. If failed entries are indeed rejected, perhaps to reduce the final number of single gel contigs, they must be noted, stored and checked subsequently against a more 'mature' database.

Summaries of results can be obtained from the View menu (and can be stored as part of the log file). 'Database information' reports the essential statistics to monitor progress of the project as a whole. 'Show relationships' lists the order of readings in contigs (answering 'yes' to 'show readings in positional order' in the show_relationships palette), or lists the readings in order of entry, supplying information on their size and relative position in contigs (answering 'no' to the same question).

The layout of templates can be seen graphically using 'show templates' (*Figure 3*). In this window, the sequence reads are colour-coded according to the primers used: red for unknown primers, cyan for forward primers, grey for

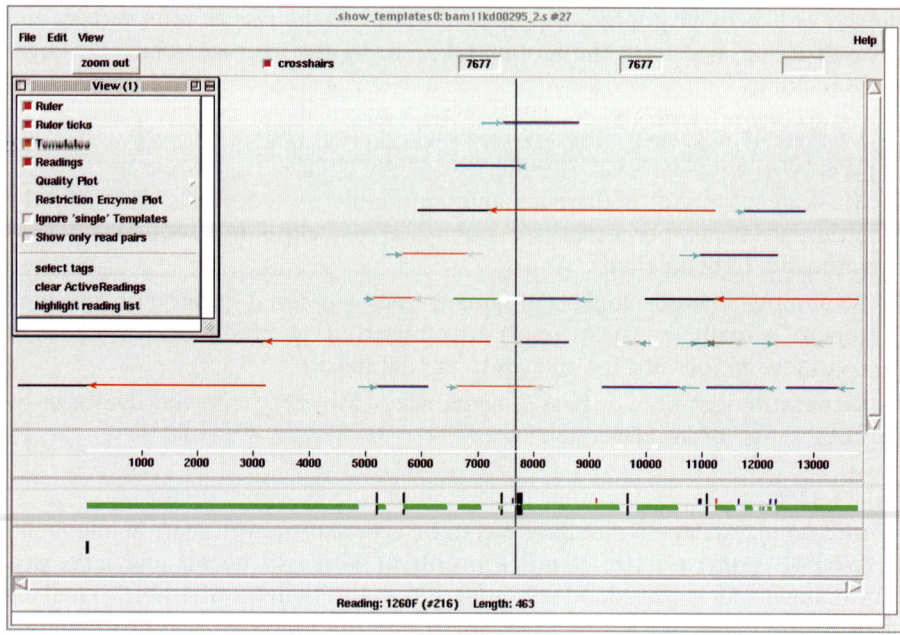

Figure 3. The 'show templates' display in GAP4 (see text).

reverse primers, dark cyan for custom forward primers, and dark grey for custom reverse primers.

Because sequence reads can be associated with their templates in GAP4, they are displayed linked to the appropriate template, represented by lines of the length specified during PREGAP (the mean of the range specified as SI in the lookup table). Templates are displayed in blue if sequenced from only one end. If they have been sequenced from both ends, they will be pink (if both ends are in the same contig), green (if the ends are in different contigs), or black (if the program identifies a problem with the template, such as a discrepancy in its size or read orientations). When the ends of templates are in different contigs and more than one contig is displayed, templates can also be yellow (if the readings are in adjacent contigs, and if the size and orientation of the template and contigs are consistent) or dark yellow (if the previous criteria are not met). In the latter case, reordering or reorientation of contigs may convert a dark yellow template to a yellow one, indicating that contigs are in the correct relative order and orientation.

It should be clear that the linkage of reads to their templates is a particularly powerful tool, particularly where two different size classes of clones have been used. Smaller clones (1 to 2 kb) can be used to build up contigs, and larger ones (up to 10 kb) can be used to order and arrange contigs. They could also be used for subsequent targeted closure of gaps by primer walking. The

2: Investigation of DNA virus genome structure

display can be configured to show only templates or only readings, to ignore templates, and to plot restriction enzyme sites and sequence quality (also colour-coded). This window is dynamically linked to the 'contig editor' window (*Figure 4*) which can also be opened from the Edit menu of GAP4.

5.1.4 Editing contigs

In the contig editor window (*Figure 4*) we can see some of the features that make STADEN such a powerful package. Flanking vector sequences and unreliable sequences at the end of readings are 'clipped' from the reading, not by permanent removal but by 'tagging'. In the contig editor display, vector sequence is marked as grey on pink, and unreliable terminal sequence (assessed by the abundance of Ns in the read) as grey. These sequences will have been ignored in the contig assembly process. They can, however, still be viewed, or used to search for joins with other contigs. They can also be partially or completely 'unmasked' during editing. The programs also make an assessment of the quality of the read for each called base. Bases failing to meet a selected quality figure (Q) are shown in red. Adjusting the consensus cut-off value (C) alters the way in which the consensus sequence reflects discrepancies between the individual reads. It can be set so that the consensus only reflects unanimity (C = 100) or to accept a majority verdict (C > 50, with Q = −1). Adjusting the Q value can then be used to exclude from the consensus determination any bases that do not meet the selected quality criterion.

The 'next problem' button moves the display to the next position where the consensus cannot be resolved according to the consensus and quality criteria

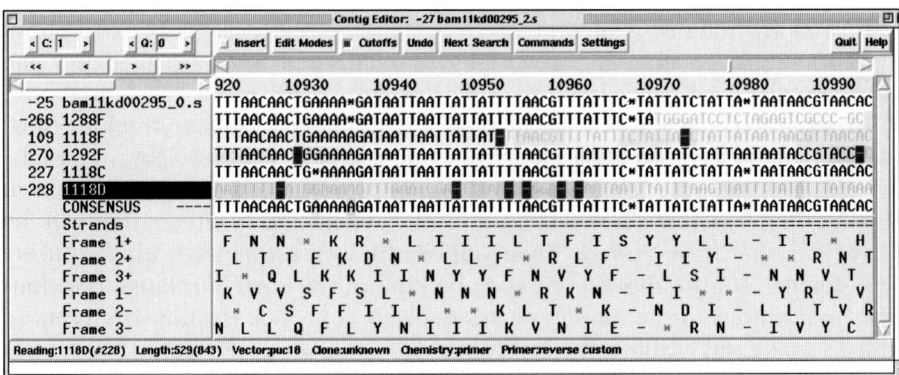

Figure 4. The 'contig editor' window of GAP4. Various optional features of this window can be observed in this example. Vector sequence is greyed-out and highlighted in pink (*e.g.* right hand end of 1288F). Low quality 3' sequence is also greyed-out (1118F & 1118D). The estimated quality of reads is also indicated by the strength of grey background shading (light is high quality). The plot also shows that readings representing both strands are present (=) and the translations of all reading frames are shown.

selected. Conflicts can be resolved by editing the 'incorrect' base or, conversely, by setting its quality to zero. The latter may be preferred, as it maintains the original call but, even in the former method, the edited base can always be identified as its quality is automatically set to 100 (it is also a good idea to put the replacement base in lowercase). By default, more 'risky' edits are not permitted. Editing of the sequences or the consensus is, however, controlled by checkboxes under the Insert and Superedit menus.

Another useful feature of GAP4 is that particular sequence features can be tagged, and translations can be viewed in all reading frames (see *Figure 4*). All of these features, and many more which cannot be described here, make the GAP4 editor a powerful and relatively safe tool. It must be remembered, though, that the greatest requirement for relatively trouble-free assembly of reliable data is good quality sequence data. Used under default parameters, GAP4 will exclude poor quality data. It would be rash to undermine the protection provided by relaxing those parameters early in an assembly project. Failed gels can always be compared to the database, and possibly even entered, at a later stage, to assist in editing or joining.

The contig editor display is, in turn, 'hot-linked' to the relevant trace displays, enabling resolution of conflicts. Double clicking with the middle mouse button on a reading will bring up the appropriate trace, centred on the selected base. Double clicking on the consensus automatically displays up to four aligned traces, which can be locked so that scrolling in one will scroll the rest (*Figure 5*).

At early stages of the project, the editor window should only be used to monitor progress of the project. Only when sufficient data have been accumulated to allow highlighting of problems should editing be undertaken. During the later stages of a sequencing project, GAP4 can be used to suggest templates for long runs or primers for extending readings, to be used either within contigs (to improve coverage), or near their ends (to facilitate gap closure).

A particularly useful feature of GAP4 is the ability to keep track of 'read pairs', that is, sequences derived from opposite ends of the same template. This is very valuable for identifying templates useful for contig extension or joining. Conversely it can also be useful to exclude read partners that will fall in regions that have already been sufficiently well sequenced, thus reducing redundancy. It also allows contigs to be kept in the correct relative positions and orientations in the database (using 'order contigs'). Results are reported in list format and graphically (see *Figure 6*).

Contigs can also be repositioned manually within the 'contig selector' (one-dimensional display) or 'contig comparator' (two-dimensional display, as shown in *Figure 6*) windows by dragging them using the middle mouse button. Within the 'contig comparator', read pairs will be seen to move closer to the diagonal as the two contigs containing the separate readings of the pair are brought closer to each other in terms of 'contig order'.

2: Investigation of DNA virus genome structure

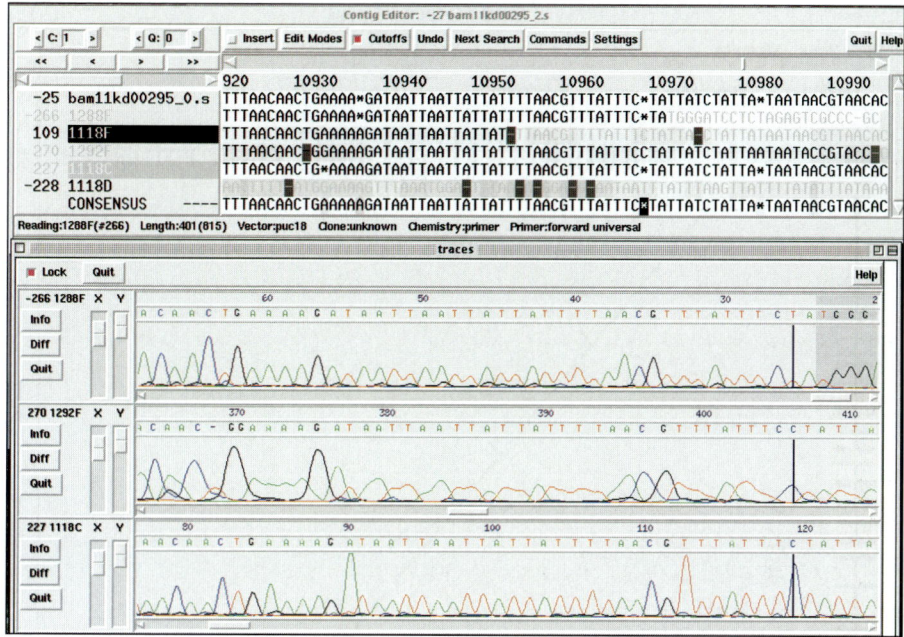

Figure 5. The 'contig editor' window of GAP4 showing associated sequence traces. The start of the vector sequence, greyed out in the trace of 1288F, is misplaced by one base. This is because the position is calculated by distance from the primer and is susceptible to sequence insertions or deletions.

5.1.5 Joining contigs

Contigs are frequently joined automatically during the assembly process as new readings are added. When the match with one contig is not sufficiently good, it is necessary to invoke the 'join editor' (*Figure 7*). This is essentially two contig editors, in which the overlapping contig segments are aligned, with a panel showing differences between the contigs. These differences should be minimized by normal editing procedures before quitting and joining the contigs.

The join editor can also be invoked when 'missed' joins are found using the 'find internal joins' function. The graphical display for the results of this function uses a 'contig comparator' window, such as that shown for the 'find read pairs' display in *Figure 6*. Crosshairs can be positioned over a reported join (the features of which are described at the bottom of the window). The positions of the contigs in the contig order are shown by two vertical bars in the upper, horizontal contig selector display. Double clicking on the highlighted join will bring up the 'join editor', as shown in *Figure 7*.

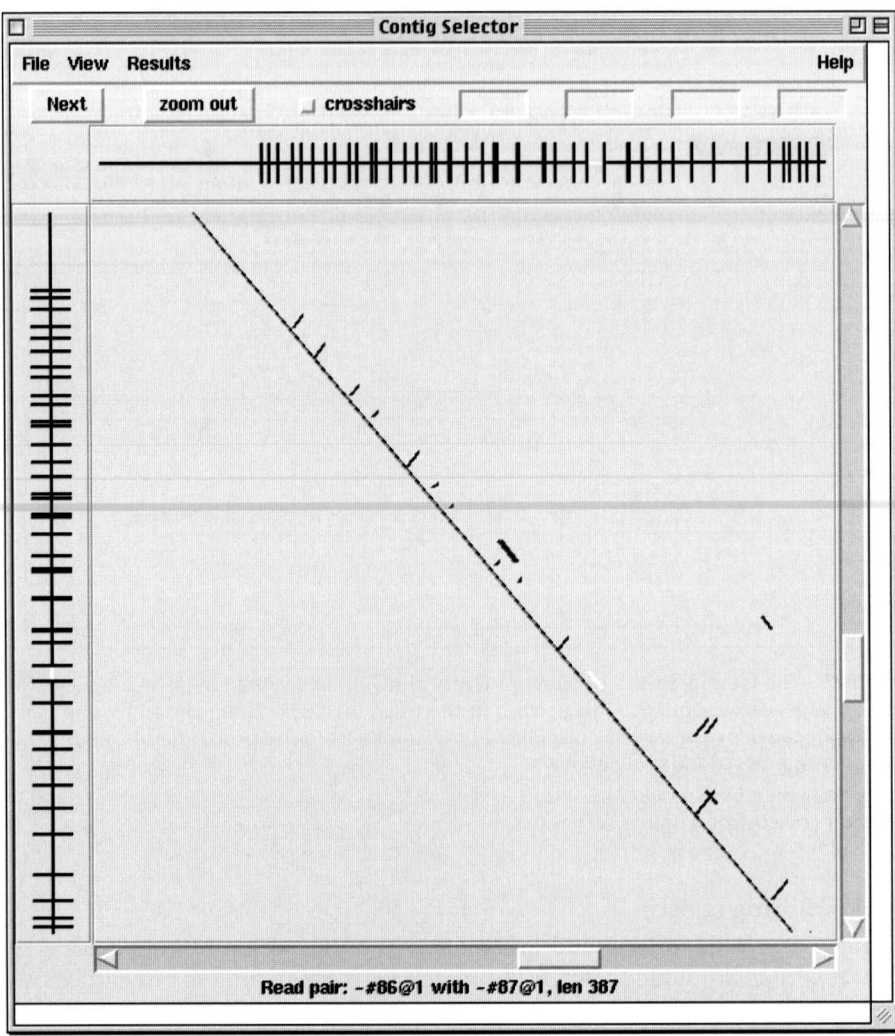

Figure 6. The 'find read pairs' plot of GAP4. Contigs (which can be single readings) are shown along each axis. Those containing the selected pair are highlighted. The diagonal indicates the location of identical positions in the database. The short (blue) lines indicate reading pairs. Their distance from the diagonal indicates the distance between the read pairs within the database (this distance is initially arbitrary, depending on the order of sequence entry). The orientation of the read pair lines indicates the relative orientation of the paired readings; for the readings to be in the correct relative orientation, the line should be parallel to the diagonal. Contigs can be complemented, but most of this will occur automatically as sequence assembly progresses.

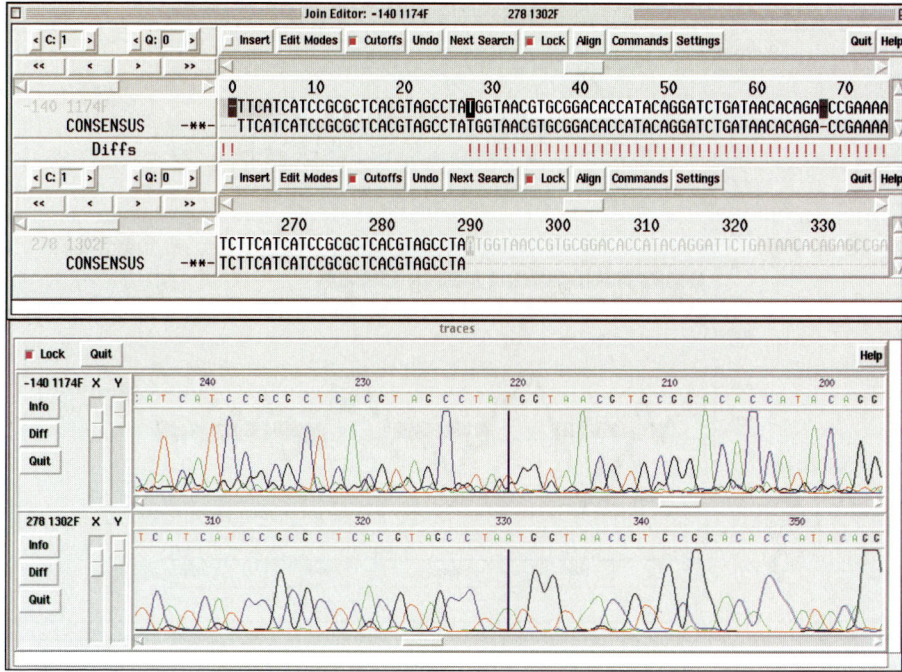

Figure 7. The 'join editor' window. The upper panel shows the alignment of two overlapping contigs. The contigs can be 'locked' together so that both scroll simultaneously. Between the contig sequences, differences are highlighted by red exclamation marks (only showing here where the sequence in one or other of the contigs is masked). Aligned traces from both contigs can be displayed and 'locked' so that clicking on a particular base displays the appropriate part of the trace (with the active base shown by a bar), facilitating editing of the join. On 'quit', the operator is asked whether or not to accept the join.

6. Sequence analysis

6.1 Prediction of open reading frames

Open reading frames can be predicted using a wide range of programs running on desktop computers and workstations. A useful program available over the world wide web is GENEMARK. It uses 'models' for a wide range of organisms, including several different herpesviruses to refine the accuracy of its predictions. Within the STADEN package, NIP offers useful analytical tools for the display of likely open reading frames by codon preference, combined with the results of signal searches (*Figure 8*).

6.1.1 Splice site prediction

Although splicing is not a consideration for poxviruses and other cytoplasmic viruses, it does need to be considered in the herpesviruses. NIP has facilities

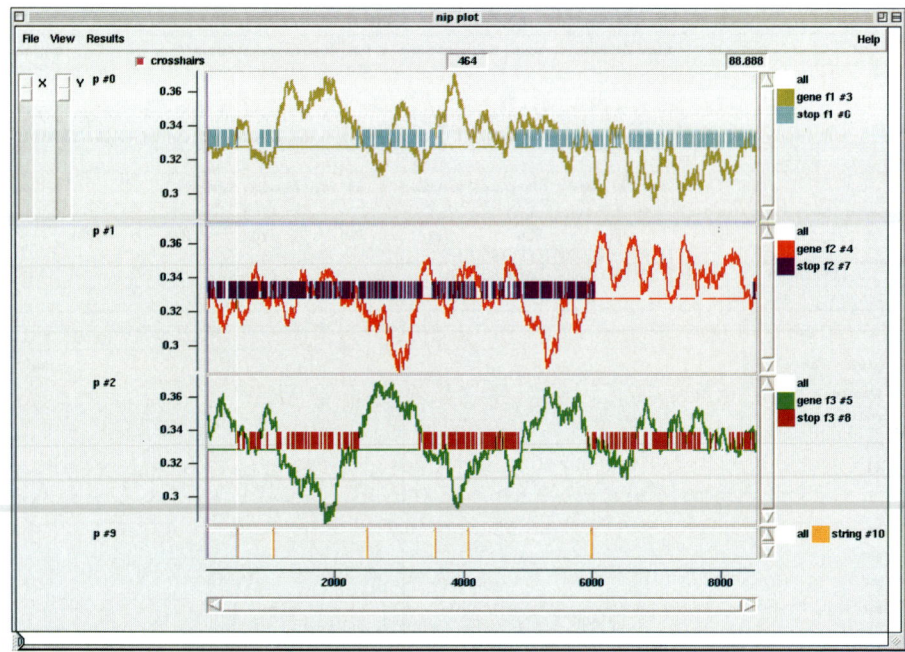

Figure 8. NIP analysis of a segment of fowlpox virus genomic DNA. The likelihood of coding for each frame (as determined by the codon usage method, (ref 32) is displayed as a graph, with the location of stop codons superimposed as vertical bars, giving a good indication of likely coding sequences. In the bottom panel, the locations of transcriptional regulator sequences are shown (in this case, poxvirus, early terminators).

for displaying likely splice sites, but programs such as GENEMARK (Section 6.1) are more useful.

6.1.2 Promoter and terminator sequences

Successful identification of coding sequences becomes more robust when combined with searches for known *cis*-acting transcriptional control sequences, such as promoters and terminators. Such analysis may also indicate the likely temporal class of a particular coding sequence. Searches can be performed manually using NIP, but such analysis is included in the GENEMARK models.

6.1.3 Initiator codons

Even when open reading frames have been identified, the likely initiator codon can only be inferred by comparison of the sequence flanking the candidate AUG initiators, using considerations discussed by Kozak (33). There may, of course be situations where more than one initiator can be used, either with different relative frequency or under different circumstances.

6.2 Homology searches
6.2.1 BLAST

Upon assembly of large stretches of genomic DNA encoding several genes, the sequence can be compared to nucleic acid and protein sequences in the databases. Combined with the results from ORF prediction software, BLAST (Basic Local Alignment Search Tool) outputs can provide useful information on the extent and integrity of gene coding sequences. General considerations relevant to database scanning have been discussed elsewhere in this series (34). The BLAST interfaces at NCBI or EBI offer flexibility in analysis of the search results. At NCBI, a graphical overview provides convenient hot links to the individual alignments (*Figure 9*).

At EBI, MVIEW (*Figure 10*) offers multiple alignments of hits for genomic nucleotide sequences or, in colour, for single proteins. A more convenient display, BLIXEM (35), is available in the X-window environment. It provides both a display (see *Figure 11*) of the distribution and strength of homology hits (as provided at the NCBI web site), and a multiple alignment display of

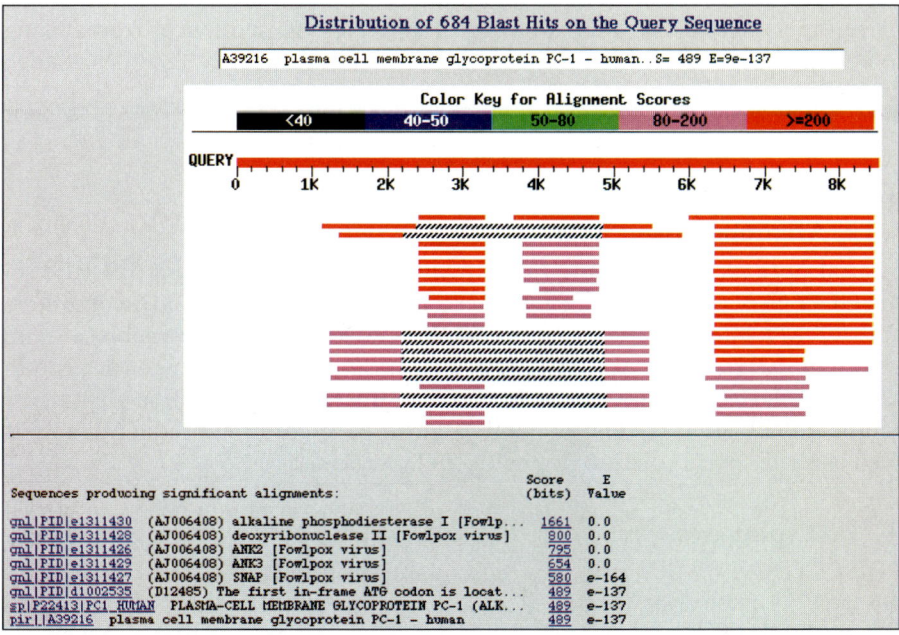

Figure 9. The results display from BLAST at NCBI. The graphical display at the top offers a useful display of the strength (colour-coded), extent, and identity of individual hits (the identity of the hit under the cursor is displayed in the window above the graphical display). Clicking on a hit moves to the relevant alignment, and thereafter it is possible to view the sequence entry and to see its Medline accession (along with a growing list of other options).

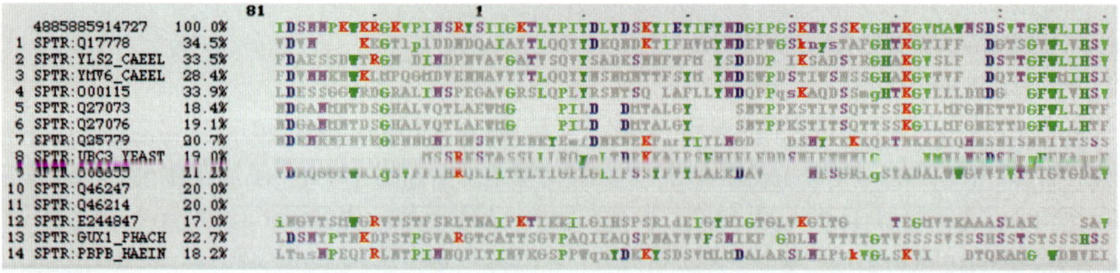

Figure 10. Part of the MVIEW display showing the alignment of a fowlpox virus protein with 'homologous' proteins in the database found by BLAST. The percentage identity/similarity between the search sequence and the hit is indicated next to the entry name. Identical residues are displayed by colour according to their properties.

the hits (which, when correctly configured, can also hot-link to the sequence database entries using a web browser).

Where large numbers of sequences need to be processed, facilities such as the BCM SEARCH LAUNCHER are useful. This provides batch processing facilities for a wide range of operations at various web sites. The results are returned as html files, viewable by a web browser, with 'hot links' therefore available. Such a search launcher avoids the sequencer from having to keep all links live, which can be time-consuming as online facilities change rapidly.

6.3 Sequence alignment

6.3.1 Alignment programs

Several options are available for sequence alignment, including PILEUP in the GCG package, and CLUSTALW (which is available for UNIX computers and for Macintosh™/PC). CLUSTALW services are also available at web sites (e.g. EBI). General considerations relevant to the alignment of protein sequences have been discussed elsewhere in this series (34). These alignments can be downloaded for display, and can also be used for subsequent analysis such as structure prediction or phylogenetic analysis.

6.3.2 Display programs

PRETTYBOX (GCG) provides high quality printouts of aligned sequences, but only as postscript format files, which cannot be read and manipulated by most graphics programs. Postscript viewers are available (such as GHOSTSCRIPT), allowing the images to be viewed and manipulated, but not without considerable loss in image quality. As shown in *Figure 10*, protein alignments can be viewed easily from BLAST at EBI using MVIEW, but the alignment is not downloadable for subsequent analysis. Display facilities for CLUSTALW are available at web sites such as EBI.

2: Investigation of DNA virus genome structure

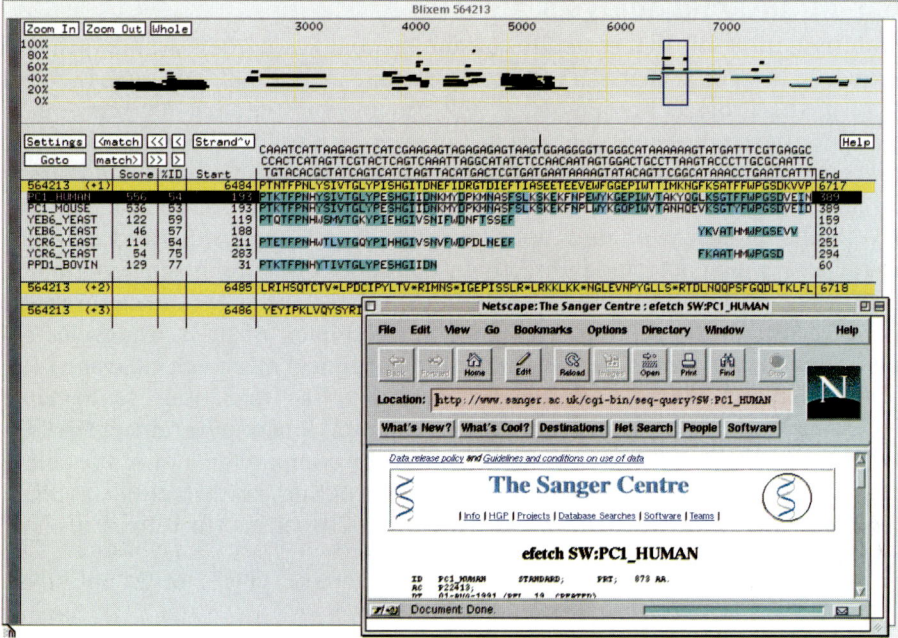

Figure 11. Output window from BLIXEM with superimposed browser window showing a selected protein sequence database entry. The upper panel shows the distribution of BLASTX hits along a DNA sequence. Their sequence identity is indicated by the y-axis. The box (blue) between bases 6000 and 7000 shows the region currently active in the lower (alignment) panel. Highlighted hits (light blue) indicate sequences from the protein currently selected in the alignment window. The alignment window shows the search sequence and all three open reading frames (yellow background); the frame showing homology to database proteins is nearest the top of the panel. Below this are arranged the hits; sequence identities and homologies highlighted by light and dark blue backgrounds, respectively). The identity of the currently selected hit is highlighted by a black background at the left side of the panel. Double clicking on the database entry name causes the database entry to be displayed in a separate browser window, as shown.

6.4 Phylogenetic analysis

The subject of phylogenetic analysis is vast and is beyond the scope of this chapter. Relatively recent reviews of the subject are available (36, 37). If the user merely requires a simple graphical display of the relative similarity between DNA or protein sequences, the PILEUP (GCG) program produces a tree file which can be viewed by FIGURE (GCG). If the user requires a more rigorous systematic approach, then programs such as PHYLIP (available for most platforms, with appropriate viewers such as TREEVIEW) or PAUP (now available under GCG) should be used.

6.5 Mutational analysis

Determination of the structure of the genome, whether at the primary sequence level or at that of a restriction enzyme map, is unlikely to be the final goal of the project. Frequently, it will provide the baseline for subsequent comparative studies between strains. Gross insertions, deletions, and genomic rearrangements can be studied by Southern blotting (Section 2.1), but there is an increasingly common need to be able to identify mutations at the genomic level. The presence of mutations in a set of otherwise identical genomic sequences can be detected indirectly by a number of methods: single-stranded conformational polymorphism analysis (SSCP, ref. 38), denaturing gradient gel electrophoresis (DGGE, ref. 39), heteroduplex analysis using electrophoresis in Hydrolink polymer (40), and by chemical mismatch cleavage (41). The genotypic changes may be studied directly by nucleotide sequencing, including fluorescent sequencing. For the latter, a new program (TRACE_DIFF) is available (42) to identify mutations by comparison, not of the called bases, but of the traces. As a baseline, the program can use either a single wild-type sequence trace, or it can generate a consensus trace from a set of aligned sequences.

7. Future directions

With the increased availability of automated sequencers, and the falling costs of primers and sequencing reagents, it is unlikely that the pace of sequencing of the large viral DNA genomes will slow. As representatives of the major groups are sequenced, the emphasis will fall on comparative genomics, requiring the sequence determination of different strains and pathotypes. Such comparative studies will help us understand the function, role and significance of the various genes, essential and non-essential, that have been (and will be) found in the large DNA viruses.

Acknowledgements

M.A.S. and S.M.L. wish to acknowledge the BBSRC and the EC (contract BIO-4-CT-960473) for financial support.

References

1. Sanger, F., Air, G. M., Barrell, B. G., Brown, N. L., Coulson, A. R., Fiddes, C. A., Hutchison, C. A., Slocombe, P. M., and Smith, M. (1977) *Nature* **265,** 687.
2. Sanger, F., Coulson, A. R., Hong, G. F., Hill, D. F., and Petersen, G. B. (1982) *J. Mol. Biol.* **162,** 729.
3. Davison, A. J., and Scott, J. E. (1986) *J. Gen. Virol.* **67,** 1759.
4. Goebel, S. J., Johnson, G. P., Perkus, M. E., Davis, S. W., Winslow, J. P., and Paoletti, E. (1990) *Virology* **179,** 247.

5. Telford, E. A., Watson, M. S., McBride, K., and Davison, A. J. (1992) *Virology* **189,** 304.
6. McGeoch, D. J., Dalrymple, M. A., Davison, A. J., Dolan, A., Frame, M. C., McNab, D., Perry, L. J., Scott, J. E., and Taylor, P. (1988) *J. Gen. Virol.* **69,** 1531.
7. Dolan, A., Jamieson, F. E., Cunningham, C., Barnett, B. C., and McGeoch, D. J. (1998) *J. Virol.* **72,** 2010.
8. Telford, E. A., Watson, M. S., Aird, H. C., Perry, J., and Davison, A. J. (1995) *J. Mol. Biol.* **249,** 520.
9. Bankier, A. T., Beck, S., Bohni, R., Brown, C. M., Cerny, R., Chee, M. S., Hutchison, C. A. d., Kouzarides, T., Martignetti, J. A., Preddie, E., and et al. (1991) *DNA Seq* **2,** 1.
10. Nicholas, J. (1996) *J. Virol.* **70,** 5975.
11. Davison, A. J. (1992) *Virology* **186,** 9.
12. Massung, R. F., Esposito, J. J., Liu, L. I., Qi, J., Utterback, T. R., Knight, J. C., Aubin, L., Yuran, T. E., Parsons, J. M., Loparev, V. N., and et al. (1993) *Nature* **366,** 748.
13. Senkevich, T. G., Bugert, J. J., Sisler, J. R., Koonin, E. V., Darai, G., and Moss, B. (1996) *Science* **273,** 813.
14. Yanez, R. J., Rodriguez, J. M., Nogal, M. L., Yuste, L., Enriquez, C., Rodriguez, J. F., and Vinuela, E. (1995) *Virology* **208,** 249.
15. Ayres, M. D., Howard, S. C., Kuzio, J., Lopez Ferber, M., and Possee, R. D. (1994) *Virology* **202,** 586.
16. Cole, S. T., Brosch, R., Parkhill, J., Garnier, T., Churcher, C., Harris, D., Gordon, S. V., Eiglmeier, K., Gas, S., Barry, C. E., Tekaia, F., Badcock, K., Basham, D., Brown, D., Chillingworth, T., Connor, R., Davies, R., Devlin, K., Feltwell, T., Gentles, S., Hamlin, N., Holroyd, S., Hornsby, T., Jagels, K., Krogh, A., Mclean, J., Moule, S., Murphy, L., Oliver, K., Osborne, J., Quail, M. A., Rajandream, M.-A., Rogers, J., Rutter, S., Seeger, K., Skelton, J., Squares, R., Squares, S., Sulston, J. E., Taylor, K., Whitehead, S., and Barrell, B. G. (1998) *Nature* **393,** 537.
17. Blattner, F. R., Plunkett, G. r., Bloch, C. A., Perna, N. T., Burland, V., Riley, M., Collado Vides, J., Glasner, J. D., Rode, C. K., Mayhew, G. F., Gregor, J., Davis, N. W., Kirkpatrick, H. A., Goeden, M. A., Rose, D. J., Mau, B., and Shao, Y. (1997) *Science* **277,** 1453.
18. Tomb, J. F., White, O., Kerlavage, A. R., Clayton, R. A., Sutton, G. G., Fleischmann, R. D., Ketchum, K. A., Klenk, H. P., Gill, S., Dougherty, B. A., Nelson, K., Quackenbush, J., Zhou, L., Kirkness, E. F., Peterson, S., Loftus, B., Richardson, D., Dodson, R., Khalak, H. G., Glodek, A., McKenney, K., Fitzegerald, L. M., Lee, N., Adams, M. D., Hickey, E. K., Berg, D. E., Gocayne, J. D., Utterback, T. R., Peterson, J. D., Kelley, J. M., Cotton, M. D., Weidman, J. M., Fujii, C., Bowman, C., Watthey, L., Wallin, E., Hayes, W. S., Borodovsky, M., Karp, P. D., Smith, H. O., Fraser, C. M., and Venter, J. C. (1997) *Nature* **388,** 539.
19. Fraser, C. M., Gocayne, J. D., White, O., Adams, M. D., Clayton, R. A., Fleischmann, R. D., Bult, C. J., Kerlavage, A. R., Sutton, G., Kelley, J. M., Fritchman, J. L., Weidman, J. F., Small, K. V., Sandusky, M., Fuhrmann, J., Nguyen, D., Utterback, T. R., Saudek, D. M., Phillips, C. A., Merrick, J. M., Tomb, J. F., Dougherty, B. A., Bott, K. F., Hu, P. C., and Lucier, T. S. (1995) *Science* **270,** 397.
20. Mewes, H. W., Albermann, K., Bahr, M., Frishman, D., Gleissner, A., Hani, J.,

Heumann, K., Kleine, K., Maierl, A., Oliver, S. G., Pfeiffer, F., and Zollner, A. (1997) *Nature* **387,** 7.
21. Sambrook, J., Fritsch, E. F., and Maniatis, T., *Molecular Cloning: A Laboratory Manual* (Cold Spring Harbor Laboratory Press, New York, ed. 2nd, 1989).
22. Mockett, B., Binns, M. M., Boursnell, M. E. G., and Skinner, M. A. (1992) *J. Gen. Virol.* **73,** 2661.
23. Schwartz, D. C., and Cantor, C. R. (1984) *Cell* **37,** 67.
24. Sanger, F., Nicklen, S., and Coulson, A. R. (1977) *Proc. Natl. Acad. Sci. USA* **74,** 5463.
25. Staden, R. (1980) *Nucl. Acids Res.* **8,** 3673.
26. Fleischmann, R. D., Adams, M. D., White, O., Clayton, R. A., Kirkness, E. F., Kerlavage, A. R., Bult, C. J., Tomb, J. F., Dougherty, B. A., Merrick, J. M., McKenney, K., Sutton, G., FitzHugh, W., Fields, C., Gocayne, J. D., Scott, J., Shirley, R., Liu, L. I., Glodek, A., Kelley, J. M., Weidman, J. F., Phillips, C. A., Spriggs, T., Hedblom, E., M. D. Cotton, Utterback, T. R., Hanna, M. C., Nguyen, D. T., Saudek, D. M., Brandon, R. C., Fine, L. D., Fritchman, J. L., Fuhrmann, J. L., Geoghagen, N. S. M., Gnehm, C. L., McDonald, L. A., Small, K. V., Fraser, C. M., Smith, H., and Venter, J. C. (1995) *Science* **269,** 496.
27. Bankier, A. T. (1993) *Methods Mol. Biol.* **23,** 47.
28. Messing, J. (1991) *Gene* **100,** 3.
29. Lander, E. S., and Waterman, M. S. (1988) *Genomics* **2,** 231.
30. Staden, R. (1982) *Nucl. Acids Res.* **10,** 4731.
31. Bonfield, J. K., Smith, K., and Staden, R. (1995) *Nucl. Acids Res.* **23,** 4992.
32. Staden, R., and McLachlan, A. D. (1982) *Nucl. Acids Res.* **10,** 141.
33. Kozak, M. (1996) *Mamm. Genome* **7,** 563.
34. Barton, G. J. (1996). In *Protein structure prediction: a practical approach* (ed. M. J. E. Sternberg). Vol. 1, p. 31. Oxford University Press, Oxford.
35. Sonnhammer, E. L., and Durbin, R. (1994) *Comput. Appl. Biosci.* **10,** 301.
36. Hillis, D. M., Mable, B. K., and Moritz, C. (1996). In *Molecular Systematics* (ed. D. M. Hillis, C. Moritz, and B. K. Mable). p. 515. Sinauer Associates, Inc., Sunderland, Mass. USA.
37. Swofford, D. L., Olsen, G. J., Waddell, P. J., and Hillis, D. M. (1996). In *Molecular Systematics* (ed. D. M. Hillis, C. Moritz, and B. K. Mable). p. 407. Sinauer Associates, Inc., Sunderland, Mass. USA.
38. Orita, M., Iwahana, H., Kanazawa, H., Hayashi, K., and Sekiya, T. (1989) *Proc. Natl. Acad. Sci. USA* **86,** 2766.
39. Sheffield, V. C., Cox, D. R., Lerman, L. S., and Myers, R. M. (1989) *Proc. Natl. Acad. Sci. USA* **86,** 232.
40. Perry, D. J., and Carrell, R. W. (1992) *J. Clin. Pathol.* **45,** 158.
41. Ellis, T. P., Humphrey, K. E., Smith, M. J., and Cotton, R. G. (1998) *Hum. Mutat.* **11,** 345.
42. Bonfield, J. K., Rada, C., and Staden, R. (1998) *Nucl. Acids Res.* **26,** 3404.

3

Mutagenesis of DNA virus genomes

KEITH N. LEPPARD

1. Introduction

1.1 The diversity of DNA viruses

DNA viruses encompass a wide range of virus types which can infect bacteria, plants, insects, and birds, as well as mammals including man. Mutant strains of these viruses have been essential to our understanding of how individual viral genes contribute to the process of infection but, for a variety of reasons, far more mutant strains have been isolated of some viruses than of others. DNA viruses which have been subject to extensive genetic characterization include various bacteriophages, and mammalian viruses such as simian virus 40, adenovirus types 2 and 5, herpes simplex virus type 1, and vaccinia virus. Although there are considerable similarities between the techniques used to isolate mutant viruses from different host systems, this chapter deals exclusively with the study of mammalian viruses.

1.2 Alternative mutagenesis strategies

The isolation and study of virus mutants depend on permissive cell-culture systems. The maintenance of cell cultures suitable for growing various DNA viruses, and the preparation and titration of virus stocks, are discussed in Section 2. Using these systems, several strategies have been employed to generate panels of virus mutants for phenotypic analysis. Broadly, these can be classified into random and site-directed approaches. Random mutagenesis involves the exposure of either virus or viral DNA to a mutagen *in vitro*, or the growth of virus in the presence of a mutagenic base analogue in cell culture (Sections 3–5). After recreating virus particles (for DNA mutagenized *in vitro*, Section 7), mutant viruses have to be identified by a phenotypic screen (Section 8), and then the genetic lesion(s) they carry must be characterized (Section 9). In contrast, site-directed mutagenesis involves the synthesis *in vitro* of DNA that has one or more specific alterations in sequence from that of wild-type viral genetic material (Section 6). Once produced, this altered DNA has to be reintroduced into virus particles (Section 7). Finally, stocks of mutant virus must be characterized to allow the phenotypic effect of the sequence change(s) to be assessed (Section 10).

An advantage of the random approach is that no presumption is made in advance about the viability of the mutants sought from the experiment: a mutagenized stock of virus or viral DNA is simply plated under appropriate conditions to allow viruses with altered growth properties to be identified. Mutant viruses in the stock that are non-viable under the growth conditions employed will simply not be seen. Although the random mutagenesis approach has this advantage, it also has serious limitations. Once a mutant strain has been identified, the position in the DNA of the mutation responsible for the phenotype must be determined. Unless a short cloned segment of the genome is chosen for random mutagenesis and fully characterized before reconstruction into a viral genome (conceptually similar to site-directed mutagenesis, since choices must be made in advance of the phenotypic characterization), finding the mutation is a significant task. Moreover, there is the possibility that genome changes other than the one responsible for the phenotype will be present in the virus. These may also have phenotypic effects under certain conditions, and may interact with the primary phenotypic mutation. A full work-up of such mutations therefore requires that they be rescued into a wild-type genetic background, to eliminate the effects of secondary mutations.

Site-directed mutagenesis has become increasingly straightforward with technical advances, and is now the method routinely chosen in any attempt to isolate mutants of mammalian DNA viruses. The advantage of knowing at the outset the site of a mutation, and hence the gene product(s) affected, outweighs the disadvantage of being unable to determine the viability of the intended mutant without some investment of time. In practice this problem is overcome by attempting the isolation of several different mutations in a gene simultaneously, so that one or more mutants is likely to be obtained.

1.3 Suiting mutagenesis strategy to the virus
1.3.1 Availability of cell culture systems

The strategy to be adopted for the isolation of virus mutants depends crucially on the virus type. The first key question is whether it is possible to grow the virus successfully in cell cultures, as little progress can be made towards the isolation of mutant virus strains in the absence of this capability. Certain types of DNA virus, notably human papillomaviruses and hepatitis B virus, have not yet been culturable *in vitro* on a significant scale. It is possible in each case to produce virus particles from certain cell types transfected with cloned DNA (1–3). However, infection with virus particles in cell culture has not been achieved. Some information about gene function in these viruses has

3: Mutagenesis of DNA virus genomes

been obtained from transfecting altered gene sequences into cell cultures in transient assays. Standard cell lines permissive for the growth of other DNA viruses are listed in *Table 1*.

1.3.2 Availability of reverse genetics

Mutagenising DNA *in vitro,* or synthesising DNA with altered sequence for reincorporation into virus particles, are known as reverse genetic approaches. These can be employed only if it is possible, for the virus in question, to recreate infectious particles which incorporate cloned genetic material. The

Table 1. Mammalian DNA viruses and cell culture systems

Virus	Particle density (g ml^{-1})	Replication cycle in culture (h)	Standard cell lines [a]	Growth medium	Maximum area split (ratio)	Duration of plaque assay[b] (days)
SV40	1.34	72	BSC-1	DMEM[c]/ 5% FBS[d]	1:5	8–14
			CV1	DMEM/ 10% NBS[e]	1:10	8–14
polyoma	1.34	72	mouse embryo cells	DMEM/ 5% FBS	primary cells	6–10
			3T3	DMEM/ 5% FBS	1:4	6–10
papilloma	1.34	?	n/a[f]	–	–	n/a
AAV	1.39–1.42	40	HeLa	DMEM/ 10% NBS	1:10	n/a
			293	DMEM/ 10% NBS	1:4	n/a
Ad 5	1.33–1.35	36	HeLa	DMEM/ 10% NBS	1:10	6–9
			A549	DMEM/ 10% FBS	1:6	6–9
			293	DMEM/ 10% NBS	1:4	6–9
HSV-1	1.27–1.29	20	Hep2	DMEM/ 5% FBS	1:10	3–4
			Vero	DMEM/ 5% FBS	1:10	3–4
vaccinia	–	8–12	Vero	DMEM/ 5% FBS	1:10	2
			BSC-1	DMEM/ 5% FBS	1:5	2

[a] Other cell types may also be used successfully.
[b] Mutant and recombinant viruses may plaque with different kinetics, even when functionally complemented.
[c] Dulbecco's modified Eagle medium.
[d] foetal bovine serum.
[e] newborn bovine serum.
[f] not available.

lack of appropriate permissive cell culture systems will preclude these, as well as more conventional mutagenesis strategies. In addition, for those viruses where one or more protein components of the virion are crucial for infectivity after penetration of the cell, naked DNA will not be infectious even when of wild-type sequence. The only major family of animal DNA viruses for which this is the case are the poxviruses. However, mutant strains of vaccinia virus can be isolated by homologous recombination between manipulated DNA sequences and a co-infecting viral genome.

1.3.3 Genome size and particle structure

A key factor in developing a mutagenesis strategy is the size of the viral genome. For mammalian DNA viruses, genome sizes range from 3 kbp to >200 kbp. Random mutagenesis of entire genomes *in vitro* is only practical if their size is small. Otherwise, cloned fragments can be mutagenized and subsequently rebuilt into intact genomes. For genomes up to the size of adenovirus (36 kbp), this can be achieved by ligating together the relevant fragments of DNA *in vitro*, prior to transfecting DNA into cells. For larger viruses, and often for adenovirus too, reconstruction is achieved by homologous recombination *in vivo* between the cloned mutagenized fragment and viral genomic DNA or suitable genomic DNA fragments.

1.3.4 The availability of complementation systems

If potentially mutant viruses are tested for phenotype in standard cell cultures which support the growth of the relevant wild-type, then only conditional mutants or those with lesions in non-essential genes can be isolated. Examples of conditional phenotypes are 'temperature sensitive' (ts), where the mutant grows normally at one temperature but abnormally or not at all at another temperature, and 'host range' (hr), where the mutant fails to grows in one or more cell types which support the growth of wild-type whilst retaining growth in others. ts mutations are generally missense changes in a gene product rather than complete deficiencies. In order to isolate virus 'null mutants', the affected function must either be dispensable for growth in standard conditions, or else its deficiency must be complemented in some way during the isolation of mutants.

Complementation of defective mutants can be achieved using a second virus (wild-type or an appropriate mutant) as a helper, but recombination between the two viruses is always a concern, and it is usually difficult or impossible to separate the desired mutant from its helper in the resulting virus stock. The helper virus will also tend to outgrow the mutant, unless it is chosen so as to be dependent on the mutant for its own growth (mutual interdependence: e.g. a mutant in gene A being complemented by a helper deficient in gene B). Complementing cell lines are more useful. These are cells derived from a standard permissive cell line which incorporate and express

3: Mutagenesis of DNA virus genomes

one or more viral genes (Section 7.1). Null mutant viruses defective in those functions provided by the cell will be able to grow specifically in that cell type.

2. Preparation and titration of virus stocks

2.1 Cell culture techniques

The availability and quality of permissive cell cultures is crucial to any study of virus mutants. Each of the DNA viruses has its own range of permissive cell types (*Table 1*) which can be easily maintained in the laboratory provided appropriate facilities are available. *Protocol 1* describes routine cell culture procedures.

Protocol 1. Maintaining monolayer cultures of mammalian cells

All parts of this protocol should be carried out in a laminar flow hood using aseptic technique. Solutions for tissue culture and virus growth should be purchased sterile, or prepared in tissue-culture grade water and sterilized by autoclaving or filtration as appropriate.

Equipment and reagents

- Class II laminar flow hood, certified for containment (or as required by local regulations)
- Humidified 37°C incubator with a 5% CO_2 in air atmosphere
- Liquid nitrogen storage space
- Tissue culture medium appropriate to cell type (Gibco Life Technologies or Flow Laboratories)
- Versene (0.2 g l^{-1}) and trypsin (2.5 g l^{-1}) solutions in buffered saline (Gibco Life Technologies)
- Newborn or fetal bovine serum, according to cell type, and heated for 60 min at 56°C prior to use (Gibco Life Technologies or Flow Laboratories)
- Sterile, tissue-culture grade plasticware (e.g. Falcon, Becton-Dickinson)
- Freezing mix: 92% newborn or fetal bovine serum, according to cell type, 8% dimethyl sulfoxide
- Cryo-storage vials (e.g. Falcon, Becton-Dickinson) [a]

A. *Passaging confluent cell monolayers*

Cell cultures should be maintained by serial passage every 3–5 days. Each cell type has its own growth requirements (type of medium, concentration and type of serum) and maximum area split, i.e. the dilution factor applied during passage (*Table 1*).

1. Take a confluent monolayer of cells, aspirate the growth medium, and wash the monolayer with versene, aspirating again.
2. Add 2.0 ml versene and 0.5 ml trypsin solution (for 10^7 cells on a 10 cm diameter culture dish; adjust volumes for other sizes of culture), and wait for 2–3 min while the cells round up and begin to detach from the surface.[b]
3. Pipette the cells into the liquid, and remove the suspension to a sterile

Protocol 1. *Continued*

tube containing 0.5 ml serum. Pellet the cells by low-speed centrifugation for 3–5 min.

4. Discard the supernatant, and resuspend the cell pellet gently in fresh growth medium containing serum at the concentration appropriate for the cell type.

5. Plate the cells onto culture dishes as required, diluting the suspension as necessary to give the cell number and growth volume appropriate for the size of the new culture and the cell type.

B. *Long-term cell storage*

Cells may be stored frozen in liquid nitrogen and viable cells recovered indefinitely, provided correct procedures are followed. It is good practice to store batches of frozen cell aliquots and to restart cultures from frozen stocks every 3–6 months.

1. Remove the cells from healthy just sub-confluent cell monolayers (five 10 cm dishes; approximately 5×10^7 cells) as described in *Protocol 1A*, steps 1–3.

2. Discard the supernatant from the cell pellet, and resuspend gently in 5.0 ml of pre-cooled freezing mix on ice. Dispense 0.5 ml aliquots into labelled cryo-storage vials.

3. Freeze the vials slowly over 24 h (this may be achieved by wrapping the vials in multiple layers of paper towel and placing at $-70\,°C$). Transfer the vials to liquid nitrogen storage.[c]

4. To recover cells from liquid nitrogen storage, take a vial and thaw immediately and rapidly in a 37 °C water bath (approximately 1 min). Pipette the cells into a 10 cm culture dish (or flask of equivalent area), pre-prepared with warmed growth medium plus serum added.

5. After overnight incubation, replace the growth medium to remove residual freezing mix, or passage immediately if required (*Protocol 1A*).

[a] Vials not specifically recommended for the purpose should not be used for liquid nitrogen storage due to the risk of explosive vial failure during storage or recovery.
[b] Monkey CV1 and BSC-1 cells require longer incubations and/or 37 °C incubation to promote detachment.
[c] The viability of a newly frozen batch should be checked after about 1 week of storage, using the procedure described in Steps 4 and 5.

2.2 Generating and titrating virus stocks

It is important to begin any mutagenesis study with a pure virus stock, and to maintain the integrity and purity of both wild-type and mutant virus stocks by use of appropriate techniques during virus passage (*Protocol 2A*). The key

3: Mutagenesis of DNA virus genomes

principles are to grow stocks from plaque-purified isolates for which the phenotype has been validated, and to infect cells at a low multiplicity (i.e. a low number of infectious units per cell) for stock preparation. Both of these precautions are to guard against the emergence in the stock of defective viruses, which arise naturally during the course of replication of many viruses, and which are sustained in culture by co-infection with the standard virus that is present; the probability of such co-infection is much greater during infections at a high multiplicity. One cycle of plaque-purification involves obtaining a series of well-separated individual plaques from a stock (*Protocol 2B*), growing small-scale stocks from each plaque, and re-checking the virus phenotype; a plaque isolate with the correct properties is then selected either to grow a large-scale stock for which the titre can be determined, or for further rounds of plaque purification.

Once a pure stock has been obtained, the concentration of infectious units in it (the infectious titre) must be determined. The most commonly used measure of infectivity is the plaque-forming unit (PFU), which is detected in a plaque assay (*Protocol 2B*). Conditions for plaque assays differ between viruses, and the times taken for plaques of wild-type virus to emerge vary greatly (*Table 1*). The general technique of the plaque assay is to establish a virtually confluent monolayer of cells prior to infection, inoculate with a diluted virus sample, and then maintain the cells in medium containing only a low concentration of animal serum so that further cell growth is minimized. The medium is solidified with agar so that virus produced by an infected cell can spread only to its immediate neighbours. After several rounds of infection, a zone of cell death known as a plaque is visible; these are revealed for quantitation by counterstaining the remaining healthy cells.

Protocol 2. Preparation and titration of virus stocks

All parts of this protocol should be carried out using aseptic technique under containment conditions appropriate to the virus concerned. All infectious waste should be sterilized before disposal.

Equipment and reagents

- Standard cell culture equipment and materials (*Protocol 1*)
- −70°C freezer storage
- Growth medium appropriate to cell type at 2× normal concentration (Gibco Life Technologies)
- Neutral red solution: 1.0 g l^{-1} neutral red (Sigma) in Earle's saline (Gibco-BRL)
- 3.7% formaldehyde solution: 100 ml per l of 37% formaldehyde solution (formalin) in phosphate-buffered saline (PBS: 140 mM NaCl, 2.7 mM KCl, 8.1 mM Na_2HPO_4, 1.5 mM KH_2PO_4, pH 7.2)
- Noble agar (Difco) at 20 g l^{-1} in water, autoclaved to dissolve and sterilize; melted at 100°C, and equilibrated to 50°C prior to use. If significant volumes of supplements are to be added to the medium for plaque assays (Step B2), increase the stock agar concentration to 28 g l^{-1} and adjust the volumes added to the 2× culture medium plus supplements accordingly, to achieve a 10 g l^{-1} final concentration
- Crystal violet solution: 1.0 g l^{-1} crystal violet (Sigma) in 20% ethanol

Protocol 2. *Continued*

A. *Growing a virus stock*

This protocol may be carried out on any scale, having regard to the amount of virus stock required and the expected titre. Typically between 5×10^7 and 5×10^8 cells is appropriate.

1. Take monolayer cultures of permissive cells on cell-culture dishes or flasks in standard growth medium at approximately 80% confluence.
2. Prepare a dilute virus inoculum from a seed stock. Use sufficient virus to give a multiplicity of infection of 0.1 plaque-forming units per cell in a volume of 0.5 ml per 10^7 cells.
3. Aspirate the growth medium, add an inoculum of virus to each culture, and immediately rock the dish to distribute the virus. Return the culture to the incubator.
4. Redistribute the virus inoculum on the cell monolayers every 15 min for 1 h.
5. Aspirate the inoculum, and add 10 ml per 10^7 cells of pre-warmed growth medium containing 2% (v/v) serum. Return the cultures to the incubator.
6. Observe the cultures daily for the appearance of cytopathic effect. Harvest the cells and growth medium together as a crude virus stock when cultures shows >80% cytopathic effect. Many viruses remain strongly cell-associated. To obtain a more concentrated stock, pellet the cells by low speed centrifugation, and discard 50–75% of the supernatant before proceeding.
7. Release virus from the cell debris by freezing and thawing the stock twice. Remove debris by low-speed centrifugation. Store the supernatant as a virus stock in aliquots at $-70\,°C$.

B. *Titration of virus stocks by plaque assay*

The apparent titre of a stock will depend on the cell type and conditions used for the assay, including incubation time; a consistent procedure should be adopted to give titres which are comparable between experiments. Virus types vary in their stability to repeated freeze–thaw cycles. Once a stock titre has been obtained, further freeze–thaw cycles should be kept to a minimum to maintain a constant titre. This protocol may also be used to obtain individual plaques as part of a cycle of plaque-purification of a virus from a mixed stock.

1. One day prior to setting up the assay, plate cells at 80–90% confluence onto either 6 cm dishes or six-well multiwell plates (smaller cultures may be used, but it is less easy to obtain accurate plaque counts). Prepare sufficient cultures for the assay of each relevant virus dilution in duplicate (see Step 4).

3: Mutagenesis of DNA virus genomes

2. Prewarm growth medium at 2× normal strength to 37°C, and supplement with bovine serum at 4% (v/v) (plus other supplements if required).[a] Melt a stock of noble agar and equilibrate it to 48°C.

3. Thaw an aliquot of the virus stock to be titred on ice. Prepare 1.0 ml serial tenfold dilutions of the stock in growth medium without serum.

4. Aspirate the growth medium from each culture, and add 0.2 ml of each diluted virus inoculum to be assayed to duplicate cultures. Choose dilutions to assay, based on the approximate anticipated titre of the stock, so that one of the dilutions tested will give between 10 and 50 plaques per culture.[b] If the titre cannot be predicted, a full range of dilutions should be assayed.

5. Redistribute the virus inoculum on the cell monolayers every 15 min for 1 h.

6. Aspirate the inoculum from a culture, mix equal volumes of 2× growth medium and molten noble agar (Step 2), and gently add 4 ml of the mixture (3 ml for six-well multiwell cultures) to the monolayer.[c] Once the overlay has spread across the surface of the monolayer, leave the dish undisturbed for 10 min so that the agar can set. Return the culture to the incubator.

7. Inspect the cultures periodically for the appearance of plaques. If the assay is likely to run for a long period, re-feed the cultures every 4 days by the addition of a further nutrient agar overlay (2–3 ml), prepared as in Step 6, on top of the existing overlay.

8. Two alternative strategies are possible once plaques are visible.

    (a) (i) Incorporate neutral red dye at 0.1 g l^{-1} into a final nutrient agar overlay, and return the cells to the incubator overnight. Protect the cells from light by wrapping the cultures in aluminium foil to avoid photoactivation of the dye. To count plaques, provide gentle illumination from below; pale pink circular zones (plaques) should be visible against a deeper red background. Cultures may be returned to the incubator for a further period and recounted; more plaques will emerge over the course of 2–3 days. Use the final plaque counts to calculate a titre (plaque-forming units per millilitre).

    (ii) To obtain plaque isolates for further characterization, they should be picked as soon as they appear. Stab through the agar overlay to the cell sheet with the tip of a Pasteur pipette, apply gentle suction, and withdraw the agar plug and underlying material. Resuspend the plug in 0.5 ml of growth medium without serum, release virus as described (see Step A6), and store at −70°C.

Protocol 2. *Continued*

(b) For viruses which produce clear plaques (e.g. herpes, vaccinia), plaques may be visualized to obtain a titre by an alternative strategy. Fix the cells by adding 3.7% formaldehyde solution on top of the agar overlays, and stand overnight (fume hood). Discard the formaldehyde solution, and remove the agar plugs from the cultures with a spatula. Stain the remaining cell material with crystal violet solution (4–5 min), then wash well with water. Count the unstained circular zones.

[a] 293 cell cultures, which are frequently used for plaque assays with adenovirus type 2 or 5, become acidic under agar. The normal concentration of bicarbonate in the growth medium (3.7 g l^{-1} in DMEM), should be doubled in the first overlay to counteract this.
[b] For example, if the expected titre is in the range 10^8–10^{10} PFU ml^{-1}, assay 10^{-6} (at 10^8 PFU ml^{-1} this gives 20 plaques per culture), 10^{-7}, and 10^{-8} (at 10^{10} PFU ml^{-1} this gives 20 plaques per culture) dilutions.
[c] The overlay mix will set within a few minutes at room temperature, and so it must be prepared in small volumes from pre-warmed components as it is required. Overlay that has set prematurely should not be re-melted for use. Avoid disturbing the cell sheet and leaving bubbles on the surface of the overlay.

3. Chemical mutagenesis of viral DNA *in vitro*

A range of chemicals is known to cause damage of various types to DNA. Nitrous acid, bisulfite, hydrazine, hydroxylamine, and alkylating agents are among the agents employed for *in vitro* mutagenesis. Generally, single-stranded DNA is far more sensitive to damage than double-stranded DNA. Techniques for the chemical mutagenesis *in vitro* of cloned DNA have been described in an earlier volume in this series (4), and are fully applicable to cloned viral DNA sequences.

4. Chemical mutagenesis of virus particles *in vitro*

The same chemicals that can be used to mutagenize isolated DNA (Section 3) will also affect DNA within virus particles. These should be purified for mutagenesis and, for the smaller viruses for which this method is applicable, this can be achieved by caesium chloride density-gradient centrifugation (*Protocol 3*). The sensitivity of virions to chemical mutagenesis will be lower than for naked single-stranded DNA, since the particle will afford some protection, and for most mammalian DNA viruses the DNA will be in double-stranded form. The duration of chemical exposure must be controlled to give an acceptable mutation frequency (*Protocol 4*). The virus arising from such an experiment will be a complex mixture of strains. Extensive plaque purification and characterization is then required to isolate those with the

3: Mutagenesis of DNA virus genomes

desired properties (Section 8). It should be borne in mind that there is a possibility of multiple mutations within any given strain.

Protocol 3. Virus purification by density gradient centrifugation

This method has been developed for adenovirus type 5, but is generally applicable to other adeno- and papovaviruses. For purification of HSV-1 particles, the densities of CsCl solutions should be reduced, reflecting the lower particle density (*Table 1*). Vaccinia virus is normally purified by velocity sedimentation on sucrose gradients. All parts of this protocol should be carried out under containment conditions appropriate to the virus concerned. All infectious waste should be sterilized before disposal.

Equipment and reagents
- Standard cell culture equipment and materials (*Protocol 1*)
- Probe sonicator with microtip
- Beckman J2 superspeed centrifuge and JA21 rotor, or similar
- Beckman ultracentrifuge and swinging-bucket rotors taking approximately 13 ml (SW41) and 5 ml (SW50.1) volume tubes, or similar
- 0.1 M Tris-HCl, pH 8.0
- TD buffer: 8 g NaCl, 0.38 g KCl, 0.1 g Na_2HPO_4, 3.0 g Tris base per litre, brought to pH 7.5 with HCl
- DNA purification reagents (*Protocol 6*)
- 1.25, 1.35, and 1.4 g ml^{-1} (density) solutions of CsCl. Add 36.16 g, 51.2 g, or 62.0 g respectively of dry CsCl to 100 ml of TD buffer, and allow to dissolve
- Virus lysis buffer: 1.0 g l^{-1} sodium dodecyl sulfate, 10 mM Tris, pH 7.5, 10 mM NaCl, 1 mM EDTA
- Virus storage buffer: 10 mM Tris-HCl pH 8.0, 100 mM NaCl, 2 mM $MgCl_2$, 50% (v/v) glycerol, 1.0 g l^{-1} bovine serum albumin, prepared from sterile components
- 10% SDS: 100 g l^{-1} sodium dodecyl sulfate
- 0.25 M EDTA
- 10 mg ml^{-1} proteinase K (Boehringer)

Method

Work with cooled reagents and apparatus whenever possible.

1. Prepare infected cultures (*Protocol 2A*) containing 10^8 cells.

2. Harvest the cells and medium when the cultures show full cytopathic effect. Pellet the cells by low-speed centrifugation, and discard the supernatant.

3. Resuspend the cell pellet in 10 ml ice-cold 0.1 M Tris-HCl, pH 8.0, and disrupt by sonication (probe sonicator, 5–10 s).

4. Pellet the cellular debris at 5000 *g* for 10 min, and collect the supernatant.

5. Prepare two CsCl step gradients in Beckman SW41 Ultraclear tubes (or similar), containing 2.0 ml of 1.4 g ml^{-1} CsCl solution overlaid with 3.0 ml of 1.25 g ml^{-1} CsCl solution.

6. Carefully overlay equal amounts of the supernatant from Step 4 on to each of the gradients, filling any remaining space in the tubes with 0.1 M Tris-HCl pH 8.0.

Protocol 3. *Continued*

7. Centrifuge the gradients for 1 h at 35 000 r.p.m. in an SW41 rotor (or similar), at 15 °C.
8. Harvest the translucent, milky white band of virus formed at the interface between the CsCl gradient steps into a Beckman SW50.1 Ultraclear tube (or similar) by puncturing the base of the tube and collecting the relevant drops. Pool the bands from the two gradients unless the yields are very large.
9. Fill the tube with 1.35 g ml^{-1} (density) CsCl solution, and centrifuge overnight to equilibrium at 40 000 r.p.m., 15 °C. Harvest the virus band as in Step 8.
10. To determine the yield of particles for adenovirus, measure the volume of virus collected, take a 16 µl aliquot, and dilute it to 400 µl with virus lysis buffer. Measure the absorbance of this sample at 260 nm. An absorbance of 1.0 is equivalent to 10^{12} adenovirus virions per millilitre in the diluted sample.
11. If infectious virus is required, dilute the virus fivefold into storage buffer, or dialyse against phosphate-buffered saline containing 20% glycerol or other buffer as appropriate for the end use of the material. Determine the infectious titre by plaque assay (*Protocol 2B*); the particle to PFU ratio for wild-type adenovirus is usually between 20 and 50.
12. To prepare viral DNA from purified virus (Step 9), add 2 volumes of water and 6 volumes of 95% ethanol, and place at −70 °C for 15 min to precipitate the virus. Collect the virus by centrifugation at 8000 g for 20 min, dry it briefly under vacuum, and gently resuspend it in 2 ml of TE buffer. Add 120 µl of 10% SDS, 40 µl of 0.25 M EDTA, and 20 µl of 20 mg ml^{-1} proteinase K, and incubate at 37 °C for 2 h. Purify the DNA as described in *Protocol 6*, Steps 8 and 9, increasing the volumes twentyfold.

Protocol 4. Chemical mutagenesis of purified virus *in vitro*[a]

N.B. Chemical mutagens present a safety hazard, and should be handled and disposed of accordingly. All parts of this protocol should be carried out under containment conditions appropriate to the virus concerned. All infectious waste should be sterilized before disposal.

Equipment and reagents

- Nitrous acid: 0.7 M $NaNO_2$ in 1 M acetate buffer (sodium acetate, acetic acid), pH 4.8
- Hydroxylamine solution: 1.0 M hydroxylamine in 0.05 M phosphate buffer (Na_2HPO_4, NaH_2PO_4), pH 7.5
- Ethyl methane sulfonate (EMS) solution: 0.5 mg ml^{-1} EMS in 0.25 M sodium acetate
- Tissue culture medium (DMEM or GMEM; Gibco Life Technologies or Flow Labs)

3: Mutagenesis of DNA virus genomes

Method
1. Take approximately 10^9 PFU of purified virus (*Protocol 3*) and dilute tenfold into either nitrous acid, hydroxylamine or EMS solution at 4°C.
2. Take samples every 1–2 min for 10–20 min (nitrous acid), every 2–5 min for 15–60 min (EMS), or every 10–20 min for 1–4 h (hydroxylamine). The larger the viral genome being mutagenized, the more rapidly it will be inactivated by mutation.
3. Dilute samples a hundredfold with cold growth medium, and dialyse against cold growth medium (2 h, 4°C) to remove the mutagen. Store at −70°C.
4. Determine the infectivity remaining at each time point by plaque assay titration under permissive conditions (*Protocol 2B*). Plot a 'kill curve' from these data, and select samples showing residual infectivity of 0.001–0.01% for the isolation of mutants (Section 8).

[a] Modified from references 5 and 6.

5. Mutagenesis through growth of virus in the presence of nucleoside analogues

Mutagenesis may be accomplished *in vivo* by growing virus in cells cultured in the presence of a mutagen, most typically a nucleoside analogue such as 5-bromodeoxyuridine. This chemical can be activated to the triphosphate form within the cell, and utilized by DNA polymerase during the course of viral replication. Incorporation opposite a normal base is potentially mutagenic because, in subsequent rounds of replication, the analogue can be mis-paired with the wrong natural base, so leading to a transition mutation (A-T converted to G-C, or vice versa). As with mutagenesis *in vitro*, the concentration of the mutagen must be controlled to give an acceptable mutation frequency, and the virus produced will be a complex mixture, from which individual isolates must be characterized further (*Protocol 5*).

Protocol 5. Mutagenesis through the use of base-analogues *in vivo* [a]

N.B. Chemical mutagens present a safety hazard, and should be handled and disposed of accordingly. All parts of this protocol should be carried out under containment conditions appropriate to the virus concerned. All infectious waste should be sterilized before disposal.

Equipment and reagents
- Standard cell culture equipment and materials (*Protocol 1*)
- 5-bromodeoxyuridine solution: 1 mg ml^{-1} in sterile distilled water

Protocol 5. *Continued*

Method

1. Take cultures of permissive cells and infect them with virus (*Protocol 2A*).
2. After adsorption of virus and removal of the inoculum, wash the monolayers three times with pre-warmed growth medium (no serum).
3. Add the appropriate volume of growth medium containing 2% serum, supplemented with either no mutagen (control) or bromodeoxyuridine from 5–30 µg ml^{-1}. Within this range, the maximum concentration that can be tolerated by the host cells without loss of viability should be used for mutant isolation. This concentration should be determined empirically for each cell type.
4. Monitor the cultures for the appearance of cytopathic effect in the non-mutagenized control, and harvest mutagenized samples when this is observed (*Protocol 2A*).
5. Dialyse the samples against cold growth medium (2 h, 4°C) to remove the mutagen. Store at −70°C, or use immediately for mutant isolation (Section 8).

[a] Modified from references 7 and 8.

6. Site-directed mutagenesis *in vitro*

The planned mutation of virus genomes, from single base changes in coding or control sequences to the gross deletion of genes, has become the method of choice for generating DNA virus mutants. As for chemical mutagenesis (Section 3), the techniques for generating mutations in viral DNA in a planned way are no different from those applicable to DNA generally. All methods involve the use of one or more mutagenic oligodeoxynucleotides as primers for DNA synthesis from a DNA template, either by the Klenow fragment of *E. coli* DNA polymerase or T4 DNA polymerase or, in a polymerase chain reaction (PCR) format, by a thermostable DNA polymerase.

The original site-directed mutagenesis methods used single-stranded DNA from clones in phage M13 as the template. A mutagenic oligodeoxynucleotide that was homologous to a target site in the template, and which contained in its sequence the mutational changes desired, flanked to both the 5′ and 3′ side by 10–15 nucleotides of sequence matching the template sequence exactly, was used to prime synthesis of the complementary strand, prior to transfection of an M13-susceptible host. The problem that had to be overcome subsequently was the low frequency of mutated DNA clones that was observed, due to the selective action of mismatch repair mechanisms on the newly synthesized, mutated, DNA strand following transfection of the *in vitro* synthesized DNA mixture into an *E. coli* host. Refinements to improve the mutation frequency by selecting against the parental strand, and to apply the procedure to double-stranded DNA templates, have since been made. A previous volume in this series considers these methods in detail (9–11). Commercially available kits for producing mutations by a site-directed approach include Altered Sites™ II (Promega) and Chameleon™ (Stratagene).

PCR protocols have become increasingly popular for site-directed mutagenesis. PCR is quick and straightforward to perform successfully, provided attention is paid to the proper design of primers. A full discussion of the practical problems which may be encountered when using PCR is presented elsewhere (12). PCR provides a means to avoid the problem of bias against an *in vitro* synthesized molecule during cloning since, by the end of the reaction, almost all the DNA present has been made *in vitro* and is virtually homogeneous. By adapting a standard PCR protocol, it is possible to produce product molecules that are mutated compared with the starting template DNA. Most simply, if one or both primers in a standard 2-primer PCR protocol contain sequence differences from the template embedded within them (at least 10 nucleotides from the 3' end to allow for proper hybridization), then the product sequence will be altered accordingly. By arranging for each primer also to incorporate naturally occurring restriction sites towards their 5' ends, the PCR product can then be cut with the relevant enzymes and substituted for the equivalent wild-type fragment in a suitable cloned DNA molecule. This strategy is, however, limited in the sequence alterations it can produce.

A more versatile method for obtaining mutated PCR products for subsequent cloning is described in *Protocol 6*. The protocol uses four primers, two non-mutagenic (A and B) and two mutagenic (C and D), in a two-stage reaction (*Figure 1*). Primers A and B should be positioned so that restriction sites suitable for sub-cloning the mutated DNA fragment (X and Y) lie within the product of PCR amplification with them. Primers C and D are of opposite polarity to each other, and overlap in the region of the desired mutation. In the two first-stage reactions, each mutagenic primer is paired with a flanking primer of opposite polarity to produce products which overlap and contain the same mutation in the overlap region. Small amounts of these two reaction products are then denatured and annealed to each other, and used as a template for amplification by primers A and B. In principle, all of the PCR product should contain the desired mutation. Using this method, any number of different mutations may be isolated in the sequence delineated by restriction sites X and Y, using primers A and B, with appropriate pairs of mutagenic primers in each case.

Several thermostable polymerases are available commercially. These differ in the frequency with which they make errors during DNA strand extension. The probability of such an error occurring increases with the size of the amplified product. A polymerase with a low error frequency is desirable to

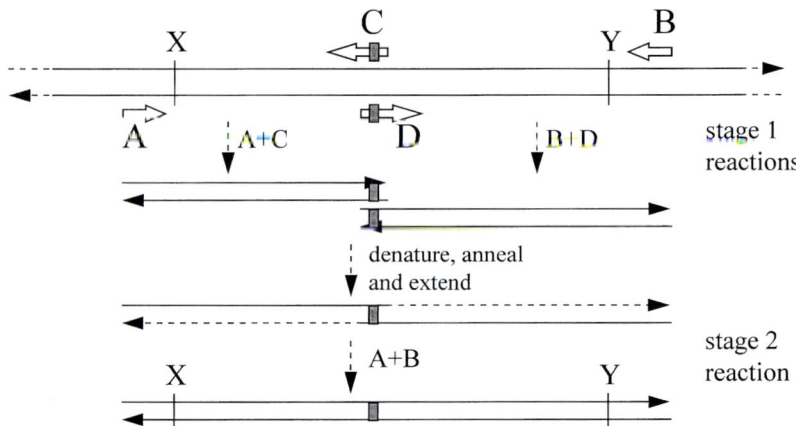

Figure 1. A two-stage PCR strategy for introducing mutations into cloned DNA. Open arrows A–D: PCR primers; X and Y, unique restriction sites for two enzymes; grey boxes, mutated DNA sequence. See reference 13.

minimize the chances of background mutations occurring. Whichever polymerase is used, it will be necessary to sequence fragment X–Y from selected clones to validate the expected mutation, and to exclude the possibility of other changes. For these reasons, the distance X–Y should be kept below 1–1.5 kbp. Non-mutagenic primers should be approximately 20 nucleotides in length, and mutagenic primers approximately 30 nucleotides. If possible, the GC content of primers should be kept between 40 and 60%.

Protocol 6. Synthesis of mutagenized DNA *in vitro* by PCR [a]

Equipment and reagents
- Thermocycler (e.g. Perkin-Elmer, Hybaid)
- Custom-synthesized oligodeoxynucleotide primers (e.g. Gibco Life Technologies)
- Thermostable polymerase and reaction buffer (e.g. New England Biolabs, Promega, Stratagene)
- Agarose gel electrophoresis
- 3M sodium acetate solution
- Phenol:chloroform: 50 % (v/v) liquefied phenol equilibrated with Tris buffer at pH 7 (Fisons), 48% chloroform (trichloromethane), 2% isoamyl alcohol (3-methyl-1-butanol, Sigma)
- TE buffer: 10 mM Tris-HCl pH 7.5, 1 mM EDTA
- 95% ethanol

Method

The PCR amplification conditions described here represent standard conditions that will work for many primer–template combinations. However, it is advisable to optimize the conditions with respect to the annealing temperature in the reaction cycle, and the concentration of Mg^{2+} ions in the reaction buffer, to maximize the yield of the desired product and to minimize the production of background artefacts.

3: Mutagenesis of DNA virus genomes

1. Set up two 25 µl first-stage amplification reactions containing: 200 ng template DNA; 200 ng each of primers A and C (or primers B and D); 0.1 mM each dNTP; 1× enzyme reaction buffer (supplied by the enzyme manufacturer); 2.5 mM $MgCl_2$ (if not included in the manufacturer's buffer); 2–4 units of thermostable polymerase.

2. Overlay the reaction mixes with paraffin oil, and incubate through 20 cycles of denaturation, annealing, and extension (94 °C, 1 min; 55 °C, 1 min; 74 °C, 3 min; extension time increased to 7 min for the final cycle).

3. Check the quality of the reaction products by analysing 2.5 µl samples on agarose gels.

4. Set up a 12.5 µl reaction containing: 2.5 µl of first stage reaction 1 (primers A and C) [b], 2.5 µl of first stage reaction 2 (primers B and D) [b], 200 ng each of primers A and B.

5. Denature at 85 °C for 2 min, and allow to cool to room temperature. Add the following additional components in a final volume of 25 µl: 0.1 mM each dNTP; 1× enzyme reaction buffer (supplied by enzyme manufacturer), 2.5 mM $MgCl_2$ (if not included in the manufacturer's buffer), 2–4 units of thermostable polymerase.

6. Overlay, and subject to thermocycling conditions as in Step 2.

7. Check the quality of the reaction product by analysing a 2.5 µl sample on an agarose gel.

8. Recover the amplification reaction mixture from the paraffin overlay, and dilute it to 100 µl with TE buffer. To clean the DNA, extract once with 100 µl phenol:chloroform, and recover the aqueous phase after brief centrifugation (microfuge). To precipitate the DNA, add 11 µl of 3.0 M sodium acetate, 250 µl 95% ethanol, mix, and incubate at −20 °C for 1 h.

9. Collect the DNA by centrifugation (microfuge, 15 min), dry it in a desiccator under vacuum, and redissolve it in 50 µl of TE buffer.

10. Cut the DNA with enzymes X and Y, according to the manufacturer's instructions, purify the fragment X–Y by agarose gel electrophoresis, and substitute for the non-mutated segment X–Y in a suitable clone of viral genomic DNA.

11. Sequence the segment X–Y of a series of candidate mutated clones. Select a clone with the desired sequence.

[a] Modified from reference 13.
[b] It may sometimes be necessary to purify the first stage reaction products by agarose gel electrophoresis and elution, in order to get a successful second stage reaction.

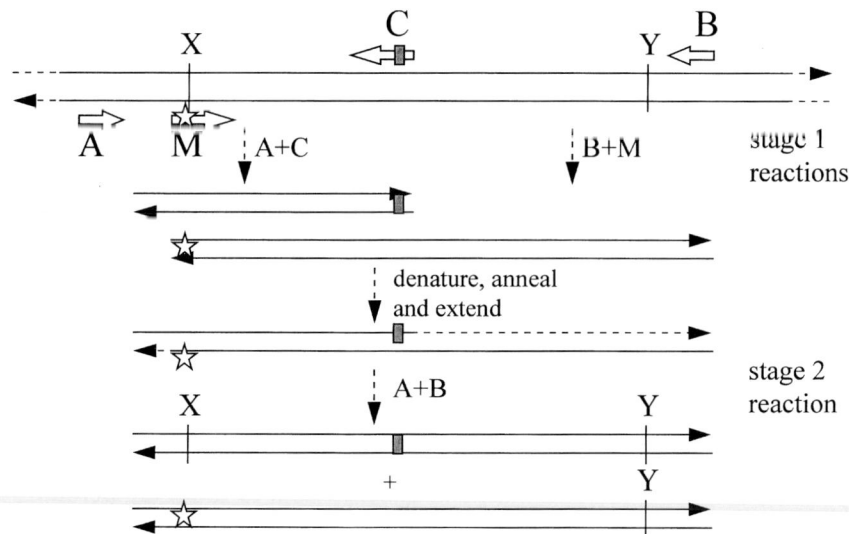

Figure 2. A modified two-stage PCR strategy for introducing mutations into cloned DNA using a single mutagenic primer. Open arrows A–C, M: PCR primers; X and Y: unique restriction sites for two enzymes; grey box, mutated DNA sequence; star, mutation removing restriction site X. See reference 14.

A further modification of this procedure allows a series of mutations to be isolated in a sequence using only one new primer in each case, plus three others (14; see *Figure 2*). This represents a cost saving where large numbers of mutations are to be created. The mutagenic primer D is replaced by primer M, which carries a mismatch with the sequence of restriction site X; DNA amplified using this primer cannot be cleaved with enzyme X. The product of the second-stage amplification will be a mixture of mutated sequences that can be cleaved with X and Y, and wild-type sequences that can be cleaved only with Y. The desired mutated product may therefore be cloned selectively as fragment X–Y.

7. Reintroducing mutagenized sequences into virus

To obtain a mutant virus strain, any DNA mutagenized *in vitro* must be incorporated into virus particles. Unlike bacteriophage λ, there is no *in vitro* packaging system for any eukaryotic DNA virus, so instead this must be accomplished *in vivo*. The experimental strategies and protocols for reconstructing mutated viruses are different for each type of DNA virus, essentially because of the great diversity in genome sizes and conformations among these viruses. However, in each case, DNA must be introduced into cells by transfection. Three different transfection methods have been used in virus

3: Mutagenesis of DNA virus genomes

reconstruction: DEAE-dextran, calcium phosphate, and lipofection (*Protocol 7*). Particular cell types may prove to be more efficiently transfected by one method than another, and the optimal method should be chosen empirically. The protocol is written for transfection of cells in 6 cm (21 cm^2) tissue culture dishes. To transfect other sizes of culture, scale the volumes quoted to suit the different surface area. Whichever transfection system is employed, success depends on the use of highly purified DNA. Lipopolysaccharide contamination of bacterial plasmid preparations is toxic to mammalian cells; purification of plasmid DNA by caesium chloride density gradient centrifugation is generally advisable. All DNA intended for transfection should be sterilized by phenol:chloroform extraction and ethanol precipitation (*Protocol 6*, Steps 8 and 9), and redissolved in sterile water or TE buffer.

Protocol 7. Introduction of DNA into cells by transfection

A. *Calcium phosphate co-precipitation*[a]

This protocol has been developed for use in recombinant adenovirus construction using 293 cells. Some cell types will tolerate longer incubations with the DNA precipitate, and transfection efficiency may be increased by modifying this protocol accordingly. For prolonged incubations, the volume of the DNA-calcium-Hepes buffer mixture (Step 3b), and the volume of medium to which it is added (step 3c), should be scaled up to prevent the cell monolayer from deteriorating. In this case, the DNA precipitate can be added directly to the culture medium in the dish, and left for up to 16 h.

Equipment and reagents

- Standard cell culture equipment (*Protocol 1*)
- 1.25 M CaCl$_2$ solution, filter-sterilized.
- Phosphate solution: 10 g l^{-1} Na$_2$HPO$_4$, filter-sterilized
- Carrier DNA: 10 mg ml^{-1} salmon or herring sperm DNA (Sigma), sterilized by phenol:chloroform extraction and ethanol precipitation (*Protocol 6*, Steps 8 and 9, but replace the −20°C incubation with a +4°C incubation), and redissolved in water (requires extended incubation at +4°C)
- Hepes buffer: 16 g l^{-1} NaCl, 0.74 g l^{-1} KCl, 2.0 g l^{-1} glucose, 10.0 g l^{-1} N-2-hydroxyethylpiperazine-N'-2-ethanesulfonic acid, pH 7.08 (Sigma), filter-sterilized[b]
- TS buffer: sterile TD buffer (*Protocol 3*) containing 0.1 g l^{-1} CaCl$_2$ and 0.1 g l^{-1} MgCl$_2$ (added from a filter-sterilized stock solution)
- TS/20% glycerol: TS buffer containing 20% (v/v) final concentration of glycerol

Method

1. Plate cells at 50% confluence[c] on 6 cm dishes in standard growth medium (*Protocol 1*).
2. 16–24 h later, prepare the DNA for transfection.
   (a) Mix the DNA for transfection (1–5 μg) with 10 μg carrier DNA[d] and 10 μl 1.25 M CaCl$_2$ solution, in a final volume of 50 μl.

Protocol 7. *Continued*

    (b) Mix 50 μl Hepes buffer and 1 μl phosphate solution,[e] and add the calcium-DNA mixture to this solution dropwise over 1 min, mixing them by blowing bubbles through the liquid with a mechanical pipettor. Allow the mixture to stand for 20–30 min to allow a fine precipitate to form.

    (c) Add 0.9 ml of growth medium containing 5% serum to the DNA precipitate.

3. Remove the growth medium from the culture, and add the DNA mixture to the cell sheet. Rock the dish to distribute the mixture across the dish, and return it to the incubator. Repeat the redistribution every 20–30 min for 3 h.

4. Apply a 'glycerol shock' to the cell sheet to promote DNA uptake.

    (a) Aspirate the inoculum, and gently add 1.0 ml TS/20% glycerol to the cell sheet. Apply the solution at one side of the dish when tilted, then bring the dish back to horizontal to spread the solution across the monolayer.

    (b) After exactly 1 min, remove the solution, and rapidly but carefully wash the monolayer twice with prewarmed TS buffer.

    (c) Either add an appropriate volume of growth medium containing serum as required for the cell type or, if plaques are required, overlay with a solidified medium (*Protocol 2B*). In some cases, it may be beneficial to delay the application of solidified medium for 24 h, maintaining the cells in liquid medium during this time to allow proper recovery from the glycerol shock.

B. DEAE-dextran method[f]

Cell monolayers are generally disrupted significantly during DEAE-dextran transfection, and are not generally amenable therefore to the direct application of an agar overlay to permit plaque formation. The toxicity of the DEAE-dextran reagent varies between cell types, and the optimal concentration in the assay should be determined in each case (useful range 50–500 μg ml^{-1}).

Equipment and reagents

- Standard cell culture equipment and materials (*Protocol 1*)
- DEAE-dextran solution: 10 mg ml^{-1} diethylaminoethyl (DEAE)-dextran (mol. wt approximately 500 000 Da; Sigma) in PBS (*Protocol 2*), filter sterilized, and stored in aliquots at 4 °C
- Chloroquine solution: 100 mM chloroquine diphosphate (Sigma) in PBS, filter sterilized, and stored at 4 °C
- DMSO solution: 10% (v/v) dimethyl sulfoxide in PBS (*Protocol 2*), filter sterilized

3: Mutagenesis of DNA virus genomes

Method

1. Prepare cells for transfection as in *Protocol 7A*.
2. 16–24 h later, prepare the DNA for transfection.
   (a) Take 2 ml of growth medium containing 2% serum, and add 2 μl of chloroquine solution. Warm the mixture to 37 °C.
   (b) Add DNA to a final concentration of 1–5 μg ml^{-1} from stock solution(s) at 0.1–1.0 mg ml^{-1} in TE buffer.
   (c) Add 20 μl prewarmed DEAE-dextran solution, and mix gently.
3. Aspirate the medium from the culture, and replace with the DNA-DEAE-dextran mixture. Incubate the culture for up to 4 h, viewing it periodically to check the health of the cells. The procedure is toxic to cells, and some cells will be lost during the incubation. It may be appropriate to terminate the incubation early if the health of the culture is deteriorating.
4. Apply a 'DMSO shock' to promote DNA uptake.
   (a) Remove the transfection mix. Add 5 ml prewarmed DMSO solution, and incubate for 2–5 min.
   (b) Aspirate the DMSO solution, and replace with 5 ml growth medium containing the normal serum concentration.
5. Change the medium on the culture after 6–16 h, to remove residual DMSO.

C. *Lipofection*

The key component in the lipofection procedure is the liposome preparation. Manufacturers provide detailed methods for use with their reagents. Different types of lipid combination, in different ratios to DNA, will be optimal for each cell type. In the absence of specific guidance about a cell type, the best reagent for the application in question, and its ideal concentration, must be determined empirically.

Equipment and reagents

- Standard cell-culture equipment and materials (*Protocol 1*)
- Polystyrene tubes[a] (e.g. Falcon, Becton Dickinson)
- OptiMEM-1 reduced serum growth medium (Gibco Life Technologies)
- Liposome reagent: e.g. Lipofectin™, LipofectACE™, LipofectAMINE™ (all Gibco Life Technologies), CLONfectin™ (Clontech), Transfectam™ (Promega), LipoTAXI™ (Stratagene)

Method

1. Prepare cells for transfection as in *Protocol 7A*.
2. 16–24 h later, prepare the DNA for transfection.

Protocol 7. *Continued*

(a) Place two 0.1 ml aliquots of OptiMEM-1 in polystyrene tubes.[g]

(b) Re-mix the liposome stock, take 15 µl (range 5–20 µl) and add this to the contents of one tube. To the other tube, add 2 µg DNA. Allow both to stand for 30–45 min at room temperature.

(c) Mix the DNA-OptiMEM and the liposome-OptiMEM preparations gently together, and allow to stand for 10–15 min at room temperature.

3. Aspirate the growth medium from the culture, wash the monolayer with serum-free growth medium, and aspirate again.

4. Dilute the OptiMEM-DNA-liposome mixture with 1.8 ml serum-free growth medium, and add gently to the cell monolayer.

5. Incubate the culture with the DNA mix for 5–24 h. At the end of this time, remove the DNA mix, and replace with an appropriate volume of normal growth medium containing serum.

[a] Originally described in reference 15.
[b] The pH of this reagent is a crucial determinant of transfection efficiency, and should not vary by more than 0.05 pH units from this figure.
[c] Optimum transfection efficiencies are obtained when cells are healthy and rapidly dividing. If the purpose of the experiment demands that the cells be overlaid with solidified medium after transfection, so that plaques can form, the cell density should be increased to 90–95%. Cells in such virtually confluent monolayers should be stimulated by refeeding with fresh growth medium containing 10% fetal bovine serum 3 h prior to transfection.
[d] Some investigators have found that this high molecular weight DNA may be substituted with more readily prepared plasmid DNA, such as an empty cloning vector containing no sequences relevant to the experiment.
[e] The volume of phosphate solution may be adjusted slightly as necessary to obtain a fine precipitate by the end of the 20 min incubation.
[f] Modified from reference 16.
[g] The use of low-protein medium and this type of plastic tube are both crucial to obtaining efficient transfection.

7.1 Complementing cell lines

Since it is usually not possible to predict whether or not a mutation created *in vitro* will be viable within the context of a viral genome, isolation of mutated virus should be attempted if possible under conditions where any gene defect which may exist is complemented. The most commonly used approach is to attempt virus isolation by transfection of DNA into specially derived cell lines which have been engineered to express, in wild-type form, the viral gene that has been mutagenized *in vitro* (*Protocol 8*). A number of useful complementing cell lines are listed in *Table 2*.

Protocol 8 describes a generic method which can be applied, in principle, to any gene product from any virus. Plasmid DNA containing a promoter-cDNA

3: Mutagenesis of DNA virus genomes

Table 2. Cell lines complementing specific DNA virus gene deficiencies

Cell Line	Parent line	Virus genes expressed	Reference
Cos1, Cos7	monkey CV1	SV40 T antigens	17
293	human embryo kidney	Ad5 E1a and E1b	18
C34	human 293	Ad5 E1a, E1b, IX	19
KB8, 16, 18	human KB	Ad2 E1a, E1a and E1b, or E1b respectively	20
–	human A549	Ad5 E1a (and E1b)	21
293-C2	human 293	Ad5 E2a	22
293-E2a	human 293	Ad5 E2a	23
293-pTP$_{14}$, -pTP$_{40}$	human 293	Ad5 E2B pTP	24
W162	monkey Vero	Ad5 E4	25
VK2-20, 10-9	human 293	Ad5 E4	26
293-Orf6	human 293	Ad5 E4Orf6	27
293-E4	human 293	Ad5 E4	23
293-E4Orf6, 6/7	human 293	Ad5 E4 Orf6 and Orf6/7	23
D6	monkey Vero	HSV-1 gB	28
L3153$_{28}$	mouse Ltk⁻	HSV-1 gC	29
L65	hamster BHKtk⁻	HSV-1 ICP4 (IE175)	30
Z4	mouse Ltk⁻aprt⁻	HSV-1 ICP4 (IE175)	31
2-6	monkey Vero	HSV-1 ICP4 (IE175)	32
G5	monkey Vero	HSV-1 U$_L$16–21	33
171	rabbit skin	HSV-1 U$_L$17, 18	34
RK13-K1L	rabbit RK13	vaccinia K1L	35

construct and a selectable marker gene is transfected into cells, and positive selection is applied for the retention of plasmid sequences. If the plasmid vector contains a functional replicon, e.g. from Epstein–Barr virus (Invitrogen), it can be maintained as a multi-copy episome, and a bulk culture of expressing cells is obtained. Otherwise, those rare cells in which the DNA has randomly integrated into the chromosome can be selected and grown up as individual clones. In practice, many viral proteins are toxic to cells, especially when expressed at high levels; attempts to isolate stable cell lines expressing such proteins will produce either no surviving cells, or else cells that grow very poorly. In such cases, the cDNA may be expressed from an inducible promoter, so that protein is produced only when required. Historically, it has been difficult to obtain eukaryotic promoters that are sufficiently tightly regulated for this purpose. Commercially available systems which have been engineered to minimize this problem are the Lac switch system (Stratagene), the Tet On/Off system (Clontech), and the ecdysone-inducible system (Invitrogen). However, there can be no certainty that a cell line expressing any given protein will be obtained, by whatever strategy is chosen.

Protocol 8. Isolation of complementing cell lines

Equipment and reagents
- Standard cell culture equipment and materials (*Protocol 1*)
- Transfection reagents (*Protocol 7*)
- Hygromycin B solution: 50 mg ml^{-1} in PBS (*Protocol 3*), sterile (Boehringer)
- G418 solution: 50 mg ml^{-1} Geneticin (G418 sulfate) (Gibco Life Technologies) in PBS (*Protocol 3*), filter-sterilized
- Plasmid expression vector containing cloning site(s) downstream of a constitutive or inducible promoter, and an expression cassette for a dominant selectable marker gene such as neomycin resistance or hygromycin resistance (Invitrogen, Stratagene, Promega, Clontech, and others)

Method
1. Clone the cDNA for the protein to be expressed under the control of a suitable promoter within a plasmid expression vector.
2. Transfect a culture of the parental cells, from which the complementing line is to be established, with the expression plasmid from Step 1 (*Protocol 7*). Control transfections with the vector alone, and with no plasmid, should also be performed.
3. One day post-transfection, passage the transfected cells into a series of replicate cultures. After the cells have replated overnight, add the appropriate selection agent to the cultures at a series of concentrations (hygromycin: 50–300 µg ml^{-1}; G418: 100–500 µg ml^{-1}).
4. Observe the cultures daily for colonies of cells growing through the selection against a background of cell death, changing the medium every 4 days, and maintaining the concentration of selective agent. Use the controls to confirm that the parental cells are killed by the drug concentration used, and that the empty vector is capable of conferring drug resistance.
5. If a vector capable of episomal persistence has been used, grow the drug-resistant cells through into a bulk culture, and confirm protein expression from the cDNA. If stable integration of the plasmid is required, pick and grow on at least ten individual colonies. When cell numbers have been expanded sufficiently, characterize each cell line for the presence and level of expression of the desired protein; the level of expression will vary between individual clones due to position effects on the expression cassette of the different sites of integration.
6. Prepare batches of aliquots of each cell line for long-term storage (*Protocol 1*).

7.2 Polyomaviruses

SV40 and polyoma virus, members of the polyomavirinae, have small circular double-stranded DNA genomes of 5–6 kbp length (36). These genomes each

3: Mutagenesis of DNA virus genomes

contain unique sites for digestion by one or more restriction enzymes, for example *Eco*R1, which can be used to clone the entire genome as a single fragment into commonly used cloning vectors. *In vitro* mutagenesis may be performed on either this clone or, if necessary, a suitable sub-clone can be mutagenized and then the viral DNA fragment re-built into the full-length clone, all using standard DNA manipulation techniques. Infectious virus can be reconstituted by liberating the full-length cloned genome from the plasmid with the relevant enzyme, religating its compatible ends *in vitro*, and then transfecting the DNA into susceptible, permissive cells (*Protocol 9*). In principle, the virus obtained from this procedure should be homogeneous. However it is good practice to plaque-purify isolates for further analysis (*Protocol 2*). All parts of this protocol should be carried out under containment conditions appropriate to the virus concerned. All infectious waste should be sterilized before disposal.

Protocol 9. Rescue of infectious SV40 from cloned DNA

Equipment and reagents

- Standard cell culture equipment and materials (*Protocol 1*)
- DNA purification reagents (*Protocol 6*)
- Transfection reagents (*Protocol 7*)

Method

1. Take 20 µg of full-length viral genomic DNA, cloned at a unique site in a plasmid vector. Cut the DNA to release the viral DNA from the vector, using the appropriate restriction enzyme in a 100 µl reaction according to the manufacturer's instructions.
2. Purify the DNA from the reaction (*Protocol 6*, Steps 8 and 9).
3. Recircularize the viral DNA in a 100 µl ligation reaction, using 10 units of T4 DNA ligase (Gibco-BRL) and reaction buffer according to the manufacturer's instructions.
4. Re-purify the DNA, redissolving it in 50 µl H_2O.
5. Transfect the DNA into permissive cells, such as CV-1 or, if early region-defective viruses are required, COS1 (*Protocol 7*).
6. Harvest the cells and medium as a crude stock after approximately 7 days, or when cytopathic effect is seen. Store at $-70\,°C$. If the culture has been overlaid with solid medium, pick individual plaques (*Protocol 2*).

7.3 Parvoviruses

Adeno-associated virus is a helper-dependent parvovirus with a single-stranded DNA genome of about 4800 nucleotides length (37). Helper functions can be provided by wild-type adenovirus or some herpesviruses. The mechanism of replication is such as to permit the rescue of infectious virus directly from a full-length cloned genome copy within a circular plasmid molecule (38), provided helper functions are present (*Protocol 10*). All parts of this protocol should be carried out under containment conditions appropriate to the virus concerned. All infectious waste should be sterilized before disposal. For larger-scale preparation of recombinant AAV, virus should be concentrated and then purified by equilibrium density gradient centrifugation (39).

Protocol 10. Rescue of infectious adeno-associated virus from cloned DNA[a]

Equipment and reagents

- Standard cell culture equipment and materials (*Protocol 1*)
- Transfection reagents (*Protocol 7*)
- Lysis buffer: 10 mM Tris-HCl pH 8.5, 100 mM NaCl

Method

1. Take 10 μg of plasmid DNA containing a full-length mutant or recombinant AAV genome, if necessary together with 10 μg AAV helper plasmid, and transfect into a just sub-confluent monolayer of HEK-293 cells on a 10 cm culture dish (*Protocol 7*); include wild-type adenovirus in the transfection mix, sufficient to give a multiplicity of infection of 2.[b]

2. Harvest the cells after 48–60 h, pellet them by low-speed centrifugation, and then resuspend in 1.0 ml of lysis buffer. Freeze–thaw the suspension four times to release the virus.

3. Heat the sample to 56 °C for 30 min to inactivate contaminating adenovirus, and store at −70 °C.

[a] Modified from reference 39
[b] The isolation of recombinant AAV without using infectious Ad as a helper has been reported recently (40).

7.4 Adenoviruses

Adenoviruses characterized to date have linear double-stranded DNA genomes of between 33 kbp and 44 kbp length; the viruses most widely studied by genetic and other approaches have been human serotypes 2 and 5, which have 36 kbp genomes (41). The genome ends are covalently linked to a protein, which must be removed before the terminal fragments can be cloned (42). Whilst clones of the entire genome of these viruses have been obtained, their size makes it inconvenient to use *in vitro* mutagenesis strategies on them.

3: Mutagenesis of DNA virus genomes

Instead, subclones are mutated, and then complete genomes are reconstituted. This may be achieved either *in vitro*, through the religation of two or more compatible DNA fragments followed by transfection of permissive cells (43), or *in vivo*, by transfection and homologous recombination between the mutagenized plasmid and an overlapping fragment, either cloned, or isolated from viral genomic DNA (44, 45; *Protocol 11*). A system in which full-length genomes are recreated by homologous recombination between plasmids in bacteria, prior to transfection into mammalian cells, has also been reported (46).

Protocol 11. Rescue of infectious adenovirus from cloned DNA

Equipment and reagents
- Standard cell culture equipment and materials (*Protocol 1*)
- DNA purification reagents (*Protocol 6*)
- Transfection reagents (*Protocol 7*)

Method

1. Prepare DNA for transfection by one of the following procedures:

   (a) (i) Cut the mutagenized DNA fragment from its plasmid, and similarly digest viral genomic DNA (*Protocol 3*) to produce 1 or 2 fragments which, when ligated to the mutated fragment, will comprise a complete genome.[a] Purify the relevant DNA fragments from these digests by recovery from agarose gels.

   (ii) Combine the fragments in a 100 μl ligation reaction (*Protocol 9*, Step 4), using a 3:1 molar excess of mutagenized fragment over the other fragment(s), and approximately 5 μg total amount of DNA. Purify the ligated DNA (*Protocol 6*, Steps 8 and 9).

   (b) (i) Prepare the mutagenized sequence in a plasmid containing viral sequences which include a genome end. Cut the plasmid to liberate the genome end using an appropriate restriction enzyme. Also digest viral genomic DNA to produce a single fragment which contains all those sequences not present in the cloned fragment, as well as sequences overlapping with it, such that recombination between homologous sequences in the two fragments will create a complete genome.

   (ii) Purify both DNA samples (*Protocol 6*, Steps 8 and 9), and combine the cloned and genomic fragments in a 3:1 molar ratio, using in total approximately 5 μg DNA.

2. Transfect the DNA into 293 cells (*Protocol 7*).

3. Recover individual plaques directly, or prepare a plate lysate from which plaques can then be obtained (*Protocol 2B*).

Protocol 11. *Continued*

4. Screen plaque isolates for the desired sequence alteration, using a PCR-based assay designed either to differentiate directly between wild-type and mutated genomes, or else to provide a fragment for DNA sequence analysis.

[a] If fragments are to be assembled by ligation in the order viral, cloned (mutated), viral, it may be useful to treat the viral DNA digest with calf intestinal alkaline phosphatase (Boehringer) to reduce further the background of wild-type virus arising in the experiment.

7.5 Herpes viruses and pox viruses

The herpes viruses constitute a diverse family, with characterized members having linear double-stranded DNA genomes of between 125 and 220 kbp length. The greatest amount of genetic analysis has been done with the related viruses, HSV types 1 and 2, which have genomes of about 152 kbp (47). Of the pox viruses, the virus subjected to the most intense analysis has been vaccinia, which has a linear genome of around 190 kbp (48). The genomes both HSV and vaccinia are too large to be handled either as single clones or even as two or three overlapping clones. Therefore, reconstruction of DNA mutated *in vitro* into virus is achieved by transfecting cells with the mutated DNA in plasmid form, and then superinfecting with wild-type virus to allow recombination *in vivo* between the cloned DNA and the viral genome to substitute the mutagenized sequences for the corresponding wild-type sequence. Some method of selection for the required recombinant is necessary, since otherwise the recombinants are hard to detect against the high background of wild-type virus. For both viruses, the general technique for generating specific null mutants is to provide a selectable marker gene in the plasmid at the site of the deletion mutation, so that recombinant viruses can be grown selectively (49, 50). As an alternative, for genes whose function is absolutely essential for infectivity, complementing cell lines may be established (Section 7.1) and used for virus isolation by *in vivo* recombination; plaque isolates must then be screened for their ability to grow only in the complementing cells (32).

Protocol 12. Rescue of infectious HSV or vaccinia virus from cloned DNA

Equipment and reagents

- Selection agents: X-gal, 40 mg ml^{-1} in dimethylformamide; mycophenolic acid; xanthine; and hypoxanthine, 10 mg ml^{-1}, in 0.1 M NaOH for gpt selection; 100x HAT for tk selection: 1.4 mg ml^{-1} hypoxanthine, 0.4 mg ml^{-1} thymidine, 0.45 mg ml^{-1} (+)amethopterin (methotrexate) (all from Sigma)
- Standard cell culture equipment and materials (*Protocol 1*)
- DNA purification reagents (*Protocol 6*)
- Transfection reagents (*Protocol 7*)

Method

1. Prepare plasmid DNA containing the mutated sequence, with an inserted marker or selectable gene if applicable, flanked by wild-type viral genomic sequences. Marker genes which can be used include β-galactosidase; selectable genes include gpt and tk.
2. For HSV, co-transfect permissive cell monolayers with 1 µg wild-type viral genomic DNA (*Protocol 3*) and 5 µg recombinant plasmid DNA (*Protocol 7*). For vaccinia virus, cells should be infected with wild-type virus at an m.o.i. of 0.05–0.2 PFU per cell, 2 h prior to transfection with plasmid DNA alone.
3. Harvest the cultures at 48 h post-transfection as crude virus stocks (*Protocol 2A*).
4. Plate out virus from these stocks to obtain plaques (*Protocol 2B*). Either apply appropriate selection conditions to obtain plaques of recombinant virus, or identify recombinant plaques by staining for a marker gene.[a]

[a] Stain for β-galactosidase using X-gal at 0.4 mg ml^{-1} in a second agar overlay. Select for gpt using 25 µg ml^{-1} mycophenolic acid, 250 µg ml^{-1} xanthine, 15 µg ml^{-1} hypoxanthine; select for tk using 1 × HAT in tk$^-$ cells.

8. Selection of mutant phenotypes in randomly mutagenized stocks

Random mutagenesis, either of DNA *in vitro* (Section 3) followed by virus reconstruction (Section 7), or of particles *in vitro* or *in vivo* (Sections 4 and 5), results in a mixed population of mutant virus strains together with a background of wild-type. Clonal populations of virus must be obtained by isolating individual plaques from the mixed stocks produced by these experiments (Section 2). These plaque-purified isolates must then be characterized phenotypically. If mutations in a particular gene are required, it may be possible to isolate non-conditional defective mutants in that gene by conducting the primary isolation on complementing cells (Section 7.1, and *Table 2*). Such isolates must then be screened for growth on non-complementing cells to exclude wild-types and those with non-defective mutations in other genes. If no complementing cells are available, then only mutants that have a conditional phenotype can be obtained (Section 1.3.4). For example, to obtain temperature-sensitive mutant strains, primary isolation is performed at a low growth temperature, typically 32°C, and individual plaque isolates are then tested alongside a wild-type control, for the ability to produce infectious progeny at a high growth temperature (39°C). Stocks obtained at the two temperatures are then titred in parallel by plaque assay at the lower, permissive, temperature (*Protocol 2B*). Using the ratio of infectious yield of wild-type at the two temperatures as a base-line, mutant isolates showing significantly reduced growth only at the higher temperature can be identified. As another example, if two or more cell lines permissive for growth of wild-type virus are available, potentially mutant plaque isolates obtained on one cell type can be screened for growth on another cell type; those failing to grow are defined as host-range mutants. Note that screens of this type give no information about which gene(s) are carrying the mutation(s) that are responsible for the phenotype. Once virus with an interesting phenotype has been identified, plaque-purification should be repeated two or three times using the phenotype as a screen to exclude any contaminating virus (*Protocol 2B*).

9. Mapping mutations in isolates obtained by random mutagenesis

The mutation(s) responsible for the phenotype of a virus obtained by random mutagenesis could lie anywhere in the genome. For viruses with small genomes, it is practical to clone and sequence completely the genome to identify differences from the wild-type parent. Since there is likely to be a background of silent mutations in addition to the changes responsible for the phenotype, each difference detected must then be separately incorporated back into a wild-type genome to test its phenotypic effect. If the number of differences is large, then 'domain-swap' experiments can be done to narrow down the part of the genome in which the significant mutation lies. In this type of experiment, both wild-type and mutant virus genomes are obtained as plasmid clones, and then part A of the wild-type genome is linked to part B of the mutant genome and vice versa. Viruses are then reconstituted from both hybrid plasmids, and tested for the mutant phenotype. If the phenotype is due to a single mutation, one virus should be wild-type and the other mutant. The mapping of the phenotypic mutation may then be further refined using additional domain swaps. The use of this strategy has been elegantly demonstrated for polyoma viruses with altered pathogenicity properties (51).

The domain-swap approach to mapping mutations is really a planned, *in vitro* version of the classic approach to mapping mutations *in vivo*, which is marker rescue. In this approach, replicate cell cultures are infected with a mutant virus under restrictive conditions (high temperature for a ts mutant, non-permissive cell line for an hr mutant) (*Protocol 2A*), and transfected with each one of a series of purified DNA fragments produced by restriction enzyme digestion, which together cover the entire wild-type viral genome (*Protocol 7*). Yield of infectious virus in each culture is then assessed by plaque assay (*Protocol 2B*). In principle, only one fragment of the series should be able to support virus growth. This fragment will 'cover' the site of the phenotypic mutation in the virus genome, and recombine with it to produce wild-type, which can then grow under the restrictive conditions.

Fragments which do not cover the site of mutation in this way can still recombine with the viral genome, but the products will still have a mutant phenotype and so will not grow. From a restriction map of the genome it is then possible to deduce the approximate location of the mutation. The use of further series of restriction fragments produced with other enzymes in similar assays allows the location to be refined with increasing precision, until DNA sequencing can be applied to pin-point the lesion. The application of this strategy to the mapping of adenovirus mutants is exemplified in reference 52.

10. Characterization of mutant viruses

The methods used to characterize a newly isolated mutant virus are diverse, and cannot be covered comprehensively in this chapter. However, certain fundamental techniques are generally useful and are described below, exemplified by their application to human adenovirus 5. In each case, the mutant is compared with wild-type in parallel infections carried out at the same multiplicity of infection (m.o.i.), under conditions where the mutant phenotype is expressed (i.e. standard cell type and growth temperature). For each experiment, cell cultures are infected at an m.o.i. sufficient to infect all the cells at the start of the experiment (typically 10 p.f.u. per cell). Since the m.o.i. used affects both the kinetics of virus growth and, in some cases, the mutant phenotype, it is crucial that this parameter is carefully controlled through the use of accurately titred virus stocks; ideally, the infectious titres of the mutant and wild-type stocks to be used should be determined in parallel, on exactly equivalent cell cultures (*Protocol 2B*). These methods may be applied to viruses other than Ad5, subject to adjustment of the host cell type used for infection and the time scale over which the infectious cycle is followed (*Table 1*).

10.1 Single step growth curve
Replicate wild-type and mutant infected cultures are harvested at times post-infection from 2 h to 72 h, and the lysates titred by plaque assay to determine the amount of infectious virus present (*Protocol 2*). This experiment can show whether progeny production by the mutant strain is reduced, delayed, or both compared with wild-type (*Figure 3*).

10.2 DNA replication assay
Total cell DNA is prepared from replicate wild-type and mutant virus-infected cultures harvested at various times post-infection. Equivalent amounts of DNA are digested with a restriction enzyme, and analysed by Southern blotting. Alternatively, samples of the uncut DNA may be immobilized on a filter using a dot- or slot-blot apparatus, and detected similarly. This experiment can reveal the time at which DNA replication begins, and the kinetic profile of DNA production thereafter (*Figure 4*). Since DNA viruses

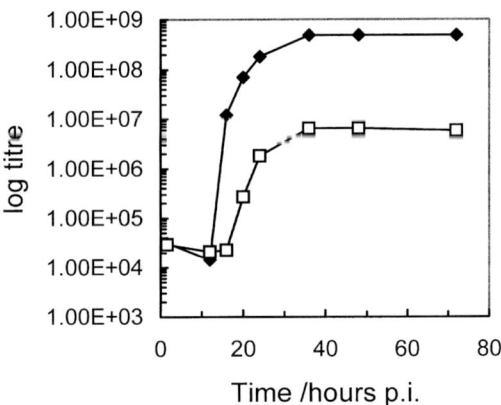

Figure 3. A comparison between the single-step growth curves of wild-type (closed symbols) and an E1b mutant adenovirus (open symbols), with virus yields (log PFU) shown against time (hours) post-infection (p.i.) at 37 °C. The mutant virus shows both a reduction in final yield of about a hundred-fold, and an eclipse period extended by about 4 h compared with the wild-type.

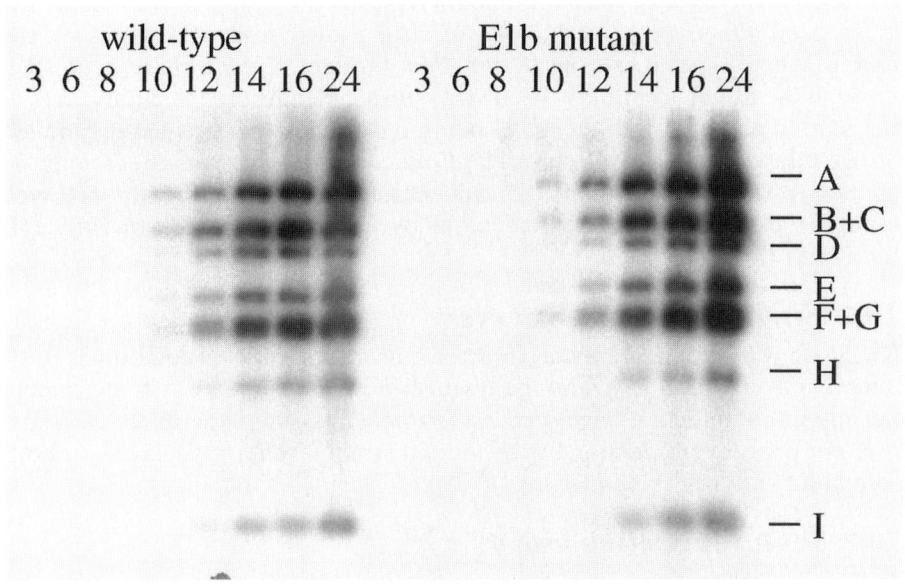

Figure 4. Viral DNA levels during the course of wild-type (left panel), or an E1b mutant adenovirus (right panel) infection. Samples of total infected cell DNA were harvested at the times indicated (hours post-infection), digested with restriction enzyme HindIII, separated by agarose gel electrophoresis, blotted, and probed with ^{32}P-labelled total adenovirus DNA produced by nick translation. In this case, the mutant virus shows an identical time of onset of DNA replication to wild-type, and similar accumulation kinetics thereafter, indicating that there is no defect in the early phase of infection.

3: Mutagenesis of DNA virus genomes

have replication cycles in which the onset of DNA replication divides gene expression into early and late phases, these data can show whether the product of the mutated gene is required early or late in the cycle.

10.3 Late protein expression

For most viruses grown in permissive cultures, the rate of synthesis of the late, mostly structural, proteins is sufficient for them to be detected in infected cell lysates after short pulse-labelling with ^{35}S-methionine, without the need to use specific antibodies. Replicate infected cultures are labelled at various times post-infection, total cell lysates are prepared, samples are analysed by SDS polyacrylamide gel electrophoresis, and proteins are detected by autoradiography. This experiment can reveal differences in the kinetics of late protein production by a mutant virus (*Figure 5*).

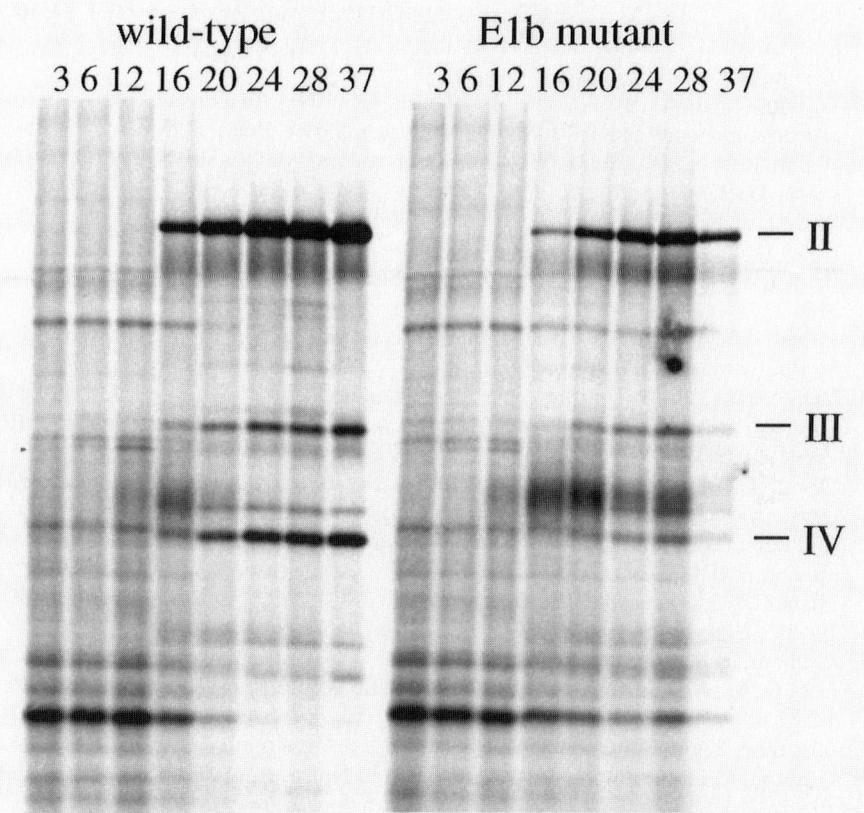

Figure 5. Protein synthesis during the course of wild-type (left panel), or an E1b mutant adenovirus (right panel) infection. Cultures were pulse-labelled with ^{35}S-methionine for 30 min prior to harvest at the times indicated (hours post-infection), and samples analysed by SDS polyacrylamide gel electrophoresis on an 8% gel. Note the delayed onset and reduced level of expression of the capsid proteins II, III, and IV in the mutant infection.

10.4 Virus assembly

If a mutant virus is defective in growth but shows no discernible defect in either replication or late gene expression, it is most likely that there is a problem with virus assembly. Viruses often assemble via intermediate particle structures which have different densities from mature particles, and which can therefore be separated from them by caesium chloride density gradient centrifugation (*Protocol 3*). Differences in the relative proportions of intermediate and mature particles between wild-type and mutant indicates an assembly defect (see reference 53 for an example of this type of analysis).

References

1. Sureau, C., Romet-Lemonne, J-L., Mullins, J. I., and Essex, M. (1986). *Cell,* **47,** 37.
2. Sells, M. A., Chen, M. L., and Acs, G. (1987). *Proc. Natl. Acad. Sci. USA,* **84,** 1005.
3. Frattini, M. G., Lim, H. B., and Laimins, L. A. (1996). *Proc. Natl. Acad. Sci. USA,* 93, 3062.
4. Walton, C., Booth, R. K., and Stockley, P. G. (1991). In *Directed mutagenesis: a practical approach* (ed. M. J. McPherson), p. 135. IRL Press, Oxford.
5. Williams, J. F., Gharpure, M., Ustacelebi, S., and McDonald, S. (1971). *J. Gen. Virol.,* **11,** 95.
6. Pringle, C. R. (1975). *Curr. Topics Microbiol. Immunol.,* **69,** 85.
7. Bader, J. P., and Brown, N. R. (1971). *Nature New Biol.,* **234,** 11.
8. Brown, S. M., Ritchie, D. A, and Subak-Sharpe, J. H. (1973). *J. Gen. Virol.,* **18,** 329.
9. Carter, P. (1991). In *Directed mutagenesis: a practical approach* (ed. M. J. McPherson), p. 1. IRL Press, Oxford.
10. Yuckenberg, P. D., Witney, F., Geisselsoder, J., and McClary, J. (1991). In *Directed mutagenesis: a practical approach* (ed. M. J. McPherson), p. 27. IRL Press, Oxford.
11. Olsen, D. B., and Eckstein, F. (1991). In *Directed mutagenesis: a practical approach* (ed. M. J. McPherson), p. 83. IRL Press, Oxford.
12. McPherson, M. J., Quirke, P., and Taylor, G. R. (1991). *PCR: a practical approach.* IRL Press, Oxford.
13. Higuchi, R., Krummel, B., and Saiki, R. K. (1988). *Nucl. Acids Res.,* **16,** 7351.
14. Ito, W., Ishiguro, H., and Kurosawa, Y. (1991). *Gene,* **102,** 67.
15. Graham, F. L., and Van der Eb, A. J. (1973). *Virology,* **52,** 456.
16. Lopata, M. A., Cleveland, D. W., and Sollner-Webb, B. (1984). *Nucl. Acids Res.,* **12,** 5707.
17. Gluzman, Y. (1981). *Cell,* **23,** 175.
18. Graham, F. L., Smiley, J., Russell, W. C., and Nairn, R. (1977). *J. Gen. Virol.,* **36,** 59.
19. Caravokyri, C., and Leppard, K. N. (1995). *J. Virol.,* **69,** 6627.
20. Babiss, L. E., Young, C. S. H., Fisher, P. B., and Ginsberg, H. S. (1983). *J. Virol.,* **46,** 454.
21. Imler, J. L., Chartier, C., Dreyer, D., Dieterle, A., SainteMarie, M., Faure, T., Pavirani, A., and Mehtali, M. (1996). *Gene Therapy,* **3,** 75.

22. Zhou, H. S., O'Neal, W., Morral, N., and Beaudet, A. L. (1996). *J. Virol.*, **70,** 7030.
23. Lusky, M., Christ, M., Rittner, K., Dieterle, A., Dreyer, D., Mourot, B., Schultz, H., Stoeckel, F., Pavirani, A., and Mehtali, M. (1998). *J. Virol.*, **72,** 2022.
24. Langer, S. J., and Schaack, J. (1996). *Virology,* **221,** 172.
25. Weinberg, D. H., and Ketner, G. (1983). *Proc. Natl. Acad. Sci. USA*, **80,** 5383.
26. Krougliak, V., and Graham, F. L. (1995). *Hum. Gene Therapy,* **6,** 1575.
27. Brough, D. E., Lizonova, A., Hsu, C., Kulesa, V. A., and Kovesdi, I. (1996). *J. Virol.*, **70,** 6497.
28. Cai, W., Person, S., Warner, S. C., Zhou, J., and DeLuca, N. (1987). *J. Virol.*, **61,** 714.
29. Arsenakis, M., Tomasi, L. F., Speziali, V., Roizman, B., and Campadelli-Fiume, G. (1986). *J. Virol.*, **58,** 367.
30. Davidson, I., and Stow, N. D. (1985). *Virology,* **141,** 77.
31. Persson, R. H., Bacchetti, S., and Smiley, J. R. (1985). *J. Virol.*, **54,** 414.
32. DeLuca, N., Courtney, M. A., and Schaffer, P. A. (1985). *J. Virol.*, **56,** 558.
33. Desai, P., DeLuca, N., Glorioso, J. C., and Person, S. (1993). *J. Virol.*, **67,** 1357.
34. Salmon, B., Cunningham, C., Davison, A. J., Harris, W. J., and Baines, J. D. (1998). *J. Virol.*, **72,** 3779.
35. Sutter, G., Ramsey-Ewing, A., Rosales, R., and Moss, B. (1994). *J. Virol.*, **68,** 4109.
36. Cole, C. N. (1996). In *Fields Virology* (3rd edn) (ed. B. N. Fields, D. M. Knipe and P. M. Howley), p. 1997. Raven Press, Philadelphia.
37. Berns, K. I. (1996). In *Fields Virology* (3rd edn) (ed. B. N. Fields, D. M. Knipe and P. M. Howley), p. 2173. Raven Press, Philadelphia.
38. Samulski, R. J., Berns, K. I., Tan, M., and Muzyczka, N. (1982). *Proc. Natl. Acad. Sci. USA*, **79,** 2077.
39. Xiao, X., Li, J., and Samulski, R. J. (1996). *J. Virol.*, **70,** 8098.
40. Xiao, X., Li, J., and Samulski, R. J. (1998). *J. Virol.*, **72,** 2224.
41. Shenk, T. (1996). In *Fields Virology* (3rd edn) (ed. B. N. Fields, D. M. Knipe and P. M. Howley), p. 2111. Raven Press, Philadelphia.
42. Rekosh, D. (1981). *J. Virol.*, **40,** 329.
43. Stow, N. D. (1981). *J. Virol.*, **37,** 171.
44. Ho, Y. -S., Galos, R., and Williams, J. (1982). *Virology,* **122,** 109.
45. McGrory, W. J., Bautista, D. S., and Graham, F. L. (1988). *Virology,* **163,** 614.
46. Chartier, C., Degryse, E., Gantzer, M., Dieterlé, A., Pavirani, A., and Mehtali, M. (1996). *J. Virol.*, **70,** 4805.
47. Roizman, B., and Sears, A. E. (1996). In *Fields Virology* (3rd edn) (ed. B. N. Fields, D. M. Knipe and P. M. Howley), p. 2231. Raven Press, Philadelphia.
48. Moss, B. (1996). In *Fields Virology* (3rd edn) (ed. B. N. Fields, D. M. Knipe and P. M. Howley), p. 2637. Raven Press, Philadelphia.
49. Post, L. E., and Roizman, B. (1981). *Cell,* **25,** 227.
50. Falkner, F. G., and Moss, B. (1988). *J. Virol.*, **62,** 1849.
51. Freund, R., Calderone, A., Dawe, C. J., and Benjamin, T. L. (1991). *J. Virol.*, **65,** 335.
52. Frost, E., and Williams, J. (1978). *Virology,* **91,** 39.
53. Hasson, T. B., Soloway, P. D., Ornelles, D. A., Doerfler, W., and Shenk, T. (1989). *J. Virol.*, **63,** 3612.

4

Interactions between viral and cellular proteins during DNA virus replication

CATHERINE H. BOTTING and RONALD T. HAY

1. Introduction

Initiation of viral DNA synthesis is preceded by assembly of a preinitiation complex at the viral origin of DNA replication. This nucleoprotein complex is composed of interacting viral and cellular proteins, which together recognize the viral origin of DNA replication and impart great specificity to the process. The initiation event is then triggered, and the pre-initiation complex dissociates as elongation proceeds. Thus the protein–protein and protein–DNA interactions involved are dynamic, with association and dissociation being induced by conformational changes, which may be triggered by one or more binding changes distal to the interaction under study. The methods described below have been used to study protein–protein and protein–DNA interactions in adenovirus DNA replication. It is therefore of value to describe briefly this replication system, to give context to the methods.

The adenovirus genome is a linear double-stranded DNA molecule of 36 000 base pairs (bp), with inverted terminal repeats (ITRs) of about 100 bp. Located within the ITRs are the *cis*-acting DNA sequences which define *ori*, the origin of DNA replication. Covalently attached to each 5' end is a terminal protein (TP), which is likely to be an additional *cis*-acting component of *ori*. Within the terminal 51 bp of the adenovirus 2 genome, four regions have been defined that are involved in initiation of replication. The terminal 18 bp are regarded as the minimal replication origin, and these sequences can direct limited initiation with just the three viral proteins involved in replication: preterminal protein (pTP), DNA polymerase (pol), and DNA-binding protein (DBP). However, two cellular transcription factors, nuclear factor I (NFI) and nuclear factor III (NFIII), are required for efficient levels of replication. In contrast, adenovirus 4 replicates efficiently without NFI and NFIII. A further cellular factor, a topoisomerase, is required for complete elongation.

The steps involved in adenovirus type 2 DNA replication can be summarized as follows. Firstly, the viral genome is coated with DBP. This protein reacts co-operatively with the cellular transcription factor, NFI, which binds to a recognition site within the origin of replication, separated from the 1–18 bp core by a precisely defined spacer region. NFIII also binds at a specific recognition site between nucleotides 39 and 48. Protein–protein interactions, between NFI and pol, and between pTP and NFIII, help to recruit the pTP–pol heterodimer into the preinitiation complex. Interaction between the heterodimer and specific base pairs 9 to 18 in the DNA sequence ensures correct positioning, and the complex is further stabilized by interactions between the incoming pTP–pol and the genome-bound TP. DNA replication is then initiated by a protein priming mechanism in which a covalent linkage is formed between the α-phosphoryl group of the terminal residue, dCMP, and the β-hydroxyl group of a serine residue in pTP, a reaction catalysed by pol. This acts as a primer for synthesis of the nascent strand. Base-pairing with the second GTA triplet of the template strand guides the synthesis of a pTP–trinucleotide, which then jumps back 3 bases, to base-pair with the first triplet (also GTA), and synthesis then proceeds by displacing the non-template strand (1). NFI dissociates as the first nucleotide binds just prior to the initiation reaction (2). Dissociation of pTP from pol begins as the pTP–trinucleotide is formed, and is almost complete by the time seven nucleotides have been synthesized (3). NFIII dissociates as the replication-binding fork passes through the NFIII binding site (4).

2. Identification and purification of proteins involved in replication

To isolate the components required for DNA replication, it is essential to have suitable assays, into which fractionated extracts of infected material can be added, to test for activity. The assay must be rapid and simple enough to allow a high throughput, so that activity can be followed during purification. As in any purification scheme, the less time spent between cell lysis and final storage of the purified protein the better.

2.1 Obtaining replication-active extracts

Adenovirus-infected HeLa cells were initially used as a source of material for the protein factors involved in viral DNA replication (5). Extracts which were active in *in vitro* assays were produced, which were used as a starting point for isolation of the replication-active components.

2.1.1 Infection of HeLa cells and preparation of cytoplasmic and nuclear extracts

Cells are infected with adenovirus, and hydroxyurea added to block replication, allowing a build up of the replication proteins. After harvesting, the cells are swollen in hypotonic buffer and lysed in a Dounce homogenizer. The

4: Interactions between viral and cellular proteins

cytoplasmic fraction is separated from nuclei by low speed centrifugation, followed by centrifugation at higher speed to remove cellular organelles. Nuclei can be treated with high salt to elute DNA binding proteins. Whether the proteins of interest are found in the cytoplasmic or nuclear fraction does not necessarily depend on their intracellular location *in vivo*. Many known nuclear proteins have been found to leach into the cytoplasmic fraction upon hypotonic swelling of cells.

Protocol 1. Preparation of cytoplasmic and nuclear extracts from adenovirus-infected HeLa cells

Equipment and reagents

- HeLa S3 spinner cells
- Earle's minimum essential medium (S-MEM) (Gibco)
- Adenovirus stock
- Hydroxyurea (Sigma)
- Isotonic buffer: 20 mM Hepes-NaOH pH 7.5, 5 mM KCl, 0.5 mM MgCl$_2$, 0.5 mM DTT, 0.2 M sucrose
- Hypotonic buffer: 20 mM Hepes-NaOH pH 7.5, 5 mM KCl, 0.5 mM MgCl$_2$, 0.5 mM DTT, protease inhibitors[a]
- Dounce homogenizer with pestle type B
- Buffer C: 25 mM Hepes-NaOH pH 7.5, 5 mM KCl, 0.5 mM MgCl$_2$, 1 mM DTT, 0.4 M NaCl, protease inhibitors[a]

Method

1. Grow HeLa S3 cells in suspension to a cell density of 4–5 × 10^5 cells ml^{-1}. Collect cells from 1 litre of culture by centrifugation at 600 *g* for 15 min, and resuspend in 100 ml serum-free medium.
2. Infect with 100 PFU per cell of adenovirus, and after 90 min add medium containing 2% calf serum plus hydroxyurea (final concentration 10 mM) to bring the volume back up to 1 litre.
3. Incubate the cells for 22 h at 37°C, and then collect by centrifugation at 600 *g* for 15 min.
4. Resuspend the cells in 20 ml isotonic buffer, collect again by centrifugation, and resuspend in 5 ml ice-cold hypotonic buffer.
5. After incubation on ice for 10 min, lyse with 10 strokes of a Dounce homogenizer using a pestle type B. All further steps should be carried out at 4°C.
6. Sediment the nuclei by centrifugation at 1400 *g* for 5 min. Remove the supernatant containing the cytoplasmic fraction, and clarify by further centrifugation at 27 200 *g* for 30 min.
7. Lyse the nuclei by incubation in buffer C for 30 min on ice, and remove cell debris by centrifugation at 27 200 *g* for 30 min.
8. Collect the supernatant containing the nuclear extract. Snap-freeze aliquots, and store at −70°C.

[a] Protease inhibitors should include: bestatin, E-64, leupeptin, phenylmethylsulfonyl fluoride (PMSF), and pepstatin A.

2.2 Assaying for activity

A number of assays have been designed to measure various aspects of the adenovirus DNA replication process. They are presented here as monitors of purification progress, but are also used to investigate the properties of the proteins, e.g. the stimulatory effect of one protein on another (6), metal-ion dependence, or template requirements (7), etc.

2.2.1 Initiation assay

This assay detects the template-dependent, pol-catalysed, covalent binding of radiolabelled dCMP to pTP (8) (*Figure 1*). Formation of the complete preinitiation complex is a prerequisite for this reaction, so the assay checks the viability of all the replication components. However, care must be taken in assuming that the protein is therefore necessarily fully functional; the DNA polymerase, for example, may be active for initiation but defective for elongation.

Protocol 2. Initiation assay

Equipment and reagents
- Buffer A: 25 mM Bicine-NaOH pH 8.0, 2 mM DTT, 1 mM $MnCl_2$, 0.15 mM ATP, 0.2 mg ml^{-1} BSA
- Disruption buffer: 20% glycerol, 5% SDS, 570 mM 2-mercaptoethanol, 33 mM Tris-HCl pH 6.7, 0.2% Bromophenol blue

Method

1. Mix together to give a final volume of 10 μl in buffer A: 500 ng DBP, 20 ng NFI, 20 ng NFIII, 2–50 ng pTP-Adpol, and 50 ng TP-DNA template. Add [α-^{32}P]dCTP (2.5 μCi, 3000 Ci $mmol^{-1}$), and incubate at 30°C for 1 h.

2. Add $CaCl_2$ to produce a concentration of 10 mM, and add 1 unit of micrococcal nuclease, to destroy the template and any DNA attached to pTP, and incubate for a further 30 min at 37°C.

3. Stop the reaction by the addition of disruption buffer to the reactions, and denature at 100°C for 2 min.

4. Analyse the reaction products by fractionation on an SDS-polyacrylamide gel, and subject the dried gel to autoradiography or Phosphor Imaging.

2.2.2 Elongation assays

The partial elongation assay requires all the components of the preinitiation complex, and again gives an easily detectable radiolabelled product (9). In contrast to the initiation assay, further unlabelled nucleotides are also added

4: Interactions between viral and cellular proteins

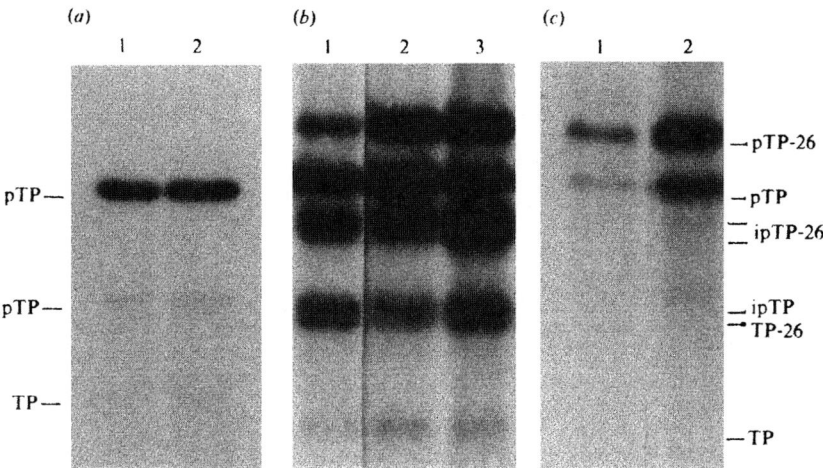

Figure 1. Formation of ^{32}P-labelled complexes after initiation (a) and partial elongation (b and c) using cellular extracts from Ad 2-infected HeLa cells. Extracts were incubated in the presence of [α-^{32}P]dCTP, using virus cores as template, and optimal concentrations of MgCl$_2$ and ATP. Polypeptides were separated by SDS-PAGE, and labelled species detected by autoradiography. Assay mixtures included the following: (a) Infected nuclear extracts: lane 1, 33 μg; lane 2, 66 μg. (b) Infected nuclear extracts: lane 1, 33 μg; lane 2, 66 μg; lane 3, 33 μg. DBP (2 μg) was added to lane 3. (c) Infected cytoplasmic extracts: lanes 1 and 2, 84 μg. DBP (2 μg) was added to lane 2. Taken from ref. 8, with permission.

to allow elongation. However, by adding the chain terminator ddGTP instead of dGTP, elongation is allowed to proceed only as far as the first dG in the sequence, at nucleotide 26 in adenovirus 2, giving the pTP-26b product (*Figure 1*). If longer elongation products are required, dGTP is used instead of ddGTP. A creatine phosphate/creatine phosphokinase ATP regeneration system is also employed. For complete elongation, an extract containing NFII, a topoisomerase, is also required. Cleavage of TP-DNA using restriction enzymes generates terminal TP-linked fragments, and also internal fragments, which can be distinguished by their difference in molecular weight. Addition of the mixture of cleavage products to the elongation assay allows the specificity of the reaction to be determined, as only the terminal fragments, and not the internal fragments, are replicated. Single-stranded DNA products are also produced, indicating that reinitiation has occurred (10).

Protocol 3. Partial elongation assay

Equipment and reagents

- Buffer A: 25 mM Bicine-NaOH pH 8.0, 2 mM DTT, 5 mM MgCl$_2$, 0.15 mM ATP, 35 μM dATP, 35 μM dTTP, 30 μM ddGTP, 0.2 mg ml^{-1} BSA
- Disruption buffer: 20% glycerol, 5% SDS, 570 mM 2-mercaptoethanol, 33 mM Tris-HCl pH 6.7, 0.2% Bromophenol blue

Protocol 3. Continued

Method

1. Mix together to give a 10 µl reaction mixture in buffer A: 500 ng DBP, 20 ng NFI, 20 ng NFIII, 2–50 ng pTP-pol, and 50 ng TP-DNA template. Add [α-^{32}P]dCTP (2.5 µCi, 365 Ci mmol^{-1}), and incubate at 30°C for 1 h.
2. Stop the reaction by adding disruption buffer, and denature at 100°C for 2 min.
3. Analyse the reaction products by fractionation in an SDS-polyacrylamide gel, and visualize by autoradiography or Phosphor Imaging of the dried gel.

Protocol 4. Elongation assay

Equipment and reagents

- Buffer A: 25 mM Hepes-KOH pH 7.5, 4 mM MgCl$_2$, 1 mM DTT, 0.1 mg ml^{-1} BSA
- Buffer B: 25 mM Hepes-KOH pH 7.5, 4 mM MgCl$_2$, 1 mM DTT, 0.1 mg ml^{-1} BSA, 80 µM dATP, 80 µM dTTP, 80 µM dGTP, 5 µM dCTP, [α-^{32}P]dCTP (0.3 µCi; 3000 Ci mmol-1), 4 mM ATP, 10 mM creatine phosphate, 10 mg ml^{-1} creatine phosphokinase
- Termination buffer: 5% SDS, 50% glycerol, 100 mM EDTA, 0.2% bromophenol blue
- Fixing Buffer: 10% acetic acid

Method

1. Mix together to give a 7 µl reaction mixture in buffer A: 500 ng DBP, 20 ng NFI, 20 ng NFIII, 2–50 ng pTP-pol, and 50 ng TP-DNA template (restriction enzyme-cleaved so that terminal TP-linked fragments and internal fragments can be distinguished), and incubate for 30 min at 30°C.
2. Add 7 µl of buffer B, and incubate at 30°C for 90 min.
3. Stop the reaction by the addition of 6 µl of termination buffer, and heat at 70°C for 5 min.
4. Resolve the reaction products on a 0.5% agarose gel containing 0.1% SDS, at 35 mA for 4 h.
5. Fix the gel in fixing buffer, and visualize the dried gel by autoradiography or Phosphor-Imaging.

2.2.3 Adenovirus DNA polymerase assay

This assay measures the ability of the DNA polymerase to elongate multiply primed ('activated') calf thymus DNA. As Ad pol is one of the few replicative

4: Interactions between viral and cellular proteins

DNA polymerases to be resistant to aphidicolin, assays can be performed in the presence and absence of this drug, to determine whether the DNA polymerase activity present in infected cell extracts is due exclusively to Ad pol (11).

Protocol 5. Adenovirus DNA polymerase assay

Equipment and reagents
- Activated calf thymus DNA, prepared as described in ref. 12
- Buffer A: 50 mM Tris-HCl pH 8.0, 5 mM MgCl$_2$, 10 mM DTT
- Solution 1: 10% trichloroacetic acid, 0.5% sodium pyrophosphate
- Solution 2: 5% trichloroacetic acid
- Ethanol
- Whatman GF/C discs
- Scintillation counter

Method

1. Incubate adenovirus DNA polymerase in buffer A with 10 μg activated calf thymus DNA, 100 μM dTTP, 100 μM dGTP, 100 μM dCTP, 20 μM dATP, 1 μCi [α-^{32}P]dATP (specific activity 3000 Ci mmol^{-1}) in a total volume of 50 μl, for 1 h at 37°C.

2. Terminate reactions by the addition of 5 ml of solution 1, and capture the acid-insoluble radioactivity on Whatman GF/C discs by filtration under vacuum.

3. Wash the disc twice with solution 1, twice with solution 2, and once with ethanol.

4. Air-dry the discs, and measure the radioactivity by scintillation counting. One unit of DNA polymerase activity is defined as the incorporation of 1 nmol of dNMP into acid-insoluble DNA in 1 h at 37°C.

2.2.4 DNA unwinding assay

The ability of DBP to unwind double-stranded DNA, as required for strand displacement DNA synthesis, can be measured using specially constructed templates. DBP can unwind short fully duplex DNA, but for longer fragments the presence of a free single-stranded end is required. The oligonucleotide to be unwound, ^{32}P-labelled at its 5' end, is annealed to a complementary sequence on recombinant single-stranded M13 phage DNA, prior to incubation with DBP (13). Unwinding is then measured by the release of free oligonucleotide. The labelled oligonucleotide bound to M13 gives a high molecular weight band at the top of an SDS-polyacrylamide gel, whereas free, unwound oligonucleotide gives a lower molecular weight band towards the bottom of the gel (*Figure 2*).

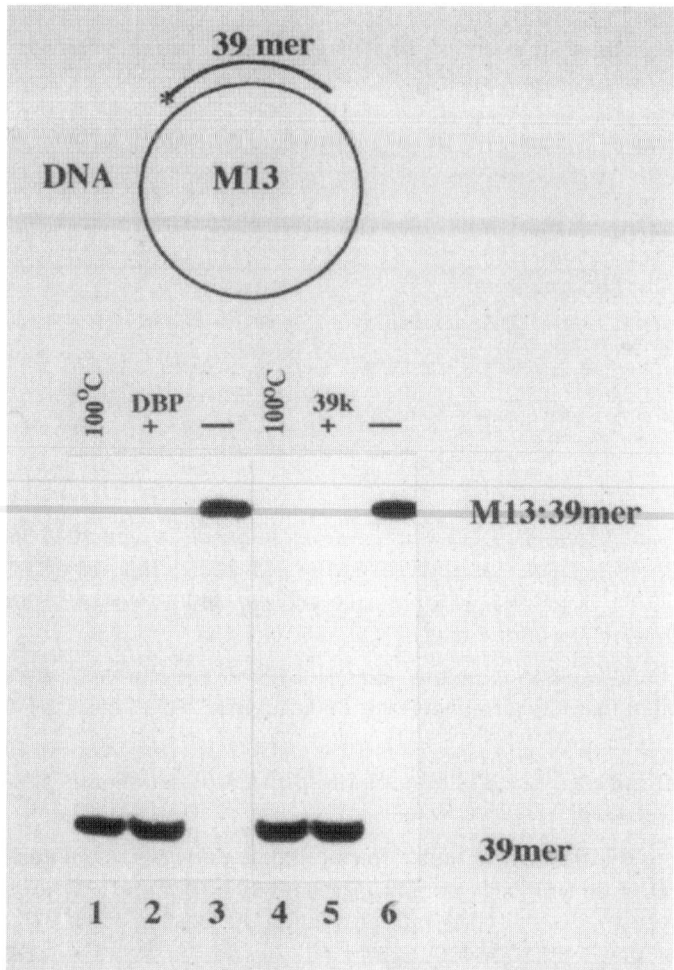

Figure 2. DNA unwinding by DBP. The assay was performed in the presence of 10 fmol M13-39mer DNA substrate, and 34 pmol of either full-length DBP (59 kDa) or the 39 kDa C-terminal fragment (39 kDa). The substrate for the assay is shown in the upper part of the figure, and consists of a 5'-^{32}P-labelled 39-mer oligonucleotide, annealed to the single-stranded form of phage M13 DNA. Lanes 1 and 4 are the heat-denatured substrate, and lanes 3 and 6 are the native substrate incubated in the absence of DBP. Lane 2 is the reaction with full-length DBP, and lane 5 is the reaction with the 39 kDa C-terminal fragment of DBP. Taken from ref. 13, with permission.

Protocol 6. DNA unwinding assay

Equipment and reagents
- Oligonucleotide, and single-stranded M13 phage DNA
- [γ-^{32}P]ATP
- Buffer A: 25 mM Tris-borate pH 8.3, 0.5 mM EDTA
- Buffer B: 10 mM Tris-HCl pH 7.5, 100 mM NaCl, 1 mM EDTA
- Buffer C: 40 mM Tris-HCl pH 8.0, 5 mM DTT, 1 mg ml^{-1} BSA
- Sepharose CL-4B
- Termination buffer: 33 mM EDTA pH 8.0, 6% SDS, 25% glycerol, 0.5% Bromophenol blue, 0.5% Xylene cyanol

Method
1. Label the oligonucleotide with [γ-^{32}P]ATP, using T4 polynucleotide kinase at 37°C for 1–2 h.
2. Isolate the labelled oligonucleotide in a 10% polyacrylamide gel containing buffer A.
3. Excise the required band, and elute from the gel by passive diffusion into buffer B, overnight at room temperature.
4. Anneal 5 pmol of 5'-^{32}P-labelled oligonucleotide with 1 pmol of complementary single-stranded M13 DNA in buffer B, by heating to 100°C for 3 min, and then slowly cooling to room temperature.
5. Remove unannealed oligonucleotide by gel filtration on a Sepharose CL-4B column equilibrated in buffer B.
6. Incubate DNA (10 fmol) and DBP (34 pmol) in buffer C, to give a total volume of 30 μl, for 30 min at 37°C.
7. Stop the reaction by the addition of 5 μl of termination buffer, and analyse the reaction products by fractionation in an SDS-polyacrylamide gel.
8. Visualize the results using autoradiography or Phosphor Imaging of the dried gel.

2.2.5 Gel-electrophoresis DNA-binding assay

This assay relies on the retardation in a native polyacrylamide gel exhibited by oligonucleotides when associated with protein (14). Retardation of the protein–DNA complex is mainly dependent on the molecular weight of the complex formed. Oligonucleotides corresponding to the origin of replication can be labelled by one of two methods:

(i) Annealing and then labelling with [α-^{32}P]dATP and [α-^{32}P]dCTP using the Klenow fragment of DNA polymerase I.
(ii) Labelling first with [γ-^{32}P]ATP using T4 polynucleotide kinase, followed by annealing.

The protein(s) of interest is then incubated with the probe, and protein–DNA complexes separated from free probe on native polyacrylamide gels.

Protocol 7. Gel-electrophoresis DNA-binding assay

Equipment and reagents
- Nucleotides and radioactive nucleotides
- Buffer I: 25 mM Bicine-NaOH, pH 8.0, 2 mM DTT, 4 mM $MgCl_2$, 20 mM KCl, 1 mg ml^{-1} BSA
- Buffer II: 50 mM Tris-borate pH 8.3, 1 mM EDTA
- Fixing solution: 20% methanol, 10% acetic acid

Method

1. Either:
   (a) anneal complementary oligonucleotides (5 pmoles) by heating to 100 °C for 3 min and then slowly cooling to room temperature, and incubate with 20 µCi [α-^{32}P]dATP, 20 µCi [α-^{32}P]dCTP, 1 mM dTTP, 1 mM dGTP, and the Klenow fragment of DNA polymerase I, for 30 min at room temperature, then add 1 mM dATP and 1 mM dCTP and incubate for a further 20 min at room temperature, or
   (b) incubate the template oligonucleotide with [γ-^{32}P]ATP and T4 polynucleotide kinase at 37 °C for 1–2 h, before annealing with the complementary oligonucleotide.

2. Isolate the labelled double-stranded DNA, as described in *Protocol 6*, Step 1.

3. Incubate the appropriately diluted protein of interest (0.1–10 ng) with 1 ng (an excess) of probe in buffer I on ice for 15 min. (It can be confirmed that the probe is in excess by checking that there is free probe, which runs towards the bottom of the gel.) Add glycerol to a concentration of 5% (**do not add any dyes**), load onto a 6% polyacrylamide gel containing buffer II, and electrophorese at 150 V for 2 h.

4. Fix the gel in fixing solution for 30 min, dry, and analyse by autoradiography or Phosphor Imaging.

2.3 Purification of proteins required for adenovirus DNA replication from HeLa cells

The original purification of the proteins involved in adenovirus DNA replication from HeLa cells, using the assays described above to follow the activity, allowed identification of the components required for activity. However, as the proteins tend to interact, it makes isolation of individual products difficult. There is also considerable variation in the expression levels of the different proteins, such that it is really only viable to isolate DBP from this source. The other proteins are expressed in too small amounts, and overexpression systems

4: Interactions between viral and cellular proteins

are required, as discussed in Section 3. The purification methods used are mainly standard biochemical techniques, but include some which are specifically useful for the purification of DNA-binding proteins. The methods should be widely applicable to other virus replication systems.

2.3.1 Purification of a replication-active fraction from adenovirus 4-infected HeLa cells

A two-column procedure is used to produce a fraction which is very active in viral DNA synthesis and contains predominantly only four polypeptide species, the most abundant being DBP (10).

Protocol 8. Isolation of a fraction from adenovirus type 4-infected HeLa cells which is active in viral DNA replication

Equipment and reagents
- DEAE-Sephacel
- Buffer A: 20 mM Hepes-NaOH pH 7.5, 5 mM KCl, 0.5 mM $MgCl_2$, 0.5 mM DTT, 50 mM NaCl
- Buffer B: 20 mM Hepes-NaOH pH 7.5, 5 mM KCl, 0.5 mM $MgCl_2$, 0.5 mM DTT, 0.2 M NaCl
- Single-stranded (denatured) calf thymus DNA-Sepharose (ssDNA-Sepharose)
- Dialysis buffer: 20 mM Hepes-NaOH pH 8.0, 5 mM KCl, 0.5 mM $MgCl_2$, 5 mM DTT

Method
1. Take the cytoplasmic extract prepared as in *Protocol 1*, and add NaCl to 50 mM. Apply the extract to a DEAE-Sephacel column equilibrated in buffer A.
2. Wash the column with 2 column volumes of this buffer, and elute with buffer B. Collect fractions, and assay for *in vitro* DNA replication activity as described in *Protocol 2*.
3. Apply the active fractions to an ssDNA-Sepharose column equilibrated with buffer B.
4. Wash the column with 2 column volumes of buffer B, and then with 5 column volumes of a linear 0.2–2 M NaCl gradient in the same buffer. Collect fractions, add PMSF to a concentration of 1 mM, and dialyse overnight.
5. Assay for *in vitro* adenovirus DNA replication activity. Two protein peaks elute from the ssDNA-Sepharose column, one at 0.5–0.8 M NaCl, and one at 1–1.2 M NaCl, but only the second peak has DNA replication activity associated with it.

2.3.2 Purification of DBP from adenovirus-infected HeLa nuclei

Due to its abundance in adenovirus-infected cells (2×10^7 molecules per infected HeLa cell), DBP can easily be purified in reasonable quantities from adenovirus-infected HeLa cell nuclei. The nucleoprotein complex is isolated and subjected to ammonium sulfate fractionation, followed by chromatography on phosphocellulose and then ssDNA-cellulose (15). This second column competes away double-stranded DNA fragments bound to the protein, as DBP binds more strongly to the ssDNA column, due to the co operative nature of DBP binding to ssDNA.

Protocol 9. Purification of DBP from adenovirus infected HeLa nuclei

Equipment and reagents

- Phosphate-buffered saline (PBS)
- Buffer A: 10 mM NaHCO$_3$ pH 8.0, 150 mM NaCl, 1 mM 2-mercaptoethanol, 0.25% NP-40, protease inhibitors (*Protocol 1*)
- Buffer B: 25 mM NaCl, 8 mM EDTA pH 8.0, 1 mM 2-mercaptoethanol, protease inhibitors (*Protocol 1*)
- Buffer C: 10 mM Tris-HCl pH 8.0, 1 mM 2-mercaptoethanol, protease inhibitors (*Protocol 1*)
- Buffer D: 20 mM Tris-HCl pH 8.0, 50 mM NaCl, 1 mM 2-mercaptoethanol, protease inhibitors (*Protocol 1*)
- 1M Tris-HCl pH 8.0
- Buffer E: 10 mM Tris-HCl pH 8.8, 1 mM 2-mercaptoethanol, protease inhibitors (*Protocol 1*)
- Buffer F: 25 mM Hepes-KOH pH 8.0, 1 mM DTT, 0.02% NP-40, 20% glycerol
- Whatman P11 phosphocellulose
- ssDNA-cellulose (Sigma)
- Heparin-Sepharose
- Mono S FPLC column

Method

1. Infect HeLa cells with 100 PFU per cell of adenovirus for 48 h, harvest, and wash with PBS, as described in *Protocol 1*.
2. Lyse the cells in buffer A on ice with stirring.
3. Collect the nuclei by centrifugation at 1400 *g* for 5 min, and wash with the above buffer without NP-40.
4. Resuspend the pellets in buffer B, centrifuge at 1400 *g* for 5 min, and wash the pellet again with buffer B.
5. Resuspend in buffer C, and wash three times with this buffer to produce a thick gel.
6. Sonicate on ice, and then adjust the salt concentration to 150 mM.
7. Clarify by centrifugation at 48 400 *g* for 15 min, and then add 1% 1 M Tris-HCl pH 8.0, to the supernatant.
8. Precipitate with ammonium sulfate at 25% and 60% saturation, and resuspend the proteins that precipitate between these saturation levels in buffer D. Dialyse overnight against this buffer.
9. Clarify the dialysed proteins, and apply to a Whatman P11 phosphocellulose column equilibrated with buffer D. Wash with the same buffer containing 150 mM NaCl, and elute with a linear gradient from 150 to 800 mM NaCl in buffer D.

4: Interactions between viral and cellular proteins

10. Take the peak fractions between 320 and 460 mM NaCl, and dilute 1:4 with buffer E and apply to an ssDNA-cellulose column. Wash the column consecutively with buffer C containing 200 mM NaCl, buffer C containing 300 mM NaCl plus 200 μg ml^{-1} potassium dextran sulfate, buffer C containing 300 mM NaCl, and buffer C containing 500 mM NaCl. Some DBP elutes with this last buffer, but purer material should be finally eluted with buffer C containing 2 M NaCl.

11. To purify further, dilute the protein 1:8 with buffer F, and apply to a heparin-Sepharose column equilibrated with buffer F containing 250 mM NaCl. Wash with buffer F containing 37 mM NaCl, and elute with buffer F containing 440 mM NaCl.

12. Dilute the salt concentration to 100 mM to apply to a Mono S FPLC column, and elute with buffer F containing 300 mM NaCl to obtain homogeneous DBP.

2.3.3 Purification of TP-DNA

Although linearized plasmid DNA can be used as the template in *in vitro* DNA replication assays, DNA isolated from virions with the terminal protein (TP) attached to the 5' end of the DNA is a more efficient template for DNA replication *in vitro* (16).

Protocol 10. Purification of TP-DNA

Equipment and reagents

- 8 M guanidinium hydrochloride
- Phenylmethylsulfonyl fluoride (PMSF)
- Buffer A: 10 mM Tris-HCl pH 8.0, 100 mM NaCl, 1 mM EDTA, 4 M guanidinium hydrochloride
- Buffer B: 10 mM Tris-HCl pH 8.0, 1 mM EDTA
- Buffer C: 10 mM Tris-HCl pH 8.0, 1 mM EDTA, 50% glycerol
- Sepharose 4B (Pharmacia)

Method

1. Incubate adenovirus virions, purified by caesium chloride gradient centrifugation (17), for 5 min on ice with an equal volume of 8 M guanidinium hydrochloride in the presence of 0.5 mM PMSF.

2. Apply the treated virions to a Sepharose 4B column equilibrated with buffer A. Elute the column with this buffer, collecting fractions.

3. Analyse the fractions by absorbance at 260 and 280 nm, to identify which fractions contain the DNA–protein complex, and dialyse them first against buffer B and then against buffer C. Store at −20 °C.

2.3.4 Cellular factors

The fraction obtained in *Protocol 8*, from infected cell cytosol, gave good *in vitro* activity for Ad 4 DNA replication. However, nuclear fractions from uninfected HeLa cells were found to augment Ad 2 DNA replication. By testing various fractions of the nuclear extract for its ability to stimulate DNA replication in the *in vitro* assays discussed above (*Protocols 2 and 4*), the nuclear factors involved were isolated. Then their sequence-specific DNA-binding activity could be utilized in their purification. The historical method given below uses sequence-specific DNA-cellulose, made by adsorbing a linearized plasmid containing repeat copies of the NFI recognition sequence onto cellulose. Nowadays an oligonucleotide, corresponding to the binding site, with a 5'-amino link would be immobilized on cyanogen bromide-activated Sepharose, as used in *Protocol 14*. The original purification of NFI is given below (18). A similar purification procedure was used to obtain NFIII (19).

Protocol 11. Purification of NFI from HeLa cells (from ref. 18)

Equipment and reagents

- Buffer A: 25 mM Hepes-NaOH pH 7.5, 5 mM KCl, 1 mM MgCl$_2$, 1 mM DTT, 0.01% NP-40, protease inhibitors (*Protocol 1*)
- Buffer B: 25 mM Hepes-NaOH pH 7.5, 10% sucrose, 1 mM DTT, 0.01% NP-40, protease inhibitors (*Protocol 1*)
- Buffer C: 25 mM Hepes-NaOH pH 7.5, 40% glycerol, 2 mM EDTA, 1 mM DTT, 0.1 mM PMSF, 0.01% NP-40
- Buffer D: 25 mM Hepes-NaOH pH 7.5, 175 mM NaCl, 20% glycerol, 1 mM EDTA, 1 mM DTT, 0.1 mM PMSF, 0.01% NP-40
- Bio-Rex 70 (Bio-Rad)
- DNA-cellulose
- Sequence-specific DNA-cellulose (made by adsorbing onto cellulose a linearized plasmid containing repeat copies of the NFI recognition sequence)

Method

1. Take HeLa nuclei purified as described in *Protocol 1*, and wash them once with buffer A and once with buffer B.
2. Resuspend in buffer B containing 350 mM NaCl, and incubate for 60 min on ice.
3. Centrifuge at 10 000 *g* for 30 min to give a crude nuclear extract.
4. Add an equal volume of buffer C to the supernatant, and apply to a Bio-Rex 70 column equilibrated with buffer D.
5. Take the flow-through, and apply to a DNA-cellulose column equilibrated with buffer D containing 200 mM NaCl, and wash with 4 column volumes of this buffer. Then elute with a linear gradient of 200–500 mM NaCl in buffer D. NFI elutes at 280 and 350 mM NaCl.
6. Dilute the peak fractions with an equal volume of buffer D containing no salt, and apply the peak fractions to a sequence-specific DNA-

4: Interactions between viral and cellular proteins

cellulose column equilibrated with this buffer. Wash the column with 3 column volumes of this buffer, and then elute with a linear gradient of 200–500 mM NaCl, followed by a step to 2 M NaCl. NFI elutes with the 2 M NaCl step.

7. Dilute the active fractions with 9 volumes of buffer D containing 200 mM NaCl, and reapply to the specific DNA-cellulose column equilibrated with this buffer. Wash the column with 2 column volumes of this buffer, and elute with a 450 mM NaCl and 2 M NaCl step. NFI elutes in the 2 M NaCl step.

8. Dialyse the eluate against buffer D containing 100 mM NaCl, and store aliquots at −70°C.

3. Overexpression of replication components

3.1 Overexpression and purification of pTP and pol

As such small quantities of pTP and pol are expressed in adenovirus-infected HeLa cells, and they are present as a tight complex, it is really only feasible to purify substantial amounts of protein from an overexpression system (14). As pTP and pol are large proteins (80 and 140 kDa respectively), bacterially expressed protein is not soluble. The alternative systems that have been used are the vaccinia virus (20) and baculovirus (21) expression systems. Purification of pTP and of pol from baculovirus-infected cells will be described here. *Spodoptera frugiperda* (SF9) cells are infected with baculovirus expressing the protein of interest, and harvested around 72 h post-infection (22, 23).

Protocol 12. Purification of pTP from baculovirus-infected SF9 cells

Equipment and reagents

- Buffer A: 25 mM Hepes-NaOH pH 8.0, 5 mM KCl, 0.5 mM $MgCl_2$, protease inhibitors (*Protocol 1*)
- Buffer B: 25 mM Hepes-NaOH pH 8.0, 1 mM DTT, 1 M NaCl, protease inhibitors (*Protocol 1*)
- Buffer C: 50 mM Tris-HCl pH 8.0, 1 mM EDTA, 10% glycerol, 10 mM 2-mercaptoethanol, 1.5 M NaCl
- Buffer D: 50 mM Tris-HCl pH 8.0, 1 mM EDTA, 20% glycerol, 0.6 M NaCl, 10 mM 2-mercaptoethanol
- Buffer E: 25 mM Hepes-NaOH pH 8.0, 1 mM EDTA, 1 mM DTT, 20% glycerol, 0.4 M NaCl, 0.05% NP-40
- Buffer F: 10 mM Hepes-NaOH pH 8.0, 50 mM NaCl, 500 mM sucrose, 1 mM EDTA, 0.2% Triton X-100
- Buffer G: 25 mM Hepes-NaOH pH 8.0, 1 mM EDTA, 0.2 M NaCl, 1 mM DTT, 10% glycerol
- Dounce homogenizer and pestle type B
- Beckman TL-100 ultracentrifuge with TLA 100.2 rotor
- Phenyl agarose
- ssDNA-cellulose

Protocol 12. Continued

Method

1. Infect 1 litre of SF9 cells (cell density 0.5–1 × 10^6 ml^{-1}) with recombinant baculovirus expressing pTP at 5 m.o.i. for 72 h at 28°C. Harvest the cells by centrifugation at 120 g for 10 min. Wash the pellet in PBS.

2. Resuspend in 5 ml ice-cold buffer A. After 5–10 min on ice, lyse the cells with 20 strokes of a Dounce homogenizer using pestle type B.

3. Collect the nuclei by centrifugation at 4000 g for 3 min, and resuspend them in 8 ml of buffer B.

4. Incubate on ice for 15 min prior to centrifugation at 60 000 r.p.m. in a Beckman TL-100 ultracentrifuge (TLA 100.2 rotor) for 20 min at 4°C.

5. Take the supernatant, and dilute 1:3 with buffer C. Apply to a 6 ml phenyl agarose column equilibrated with the above dilution buffer. Wash with 2 column volumes of this buffer, and elute with buffer D.

6. Identify the pTP-containing fractions by SDS-PAGE, and dialyse the pooled fractions against buffer E. Take the dialysed protein, and slowly dilute it 1:2 in buffer F. Then dilute this 1:2 in buffer G. This serial dilution procedure is required as a precaution against precipitation.

7. Apply to an ssDNA-cellulose column equilibrated with buffer G, wash with 4 column volumes of this buffer, and elute pTP with buffer G containing 0.6 M NaCl.

8. Snap freeze pTP-containing fractions, and store at −70°C.

Protocol 13. Purification of adenovirus DNA polymerase from baculovirus-infected SF9 cells

Equipment and reagents

- Buffer A: 25 mM Hepes-KOH pH 8.0, 5 mM KCl, 0.5 mM $MgCl_2$, 0.5 mM DTT, protease inhibitors (*Protocol 1*)
- Buffer B: 25 mM Hepes-NaOH pH 8.0, 0.2 M NaCl, 1 mM EDTA, 2 mM DTT, 10% glycerol
- Buffer C: 5 mM potassium phosphate pH 7.0, 0.4 M NaCl, 1 mM DTT, 10% glycerol
- Buffer D: 100 mM potassium phosphate pH 7.0, 0.4 M NaCl, 1 mM DTT, 10% glycerol
- Dounce homogenizer and pestle type B
- ssDNA-Sepharose
- Hydroxyapatite

Method

1. Infect SF9 cells with recombinant baculovirus expressing pol, as in *Protocol 12*, Step 1.

2. Resuspend the harvested pellet in 5 ml ice-cold buffer A, and incubate on ice for 5–10 min.

4: Interactions between viral and cellular proteins

3. Lyse the cells with 20 strokes of pestle type B in a Dounce homogenizer, on ice.
4. Add NaCl to a concentration of 0.2 M, and incubate the extract on ice for 30 min. Remove cell debris by centrifugation at 15 000 g for 5 min, and clarify the extract by further centrifugation at 100 000 g for 15 min.
5. Apply the clarified cell extract to an ssDNA-Sepharose column equilibrated with buffer B. Wash the column extensively with the same buffer, and elute pol with buffer B containing 0.6 M NaCl.
6. Identify the pol-containing fractions by SDS-PAGE. Pool the protein-containing fractions, and dialyse against buffer C.
7. Apply the dialysate to a hydroxyapatite column equilibrated in the same buffer. Wash extensively in the same buffer, and elute with buffer D.
8. Snap freeze the pol-containing fractions, and store at −70°C.

3.2 Expression and purification of the cellular factors

It has been shown that only the DNA-binding domains (DBDs) of the NFI and NFIII are required for adenovirus DNA replication (24, 25). Hence it is not necessary to purify the full-length proteins for replication studies. However, GST fusions of just the NFI_{DBD} are insoluble when expressed in *Escherichia coli*. Although a small proportion can be denatured and refolded into an active form (2), it is more convenient to produce the protein using the baculovirus expression system, which yields soluble protein. The $NFIII_{DBD}$, also known as the POU domain, is soluble when expressed in *E. coli*.

3.2.1 Purification of NFI_{DBD} from baculovirus-infected SF9 cells

The purification of NFI_{DBD} again utilizes sequence-specific DNA affinity chromatography (24).

Protocol 14. Purification of NFI_{DBD} from baculovirus-infected SF9 cells

Equipment and reagents
- Buffer A: 25 mM Hepes-NaOH pH 8.0, 0.4 M NaCl, 1 mM EDTA, 2 mM DTT, 0.5% NP-40, protease inhibitors (*Protocol 1*)
- Buffer B: 25 mM Hepes-NaOH pH 8.0, 0.4 M NaCl, 1 mM EDTA, 2 mM DTT, 10% glycerol
- Bio-Rex 70 (Bio-Rad)
- poly(dI-dC) (Pharmacia)
- NFI DNA binding-site oligonucleotide, immobilized on cyanogen bromide-activated Sepharose via a 5'-amino link (NFIbs-CNBr-Sepharose)

Method
1. Infect SF9 cells with recombinant baculovirus expressing NFI_{DBD}, as in *Protocol 12*, Step 1.

Protocol 14. *Continued*

2. Resuspend the harvested pellet in 5 ml ice-cold buffer A. After 30 min on ice, remove the nuclei by low-speed centrifugation, followed by clarification at 45 000 g for 30 min.
3. Apply the clarified extract to a Bio-Rex 70 column equilibrated with buffer B. Wash with 2 column volumes of this buffer, and elute with buffer B containing 0.7 M NaCl.
4. Perform a gel electrophoresis DNA-binding assay (*Protocol 7*) to determine which fractions are active for NFI.
5. Dilute NFI-containing fractions to give an NaCl concentration of 250 mM, and add 6 μg poly(dI-dC) per mg of protein.
6. Apply to a NFIbs-CNBr-Sepharose column equilibrated with buffer B containing 250 mM NaCl. Wash the column with 5 column volumes of this buffer, and elute NFI_{DBD} with buffer B containing 1 M NaCl.
7. Snap-freeze NFI_{DBD}-containing fractions, and store at −70°C.

3.2.2 Purification of POU ($NFIII_{DBD}$) from *E. coli*

GST-POU overexpressed in *E. coli* is purified using a standard protocol for the purification of a GST-fusion protein, after first allowing it to pass through a DEAE-cellulose column to remove any associated DNA. It was found that an excess of glutathione-agarose beads was required to get efficient binding of this particular protein, hence the large column volume used (26).

Protocol 15. Purification of POU ($NFIII_{DBD}$) from *E. coli*

Equipment and reagents

- Isopropyl-β-D-thiogalactopyranoside (IPTG)
- Buffer A: 50 mM Tris-HCl pH 8.0, 150 mM NaCl, 1 mM EDTA, 5 mM DTT, 5 mM sodium metabisulfite (SMBS)
- Buffer B: 50 mM Tris-HCl pH 8.0, 150 mM NaCl, 1 mM DTT, 5 mM SMBS, 15% glycerol
- Buffer C: 25 mM potassium acetate pH 6.8, 50 mM NaCl, 1 mM DTT, 10% glycerol, 0.1% Triton X-100, 10 mM glutathione
- Buffer D: 25 mM potassium acetate pH 6.8, 50 mM NaCl, 1 mM DTT, 10% glycerol
- Lysozyme
- Triton X-100
- DEAE-cellulose
- Glutathione-agarose
- Thrombin
- Mono S FPLC

Method

1. Grow 1 litre of *E. coli* B834 containing the plasmid encoding for GST-POU to an OD of 0.6 in Luria broth, supplemented with ampicillin to a concentration of 0.1 mg ml^{-1} at 37°C.
2. Induce protein expression by the addition of IPTG at room temperature for 3 h.

4: Interactions between viral and cellular proteins

3. Collect the bacteria by centrifugation at 9000 *g* for 15 min, and resuspend the pellet in 15 ml of buffer A.
4. Treat with lysozyme (1 mg ml^{-1}) for 5 min at room temperature, sonicate in the presence of protease inhibitors, and add Triton X-100 to 1%.
5. Pass the supernatant obtained upon centrifugation through a 7 ml DEAE-cellulose column equilibrated with buffer A, and then apply it to a 20 ml glutathione-agarose column equilibrated with buffer B. Wash with 5 column volumes of this buffer, and elute with buffer C.
6. Pool the GST-POU-containing fractions, and digest with thrombin (1 unit per 0.5 mg protein) at room temperature for 4 h, to obtain the free POU.
7. Separate from GST and other contaminants on a Mono S FPLC column, equilibrated with buffer D, and elute with a linear salt gradient from 50 mM to 1 M NaCl. POU elutes at 230 mM NaCl.

4. Mapping interactions

Once the components of the replication system have been identified and purified, their roles in the replication process can be explored. In the adenovirus there are many complex protein–DNA and protein–protein interactions which bring about the formation of the preinitiation complex. Although two components of the adenovirus replication system have been crystallized individually, namely DBP (27) and POU (NFIII$_{DBD}$) (28), we are still a long way from having a structural model for the whole preinitiation complex. However, regions of proteins which contact the DNA or each other can be identified by other methods. If epitopes important for binding are identified, they can be targeted for drug design. However the success of the methods available depends a lot on the strength of the interaction being studied.

4.1 Protein–DNA interactions

Several methods for mapping protein–DNA interactions are described below. The first two methods map on the DNA, and the third on the protein. The source of DNA may be chemically synthesized oligonucleotides, or a plasmid containing the sequence of interest.

4.1.1 DNase I footprinting

The protein of interest is allowed to interact with DNA containing the putative binding site that has been ^{32}P-labelled on one strand only, before partially digesting with DNase I. A comparison of the separated fragment patterns obtained in the presence and absence of the binding protein allows sites of undigested, and therefore protected, DNA to be identified (29).

Protocol 16. DNase I footprinting

Equipment and reagents
- Buffer A: 25 mM Hepes-NaOH pH 8.0, 100 mM NaCl, 5 mM $MgCl_2$, 1 mM DTT, 20 μg ml^{-1} BSA
- Buffer B: formamide (98%), 20 mM NaOH, 1 mM EDTA
- Termination buffer: 0.3 M sodium acetate, 20 mM EDTA, 100 μg ml^{-1} tRNA
- 10% acetic acid

Method

1. Cleave the Ad 2 ITR from plasmid pHR1 using the restriction enzymes *Eco*RI and *Pst*I, which generate a 5' overhang and a 3' overhang, respectively. Specifically label the *Eco*RI site by incubation with [α-^{32}P]dATP (specific activity 3000 Ci $mmol^{-1}$), the other unlabelled dNTPs, and the Klenow fragment of DNA polymerase I.

2. Purify the labelled DNA on a polyacrylamide gel, as described in *Protocol 6*.

3. Incubate the labelled DNA (10 000 c.p.m.) for 30 min at room temperature in buffer A, with varying amounts of the protein of interest.

4. Add 0.25 units of DNase I, and incubate for a further 60 s at room temperature, before stopping the reaction by the addition of termination buffer.

3. Extract the DNA with phenol-chloroform, and precipitate with ethanol, wash with 70% ethanol, and resuspend in buffer B.

4. Denature at 100°C for 2 min, cool on ice, and fractionate on a 50 cm 8% polyacrylamide gel containing 50% urea.

5. Fix the gel in 10% acetic acid, and expose the dried gel to X-ray film or a Phosphor Imager screen.

4.1.2 Methylation protection

In this technique, guanidine bases that are not protected by interaction of the protein with DNA are methylated, thus making them susceptible to cleavage by piperidine. Hence uncleaved, protected guanidines can be identified. Each DNA strand is mapped in turn by introducing a ^{32}P-label into the appropriate strand of the DNA fragment, using one of the labelling methods detailed in the following protocol (*Protocol 17*). The converse experiment may be performed, where DNA, methylated such that only one base is modified per molecule, is tested for protein binding (29, 30). The DNA molecules methylated at sites which prevent interaction with the protein are identified. These include guanidines where methylation on N7, which projects into the major groove, directly interferes with protein binding. However, molecules with methylation of adenine residues on N3, which projects into the minor groove, also do not bind, suggesting contacts here too. However, it seems likely that this methylation may act indirectly, to prevent hydrogen bonding between bases in the major groove, and not through a direct contact.

Protocol 17. Methylation protection

Equipment and reagents
- DMS stop buffer: 1.5 M 2-mercaptoethanol, 0.3 M sodium acetate, containing 100 μg ml^{-1} tRNA
- Dimethyl sulfide (DMS)
- 1 M piperidine

Method

1. 5'-label the top strand of the Ad ITR, exposed by cleavage with *Eco*RI, using [γ-^{32}P]ATP and T4 polynucleotide kinase, as in *Protocol 6*.

2. Repair the bottom strand of another portion of plasmid, exposed as above with the Klenow fragment of DNA polymerase I in the presence of [α-^{32}P]dATP to 3'-label as in *Protocol 7*. Recut with *Bam*HI (both methods) to release the labelled ITR-containing fragments.

3. Incubate each labelled DNA (1.5 ng) in turn with the protein of interest for 1 h at 0°C.

4. Treat with DMS (1 μl) for 60 s at room temperature, and stop the reaction with 200 μl DMS stop buffer.

5. Extract the DNA with phenol-chloroform, precipitate it with ethanol, and dissolve in 50 μl 1M piperidine.

6. Incubate for 30 min at 90°C in a tightly sealed tube, lyophilize the samples, and separate the DNA on a 12% polyacrylamide sequencing gel.

7. Expose the dried gel to X-ray film.

4.1.3 UV cross-linking of protein to DNA

The protein of interest is allowed to interact with labelled DNA and then irradiated with UV light to cross-link the protein to the DNA. This procedure can be performed using oligonucleotides labelled only with ^{32}P (14), or using DNA into which bromodeoxyuracil has been incorporated, instead of thymidine, using synthesis catalysed by the Klenow fragment of DNA polymerase I, and using single-stranded M13 recombinant DNA as the template (31). Bromodeoxyuracil is very susceptible to UV light, and therefore easily cross-links to protein. The protein can then be digested, and identification of the radioactively labelled fragments allows sites within the protein which interact with the DNA to be identified (14). Alternatively, identification of a protein exhibiting specific binding to the DNA can be made by digesting the DNA to give a labelled protein product (31).

Protocol 18. UV cross-linking of protein to DNA

Equipment and reagents
- Buffer A: 25 mM Bicine-NaOH pH 8.0, 2 mM DTT, 1 mM EDTA, 20 mM KCl, 1 mg ml^{-1} BSA
- UV transilluminator

Method

1. Incubate 100 ng of the protein of interest with 10 ng of ^{32}P-labelled double-stranded oligonucleotide, or ^{32}P- and bromodeoxyuracil-labelled M13 DNA (digested with appropriate restriction enzymes to give the fragment of interest, and purified on a polyacrylamide gel) on ice for 15 min in buffer A.

2. Transfer the samples to the surface of a siliconized glass plate, or, if evaporation is a problem, to a microtitre plate (lidded and sealed), or to sealed glass capillary tubes, and place on ice 2 cm below a UV transilluminator (λ = 300 nm) and irradiate for 45 min (normal DNA) or 10 min (bromodeoxyuracil-containing DNA).

3. Either:

   (a) Incubate samples with the required protease before reducing and denaturing them, and separating the fragments on an SDS-polyacrylamide gel. Fix and dry the gel, and analyse by autoradiography or Phosphor Imaging, or

   (b) Add MgCl$_2$ and CaCl$_2$ to final a concentration of 3 mM, and digest with 0.17 units ml^{-1} of DNase I and/or 0.07 units ml^{-1} of micrococcal nuclease for 30 min at 37°C. Electrophorese the reaction products on an SDS-polyacrylamide gel, and visualize by autoradiography or Phosphor Imaging.

4.2 Protein–protein interactions

There are many methods for investigating protein–protein interactions. Some of those which have proved most useful in investigating the interactions involved in the adenovirus DNA replication complex are discussed below. The interaction can be studied using the whole protein, domains of interest can be expressed separately, or deletion mutants can be constructed to narrow down the area of interaction. It has generally been found that there are multiple sites of contact between the individual proteins involved in adenovirus DNA replication (14, 26, 32). This may allow the binding affinity between proteins to change to accommodate the changing roles of the

4: Interactions between viral and cellular proteins

proteins as replication progresses from preinitiation through initiation to elongation.

4.2.1 Gel filtration

A protein–protein interaction should remain intact during gel filtration, giving rise to a peak of higher molecular weight than the interaction constituents. Depending on the relative molecular weights, information about the stoichiometry of the interaction can be gleaned by peak integration, and calibration using standard proteins. The separation of pTP, pol, and the pTP–pol heterodimer was successfully shown by application to a Zorbax GF250 gel filtration column mounted in a Waters HPLC system (23).

4.2.2 GST pull-downs

This is a straightforward experiment to perform if one of the components is available as a GST fusion. One component is immobilized as the GST fusion on glutathione-agarose beads and, after blocking with BSA, the interacting protein is introduced (33). Extensive washing of the agarose beads removes non-interacting protein. The choice of wash buffer depends on the non-specific binding properties of the interacting protein and the nature and strength of the interaction. Thus 50 mM Tris-HCl (pH 8.0), 0.4 M NaCl, 1 mM EDTA, 0.05% NP-40 would be considered a stringent buffer. However, we have found potassium acetate to be an effective wash buffer, with no binding of the 'sticky' pTP to the negative control under the conditions described below.

Protocol 19. GST pull-downs

Equipment and reagents

- Buffer A: 1% BSA in 25 mM potassium acetate pH 7.5
- Buffer B: 25 mM potassium acetate pH 7.5
- Glutathione-agarose
- Tumbling wheel
- Disruption buffer: 4x stock containing 2 ml 20% SDS, 1 ml 2-mercaptoethanol, 1 ml 1 M Tris-HCl pH 7.0, 1 ml 50% glycerol, enough Bromophenol blue to give a blue colour

Method

1. Bind the GST fusion (400 ng), or GST as a negative control, to glutathione-agarose (5 µl) at 4°C for 30 min on a tumbling wheel.
2. Wash with buffer A, and add the interacting protein (150 ng).
3. Incubate for a further 1 h at 4°C on a tumbling wheel.
4. Wash once with buffer A, and 3 times with buffer B.
5. Transfer the beads to a fresh tube, and boil in disruption buffer for 2 min, before separating by SDS-PAGE and Western blotting if required.

4.2.3 Affinity chromatography

In this technique, one protein is immobilized on a small column, for example, as a GST fusion if available (4), or by covalent cross-linking, for example, onto cyanogen bromide-activated Sepharose (6) or activated thiol-Sepharose. The second protein which is thought to interact is applied to the column, and after washing is eluted. Unpurified extract makes a convenient source of protein, with the advantage that if the second protein can be pulled out from such a cocktail of proteins, the interaction must be specific. However, it is important to show that the same results can be obtained with purified protein, to determine that a third, bridging protein is not involved in manifesting the interaction. The elution conditions required to dissociate the complex can also be investigated, giving information about the strength of the interaction.

Protocol 20. Affinity chromatography

Equipment and reagents

- Buffer A: 25 mM Hepes-KOH pH 7.5, 1 mM DTT, 0.5 mM $MgCl_2$, 50 mM KCl, 2 mM SMBS, 0.05% NP-40, 10% glycerol
- Glutathione-agarose, CNBr-activated Sepharose 4B, or activated thiol-Sepharose 4B

Method

1. Couple one protein of interest to a column. Also prepare a control column to which GST or BSA is coupled.
2. Apply the second purified protein or extract to the columns, equilibrated in a suitable buffer, e.g. buffer A.
3. Remove the unbound material by washing the column in the above buffer, and elute the bound protein, e.g. by increasing the salt concentration of the buffer, by the addition of urea or 3 M $MgCl_2$, or by varying the pH.
4. Analyse fractions by SDS-polyacrylamide gel electrophoresis.

4.2.4 Use of antibodies

The importance of having antibodies against the protein of interest goes without saying. Simple uses include confirming the identity of the band of interest in the early steps of a protein purification procedure, by Western blotting. To detect protein–protein interactions, far-Western blots can be performed (32). In this procedure one protein is denatured, run on an SDS-polyacrylamide gel, and transferred to a polyvinylidine difluoride (PVDF) membrane. After blocking, the membrane is incubated with a second protein, and probed for binding of this second protein. Thus it can be determined whether the interaction relies exclusively on conformational interactions

4: Interactions between viral and cellular proteins

(destroyed in the denatured protein) or on recognition of short linear sequences.

Protocol 21. Far-Western blots

Equipment and reagents
- Polyvinylidine difluoride (PVDF) membrane
- Blocking buffer: PBS containing 10% non-fat milk, 0.05% Tween 20
- Purified proteins or extracts
- Primary antibody and secondary antibody, e.g. anti-mouse Ig–horseradish peroxidase conjugate (Amersham)
- Interaction buffer: buffer in which interacting proteins are known to interact, e.g. 50 mM potassium phosphate pH 7.0, 0.4 M KCl, 0.05% Tween 20, containing 10% non-fat milk
- Enhanced chemiluminescence (ECL) system (Amersham)

Method
1. Resolve the first protein and a control protein by SDS-PAGE, and transfer electrophoretically on to PVDF membrane. Incubate the membrane in blocking buffer for 1 h at room temperature on a rocking platform.
2. Wash the membrane in interaction buffer before incubating with a source of the second protein (or BSA as a control protein for a second control membrane) for 1 h at room temperature.
3. Wash the membrane extensively with interaction buffer before incubating with a primary antibody against the second (native) protein.
4. Wash again, then incubate with secondary antibody.
5. Wash and detect using the ECL kit (follow the manufacturer's instructions).

The protein of interest may be immunoprecipitated, by an antibody raised against it that has been immobilized on protein A- or protein G-agarose. The technique can be expanded to determine whether a second protein interacts with the first by investigating whether this protein too can be precipitated by the immobilized antibody–protein complex. Controls to show that the second protein is binding via the first protein, and not by non-specific interaction with the antibody or agarose beads, are required.

The methods discussed so far can be performed using polyclonal antibodies. However, the preparation of a panel of monoclonal antibodies against the protein of interest can be even more valuable. Immediately upon mapping the epitopes to which the monoclonal antibodies bind, structural information is gained about exposed sequences in the protein. It is likely that the exposed regions recognized by the antibodies are also, due to their exposed nature, involved in the functional protein–protein interactions being investigated. Hence, when performing immunoprecipitations using monoclonal antibodies, a second protein can only be immunoprecipitated with the first as long as the monoclonal antibody does not bind at the site of interaction with the second protein on the first. Thus sites of interaction may be defined (14).

The antibody can be cross-linked to protein A-agarose beads using dimethylpimelimidate (DMP). This helps to reduce the background caused by interaction of the secondary antibody with the heavy and light chains of the precipitating antibody, if a Western blot is used to analyse the end result (34).

4.2.5 Partial proteolysis

Although a protein may have a number of cleavage sites for a protease within its linear sequence, when the protein is in its native conformation only a proportion of these sites will be accessible to the protease. Hence partial proteolysis followed by N-terminal peptide sequencing of the fragments produced can be used to give structural data about exposed areas of the protein. Comparison of the digestion profile obtained for the protein proteolysed in the presence and absence of the interacting protein can indicate sites protected from proteolysis by the interaction (14, 32). Care must be taken in interpreting the results, as protection may be due to steric hindrance preventing the protease from cleaving a site, and not due to an actual contact between the proteins being studied. Determining whether individual cleavages can occur independently of each other allows one to ascertain if protection at one site may automatically result in a lack of cleavage at another site, without actual protection at this second site. Experiments can be set up in various ways, for example, as a time course of digestion, by addition of varying amounts of protease with a fixed digestion time, or by increasing titrations of the second protein with a fixed digestion time. Proteases which may prove useful include trypsin, chymotrypsin, endoproteinase Asp-N, endoproteinase Lys-C, and V8 protease. Although visualizing the digestion products by Western blotting, probing with antibodies against one of the proteins being studied, gives a picture which is not confused by bands from the second protein, it is also important to check that no major bands are being missed because they are not recognized by the antibodies available, and to verify that bands that are strongly detected by the antibodies really are significant, using a staining method. Control digestions of the second protein alone can be used to identify bands that have come from this source.

Protocol 22. Proteolytic digestions

Equipment and reagents
- Protease (Sigma or Boehringer Mannheim)
- Protease-free BSA
- Protease inhibitor

Method
1. Preincubate the two interacting proteins together for 30 min at room temperature. Also include control incubations where the protein

4: Interactions between viral and cellular proteins

> whose digestion will be followed is incubated with an equivalent amount of BSA.
> 2. Add the protease (protein:protease ratio 15:1), and incubate at room temperature; remove aliquots at various time points, and quench with the appropriate protease inhibitor.
> 3. Analyse by SDS-PAGE, followed by silver staining and also by Western blotting.

5. Investigating the dynamics of the replication process

The proteins involved in adenovirus DNA replication come together to form the preinitiation complex, the interactions involved in which are being mapped. However, this is only one snapshot of a multi-step process; as soon as the initiation reaction occurs the complex begins to dissociate. It is understanding how this dynamic process takes place and how it is triggered that will really allow the replication process to be fully understood. Progress is being made towards this aim, and some of the techniques that have been used are discussed here.

5.1 Immobilized replication assay

In this assay one component of the pre-initiation complex (normally NFI_{DBD}) is used as the GST fusion (2). The pre-initiation complex is allowed to form in solution, and then the complex is immobilized by the addition of glutathione-agarose beads. The addition of nucleotides allows the reaction to proceed, and the products are then analysed on an agarose gel. It is thus possible to measure the dissociation of species from the immobilized entity. Using this assay, it was determined that NFI dissociates from the replication complex as the first nucleotide is bound, prior even to the initiation event (2).

5.2 Glycerol gradient centrifugation

Glycerol gradient centrifugation in the presence of heparin allows free pTP to be distinguished from the pTP-pol heterodimer, as they sediment differently. Taking advantage of the DNA template sequence, King *et al* (3) were able to show, by the addition of only the nucleotide dCTP to an elongation assay, that pTP.C was still associated with pol, as was pTP.CA. However, when three nucleotides were added, with dCTP at low concentrations so that pTP.CAT but not the pTP.26 product was formed, it was shown that 40% of the pTP.CAT still cosediments with pol, but that 60% is free. By the time seven residues have been added, the pTP-oligonucleotide is all free of pol.

References

1. Hay, R. T. (1996). In *DNA replication in eukaryotic cells* (ed. M. L. DePamphilis), p. 699. Cold Spring Harbor Laboratory Press, NY.
2. Coenjaerts, F. E. J., and van der Vliet, P. C. (1994). *Nucl. Acids Res.*, **22**, 5235.
3. King, A. J., Teertstra, W. R., and van der Vliet, P. C. (1997). *J. Biol. Chem.*, **272**, 24617.
4. Van Leeuwen, H. C., Rensen, M., and van der Vliet, P. C. (1997). *J. Biol. Chem.*, **272**, 3398.
5. Challberg, M. D., and Kelly, T. J. (1979). *Proc. Natl. Acad. Sci. USA*, **76**, 655.
6. Bosher, J., Robinson, E.C., and Hay, R.T. (1990). *New Biol.*, **2**, 1083.
7. Temperley, S. M., Burrow, C. R., Kelly, T. J., and Hay, R. T. (1991). *J. Virol.*, **65**, 5037.
8. Leith, I. R., Hay, R. T., and Russell, W. C. (1989). *J. Gen. Virol.*, **70**, 3235.
9. Webster, A., Leith, I. R., Nicholson, J., Hounsell, J., and Hay, R. T. (1997). *J. Virol.*, **71**, 6381.
10. Temperley, S. M., and Hay, R. T. (1991). *Nucl. Acids Res.*, **19**, 3243.
11. Monaghan, A., and Hay, R. T. (1996). *J. Biol. Chem.*, **271**, 24242.
12. Fansler, B. S., and Loeb, L. A. (1974). In *Methods in enzymology*. (ed. L. Grossman and K. Moldave). Vol. 29, p. 52. Academic Press, London.
13. Monaghan, A., Webster, A., and Hay, R. T. (1994). *Nucl. Acids Res.*, **22**, 742.
14. Webster, A., Leith, I. R., and Hay, R. T. (1997). *J. Virol.*, **71**, 539.
15. Coenjaerts, F. E. J., and van der Vliet, P. C. (1995). In *Methods in enzymology*. (ed. J. L. Campbell). Vol. 262, p. 548. Academic Press, London.
16. Goding, C. R., and Russell, W. C. (1983). *EMBO J.*, **2**, 339.
17. Russell, W. C., Valentine, R. C., and Pereira, H. G. (1967). *J. Gen. Virol.*, **1**, 509.
18. Rosenfeld, P. J., and Kelly, T. J. (1986). *J. Biol. Chem.*, **261**, 1398.
19. O'Neill, E. A., and Kelly, T. J. (1988). *J. Biol. Chem.*, **263**, 931.
20. Mul, Y. M., van Miltenburg, R. T., de Clercq, E., and van der Vliet, P. C. (1989). *Nucl. Acids Res.*, **17**, 8917.
21. Watson, C. J., and Hay, R. T. (1990). *Nucl. Acids Res.*, **18**, 1167.
22. Richardson, C. D. (ed.) (1995). *Baculovirus expression protocols: methods in molecular biology.* Vol. 39. Humana Press, Totowa, NJ.
23. Temperley, S. M., and Hay, R. T. (1992). *EMBO J.*, **11**, 761.
24. Bosher, J., Leith, I. R., Temperley, S. M., Wells, M., and Hay, R. T. (1991). *J. Gen. Virol.*, **72**, 2975.
25. Verrijzer, C. P., Kal, A. J., and van der Vliet, P. C. (1990). *EMBO J.*, **9**, 1883.
26. Botting, C. H., and Hay, R. T. (1999) *Nucl. Acids Res.*, **27**, 2799–2805.
27. Tucker, P. A., Tsernoglou, D., Tucker, A. D., Coenjaerts, F. E. J., Leenders, H., and van der Vliet, P. C. (1994). *EMBO J.*, **13**, 2994.
28. Klemm, J. D., Rould, M. A., Aurora, R., Herr, W., and Pabo, C. O. (1994). *Cell*, **77**, 21.
29. Cleat, P.H., and Hay, R.T. (1989). *EMBO J.*, **8**, 1841.
30. Clark, L., Nicholson, J., and Hay, R. T. (1989). *J. Mol. Biol.*, **206**, 615.
31. Clark, L., Pollock, R. M., and Hay, R. T. (1988). *Genes Devel.*, **2**, 991.

32. Parker, E. J., Botting, C. H., Webster, A., and Hay, R. T. (1998). *Nucl. Acids Res.,* **26**, 1240.
33. Jaffray, E., Wood, K. M., and Hay, R. T. (1995). *Mol. Cell. Biol.,* **15**, 2166.
34. Harlow, E., and Lane, D. (1988). *Antibodies: a laboratory manual.* Cold Spring Harbor Laboratory Press, NY.

5

Analysis of transcriptional control in DNA virus infections

S. K. THOMAS and D. S. LATCHMAN

1. Introduction

Transcriptional control in DNA viruses is highly regulated, showing a high degree of both tissue-specificity and temporal control of gene expression. During the replication of DNA viruses, transcription occurs first from the immediate-early (if applicable) and early genes, followed by the late genes, with these two phases of transcription being separated by DNA replication. In general, genes expressed in the early phase encode regulatory and other non-structural viral proteins, while most of the genes expressed in the late phase encode virion components. Although this highly regulated cascade is controlled by virus-encoded factors, it is now well established that expression is also controlled by host-encoded factors. For example, although the herpes simplex virus (HSV) virion protein Vmw65 (VP16) is essential for the transactivation of the immediate-early genes, it has to be complexed with the cellular transcription factors Oct-1 and host cell factor (HCF) in order to function (1). The presence or absence of these necessary cellular transcription factors can obviously determine the outcome of infection, thus (at least in part) determining the cellular tropism of the virus. An understanding of the control of these interactions is therefore clearly important both for an understanding of the cellular tropism of the virus, and also to aid the development of antiviral agents.

This chapter is divided into two main sections. The first will describe the most commonly used methods for defining the pattern of viral gene expression during infection, either in tissue culture cells or *in vivo*. The second section describes techniques that can be used to analyse the interactions between viral promoters and viral and cellular transcription factors, using the example of methods we have used in our laboratory to dissect the role of the cellular Oct-2 transcription factor in the regulation of herpes simplex virus (HSV) immediate-early (IE) gene expression. It has been demonstrated that the protein products from these IE genes are essential for the progression of the HSV lytic (productive) replication cycle. Interestingly, HSV is able to productively infect most cell types, except for sensory neurones, where it instead establishes a life-long latent infection. Although asymptomatic, these latently infected cells provide a reservoir of virus that can be periodically reactivated (by factors such as stress), allowing virus to migrate to susceptible epithelial cells where it can establish a lytic infection (2).

The major characteristic of these latent infections is that except for a small region of the genome encoding the latency-associated transcript (LAT), the viral genome (including the regions encoding the IE genes) is completely silent. Further evidence suggests that the establishment of latency itself may be due to the failure of IE gene expression following infection (3). As mentioned previously, transcription of the IE genes is controlled by a complex composed of Vmw 65, Oct-1, and HCF, which binds to the TAATGARAT (R = purine) and related overlapping octamer–TAATGARAT motifs present in the IE promoters. Silencing of these genes could thus occur if another cellular transcription factor (such as an octamer-binding protein) bound to these motifs, preventing interaction with the Vwm65 transcription complex. Although the precise nature of the factors which might be involved in this inhibition of gene expression remain to be determined, we have demonstrated (by a variety of means) that the neuronally expressed isoforms of Oct-2 (Oct 2.4 and 2.5) can inhibit viral growth and activity of the IE promoters. These and other methods for analysing the control of viral gene expression, including the use of reporter constructs, strategies for mutagenesis, and methods for analysing interactions between viral and cellular factors, will be discussed in the second half of this chapter.

2. Analysis of viral gene expression during infection

This section describes techniques that can be used to analyse the pattern of viral transcription following infection in tissue culture cells or *in vivo*. This type of analysis can yield valuable information about the tissue-specific and temporal pattern of expression of a particular transcript. In the former type of analysis, RNA is prepared from various tissues from the infected host, and then monitored for the expression of the gene(s) of interest. Evidence of aborted infections detected in this way can provide clues to the host–viral interactions which occur during the virus replication cycle. Similarly, comparing the patterns of gene expression in permissive and non-permissive tissue-culture cell lines can also be informative. Analysis of the temporal pattern of gene expression is usually carried out at various time points following infection of tissue culture cells, enabling the pattern of gene expression of individual genes to be defined throughout the replication cycle.

As well as yielding information on the pattern of gene expression, RNA analysis enables the fine mapping of the gene of interest. Thus, as well as being able to identify the 5' and 3' termini of a gene, the use of alternative transcription initiation or splice donor and acceptor sites can also be determined. This information is important, since the differential splicing of a single primary transcript is used by some DNA viruses (e.g. adenovirus) as a means of controlling the expression of different open reading frames during the course of infection (4). Methods for this type of analysis are not included in this chapter, but can be readily found elsewhere (e.g. Sambrook *et al.*, ref. 5).

2.1 Preparation of RNA samples

Many methods are available for preparing RNA from tissues or cells, most of which are either based on the use of ribonuclease inhibitors (such as RNasin or guanidinium isothiocyanate), or on the use of proteinase K to degrade ribonucleases, or extraction with organic solvents such as phenol. Examples of methods for preparing total and polyA+ RNA (*Protocols 1 and 2*) which are suitable for most types of analysis are given here, and alternative methods for preparing these and cytoplasmic or nuclear RNA (useful for examining precursor transcripts and RNA splicing intermediates) can be found in Sambrook *et al* (5). Alternatively, many kits designed for isolating total or polyA+ RNA from cells and tissues are commercially available. Whilst these are costly for large numbers of samples, they do have the advantage of being quick and easy to use.

The success of the analysis of RNA transcripts is dependent on the quality of the RNA preparations, since RNA is very susceptible to degradation. It is therefore important to use 'RNase-free' reagents and equipment, and to wear gloves for all work involving RNA. It is also advisable to treat all aqueous solutions with DEPC as described in *Protocol 1* (except for those containing Tris, which should instead be made up using RNase-free water, then autoclaved). Once prepared, RNA samples should be protected from nuclease contamination during analysis. It is advisable to store the RNA samples at –70°C, either resuspended in water, or precipitated in the presence of ethanol and salt.

Protocol 1. Isolation of total RNA[a]

Equipment and reagents

- Polytron high speed homogenizer (or similar)
- DEPC-water: Add diethyl pyrocarbonate (DEPC) to a final concentration of 0.1%. Autoclave for 1 h, then leave overnight at 65°C (with the lid loose) to allow all traces of residual DEPC to decompose. **N.B.: DEPC should be handled in a fume hood.**
- Solution D: dissolve 250 g guanidinium isothiocyanate (in the manufacturer's bottle) in 293 ml water (preheated to 65°C), then add 17.6 ml 0.76 M Na citrate, pH 7, and 26.4 ml 10% (w/v) Na sarcosinate. Aliquot into 50 ml tubes, and store at –20°C. Before use, add 360 μl β-mercaptoethanol per 50 ml stock

Method

1. Prepare cells as follows:
   (a) Adherent cell monolayers: wash cells twice with ice-cold PBS. Add 1.8 ml[b] solution D and scrape cells. Transfer to a 15 ml conical tube.

Protocol 1. *Continued*

   (b) Suspension cells: transfer cells to a 15 ml conical tube, then wash cells twice with ice-cold PBS by spinning in a bench-top centrifuge at 350 *g*, 4 °C, for 3–5 min. Remove all traces of PBS, then add 1.8 ml solution D.

   (c) Tissues: freeze tissues in liquid nitrogen. Once frozen, homogenize (e.g. in a Polytron homogenizer) at high speed (according to the manufacturer's instructions) with 8 ml solution D.[c] For the following steps, add the volumes indicated in brackets.

2. Add 180 µl (800 µl) 2 M sodium acetate pH 4, 1.8 ml (8 ml) phenol, and 360 µl (1.6 ml) chloroform:IAA (24:1), mixing thoroughly between the addition of each reagent.

3. Vortex, or shake vigorously for 15–20 s, then incubate on ice for 15 min.

4. Spin at 8000 r.p.m., 4 °C, for 20 min in an JA20 rotor.

5. Transfer the supernatant to a fresh tube, then add 1.8 ml (7 ml) isopropanol. Mix well, then precipitate RNA overnight at −20 °C.

6. Pellet the RNA by spinning at 8000 r.p.m., 4 °C, for 20 min.

7. Thoroughly remove the supernatant, and dissolve the pellet in 300 µl of solution D. Transfer to a microfuge tube, and precipitate with 600 µl ethanol. Mix well, and incubate overnight at −20 °C.

8. When required, spin at maximum speed in a microcentrifuge, 4 °C, for 10 min to pellet the RNA. Remove the supernatant, and wash the pellet in 70% (v/v) ethanol.

9. Briefly dry the pellet,[c] and dissolve in 50 µl (100 µl) DEPC-water. Quantify by spectrophotometry at 260 nm (where one OD unit = 40 µg ml^{-1} of RNA, ref. 5).

[a] Based on the method of Chomczynski and Saachi (6).
[b] Volumes given here are those required to prepare RNA from cells grown on a 10 cm tissue culture dish (10^7 cells). Preparation of RNA from other size dishes will require the volumes to be adjusted accordingly.
[c] A maximum of 0.4 g of tissue should be homogenized in 8 ml buffer D.
[d] Care should be taken not to over-dry the RNA pellet, as this can lead to difficulties in resuspending it.

Ribosomal RNA (rRNA) is, by far, the most abundant RNA species in eukaryotic cells, representing 80–90% of the total cellular RNA. In contrast, messenger RNA (mRNA) represents only 1–5% of the total cellular RNA. Most mRNA transcripts are polyadenylated at the 3' end, allowing them to be affinity purified on an oligo(dT)-cellulose column. The method described below can be used to isolate polyA+ RNA either from cell lysates, or from pre-prepared total RNA samples. Although isolating polyA+ RNA from cell

5: Analysis of transcriptional control in DNA virus infections

lysates is the most rapid approach, it is not recommended for cells which possess a high ribonuclease content.

Protocol 2. PolyA+ RNA isolation

Equipment and reagents

- Polytron high speed homogenizer (or similar)
- Solution D: see *Protocol 1*
- STE: 0.1 M NaCl, 20 mM Tris-HCl pH 7.4, 10 mM EDTA pH 8
- Oligo(dT)-cellulose (Sigma)
- Proteinase K: 10 mg ml^{-1} stock made up in water. Store at –20°C
- Loading buffer: 0.4 M NaCl, 20 mM Tris-HCl pH 7.4, 10 mM EDTA pH 8, 0.2% SDS
- Wash buffer: 0.1 M NaCl, 10 mM Tris-HCl pH 7.4, 1 mM EDTA pH 8, 0.2% SDS
- Elution buffer: 1 mM Tris-HCl pH 7.4, 1 mM EDTA pH 8, 0.2% SDS
- DEPC-water: see *Protocol 1*

Method

1. Prepare tissues and cells as follows:
   (a) Adherent and suspension cells: harvest cells and transfer to a 50 ml polycarbonate tube. Wash cells twice with PBS, spinning each time at 350 *g* for 5 min to pellet the cells. Resuspend cells in 5 ml STE, then add 200 µl 10 mg ml^{-1} proteinase K, and 250 µl 10% (w/v) SDS. Homogenize for 30 s at high speed in a Polytron homogenizer (or similar, according to manufacturer's instructions). Incubate at 37°C for 30 min. Alternatively prepare total RNA according to *Protocol 1*. Resuspend the RNA pellet in 5 ml STE.
   (b) Tissues: prepare total RNA as detailed in *Protocol 1*, then resuspend the RNA pellet in 5 ml STE.

2. Resuspend the oligo-dT cellulose[a] at 50 mg ml^{-1} in loading buffer.

3. Add 450 µl of 5 M NaCl to the RNA samples, then add 2 ml of the oligo(dT)-cellulose slurry. Incubate the mixture (with slow shaking) for 3 h at room temperature.

4. Spin at 500 *g* for 5 min to pellet the oligo(dT)-cellulose. Wash twice with 2 ml loading buffer, spinning as before.

5. Resuspend the mixture in 10 ml loading buffer, and add to a pre-prepared column.[b]

6. Wash the column once with 10 ml loading buffer, then once with 10 ml wash buffer.

7. Place a 30 ml glass Corex tube under the column, then elute the polyA+ RNA with 5 ml elution buffer (pre-warmed to 37 °C).

8. Mix the eluted solution, remove 350 µl, and quantify by spectrophotometry at 260 nm (where one OD unit = 40 µg ml^{-1} of RNA, ref. 5).

Protocol 2. *Continued*

9. Add 30 μg tRNA, 880 μl 2 M sodium acetate, and 12 ml ethanol. Mix thoroughly, then incubate overnight at −20°C to precipitate the polyA+ RNA.
10. Spin at 10 000 r.p.m. in a JA20 (or similar), for 45 min, to pellet the RNA. Air-dry the pellet, then dissolve in DEPC-water at 0.5 μg ml^{-1}.

[a] The oligo(dT)-cellulose can be reused up to four times. To prepare, resuspend in 0.1 N NaOH and shake for 30 min. Spin at 500 *g* for 5 min to pellet the oligo(dT), then wash three times with loading buffer (check the pH of the final wash; it should be pH 7.4).
[b] Suitable commercially prepared columns are readily available (e.g. from Bio-Rad), or columns can be prepared either from a 10 ml plastic, or siliconized (5) glass, pipette.

2.2 Analysis of transcripts by Northern blotting

Northern blots are used to analyse specific RNA molecules in a complex RNA population. The RNA is separated by electrophoresis under denaturing conditions, transferred to a nylon (or nitrocellulose) filter, then hybridized with a specific, radioactively labelled probe. The distance migrated by the RNA species can be used to determine its molecular weight, and the intensity of the autoradiographic signal gives a measure of its abundance. Two methods of preparing denaturing agarose gels are commonly used to analyse RNA, based on the use of either glyoxal (7) or formaldehyde (8) to denature the RNA. The method given here (*Protocol 3*), which uses formaldehyde, has been found to be easy and to give good resolution.

Total, cytoplasmic, or polyA+ RNA preparations can all be used in Northern blot analysis. The amount of RNA loaded per lane will depend on the abundance of the transcript of interest. Abundant species (>0.1% of the mRNA population) can readily be detected in 20 μg of total RNA, whereas rare RNA species are best analysed using polyA+ RNA (typically 5–10 μg). It is also advisable to use polyA+ rather than total RNA if the filter is to be stripped and repeatedly hybridized with different labelled probes.

To compare the amount of an RNA species in different samples, it is necessary to ensure that equal amounts are loaded in each lane. The best way to do this is to hybridize the blot with a probe which detects an RNA species which would be expected to be present at similar levels in the different samples (e.g. actin or glyceraldehyde-3-phosphate dehydrogenase (GAPDH)). If the test and control RNA species are sufficiently different in size, the two probes can be hybridized to the blot at the same time. However, if the species are similar in size, hybridization with the control probe should be carried out after the test probe has been stripped from the blot (*Protocol 6*). The signals from the test RNA are then normalized to those from the control RNA by densitometry.

5: Analysis of transcriptional control in DNA virus infections

Protocol 3. Electrophoresis of RNA in denaturing gels, and Northern transfer

Equipment and reagents
- 50 × MOPS: 1 M MOPS, 0.25 M sodium acetate, 0.05 M EDTA, adjust to pH 7
- 1 M NaPi: Dissolve 89 g $Na_2HPO_4.2H_2O$ in 700 ml water. Add 4 ml concentrated H_3PO_4, and adjust the volume to 1000 ml
- 0.05% (w/v) Bromophenol blue (in water)
- 20 × SSC: 3 M NaCl, 300 mM sodium citrate
- Nylon membrane: e.g. Gene screen (NEN DuPont) or Hybond N+ (Amersham).

Method

1. Wash all equipment (gel tanks, combs, etc.) with 0.1 N NaOH before use. Rinse thoroughly with DEPC-water.

2. Prepare a horizontal 0.9% agarose gel: for a 250 ml gel, dissolve 2.25 g agarose in 5 ml 50 × Mops and 204 ml DEPC-water. Allow to cool to 65°C, then add 42 ml formaldehyde, and pour the gel immediately (**this should be carried out in a fume hood, as formaldehyde vapours are toxic**).

3. Once set, place the gel in an electrophoresis tank, and immerse in 1 × MOPS buffer.

4. Prepare the RNA samples by adding: 20 µl RNA in water (20 µg total RNA, or 5–10 µg polyA+ RNA), 20 µl formamide, 7 µl formaldehyde, 1 µl 50 × MOPS, 1 µl 1mg ml^{-1} ethidium bromide (**care: ethidium bromide is a known carcinogen**), 2 µl 0.05% (w/v) Bromophenol blue. Mix the samples, then heat to 70°C for 5 min. Incubate on ice until ready to load the gel.

5. Load the samples into the appropriate wells. Run the gel at 100 V until the marker dye has migrated ~10 cm (~ 4.5 h). During this time the buffer should be slowly pumped from the cathode to the anode end of the gel tank.

6. Carefully remove the gel from the tank, and rinse briefly in water to remove excess formaldehyde. Carefully cut off one corner to allow orientation of the gel.

7. Visualize the RNA bands under UV transillumination (**care: suitable eye protection should be used**).

8. Wash the gel in 10 × SSC for 20 min at room temperature (with shaking).

9. Using 10 × SSC, transfer the RNA overnight onto a suitable nylon membrane according to the manufacturer's instructions.

10. The next day, dismantle the transfer apparatus, and mark the positions of the wells on the filter with a pencil. At this stage, transfer of the

Protocol 3. *Continued*

RNA can be monitored using a hand-held UV transilluminator. Mark the position of the 18S and 28S rRNA species.[a] Float the blot on 50 mM NaPi for 10 min at room temperature, then wash with shaking for a further 10 min.

11. Cross-link the RNA to the blot according to the manufacturer's instructions.

[a] Although RNA markers are readily available (e.g. from Promega), many workers use the positions of the 18S and 28S rRNA species (~1.8 and ~4.8 kb respectively) to calculate the size of the transcript of interest.

Several methods exist for radiolabelling DNA probes, including nick translation (9) and random priming (10). In general, it is advisable to prepare DNA probes using sequences between 150 bp and 1.5 kb in length (although longer sequences can often be used, since both nick translation and random priming produce smaller DNA fragments). In our laboratory we have found that the random-priming method given in *Protocol 4* routinely generates probes with a high specific activity. In this method, the DNA template is hybridized with random hexanucleotides, which then act as primers for the synthesis of a radiolabelled second strand by *E. coli* DNA polymerase 1 (Klenow fragment). There are also many kits commercially available, such as the Ready-to-Go kit (Pharmacia) and Prime-a-gene (Promega), which also use random priming to produce probes with high specific activity. In either case, the probes can be prepared from DNA in solution or in a low melting point (LMP) gel slice. *Protocol 5* gives the spun column chromatography method, which we routinely use in our laboratory to purify the probe before hybridization.

Protocol 4. Oligo-labelling of the DNA probe

Equipment and reagents

- 10 × HEPES buffer: 50 mM $MgCl_2$, 40 mM HEPES pH 6.6
- 100 μM solutions of dATP, dGTP, and TTP. Mix 1 μl of each to give stock dNTPs
- DNA polymerase 1 (Klenow fragment), 5 units ml^{-1} (Promega)
- [α-^{32}P]dCTP at 10 mCi ml^{-1} (Amersham Life Sciences)
- Random hexanucleotide primers (500 OD units ml^{-1} in water) [p(dN)$_6$] sodium salt (Pharmacia Biotech)

Method

1. Add the following to a microcentrifuge tube: HEPES buffer 3 μl, stock dNTPs 1 μl, random hexamers 3 μl, 100 ng probe DNA[a], water to 30 μl.
2. Denature the DNA by incubating the tube at 100°C for 5 min.

5: Analysis of transcriptional control in DNA virus infections

3. Allow the tube to cool to room temperature, then add 1 μl DNA polymerase 1 (Klenow fragment) and 3 μl [α-^{32}P]dCTP.

4. Mix gently and incubate for 1 h at 37 °C.

[a] DNA, either in solution, or in an LMP agarose gel slice, is suitable. If using DNA in a gel slice, first incubate the DNA at 80 °C for 5 min to melt the gel, then add to the mix as above.

Protocol 5. Purification of the probe using Sephadex G50 spun column chromatography

Equipment and reagents

- Plastic 1ml syringe
- TE Buffer: 0.01M Tris HCl pH 7.5, 0.001M EDTA
- Sephadex G50 (Sigma G50-150) that has been swollen in TE buffer (follow the manufacturer's instructions)

A. Preparation of column

1. Take a 1 ml syringe without the plunger, and plug the bottom with glass wool.

2. Fill the syringe to the top with a slurry of Sephadex G50 in TE buffer.

3. Place the column in a 15 ml Falcon tube, together with a 1.5 ml microcentrifuge tube (with the lid removed) to catch the waste.

4. Centrifuge at 1500 *g* for 3 min. The column should now contain about 0.7 ml of dry Sephadex G50. Discard the eluate.

5. Add more Sephadex G50 slurry to the column, and spin as before.

6. Add 100μl TE to the column, and spin as before. Repeat this step until the eluted volume equals 100μl.

B. Purification of the probe

1. Replace the microcentrifuge tube beneath the column.

2. Load the probe-labelling reaction onto the column, and spin as before. The microcentrifuge tube should contain the labelled DNA in a volume equal to the volume loaded, while the unincorporated radioactivity should remain in the column.

Protocol 6. Hybridization of Northern blots

Equipment and reagents

- Hybridization oven (e.g. Hybaid Maxi Hybridization oven)
- 1 M NaPi pH 7.2: see *Protocol 3*
- Church's Buffer: 0.5 M NaPi pH 7.2, 7% SDS, 1 mM EDTA pH 8
- Wash solution: 40 mM NaPi, 1% SDS

Protocol 6. *Continued*

Method

1. Prehybridize the blot in 50 ml Church's buffer for at least 1 h at 65 °C. Preferably this should be carried out in a rotary oven, although heat-seal bags (used double) are also suitable.
2. Denature the probe (*Protocol 4*) by adding 0.25 volumes 2 N NaOH, and incubating at room temperature for 5 min. Alternatively, the probe can be denatured by incubation at 95 °C for 5 min, after which it should immediately be put on ice.
3. To hybridize the blot, pour off the prehybridization solution, and replace with the probe diluted in Church's buffer to at least 1.5×10^6 c.p.m. ml^{-1} (in ~15 ml). Incubate overnight at 65 °C.
4. After hybridization, wash the filter with 40 mM wash solution as follows: 2×5 min at 55 °C; 15 min in a 65 °C shaking water bath; 45 min in a 65 °C oven.
5. Remove the filter from the wash solution, and wrap in cling film (do not allow the filter to dry, or the probe will be irreversibly bound to the filter). Expose the filter either to X-ray film, or to a Phosphor-Imager (Bio-Rad) screen according to the manufacturer's instructions.
6. To reuse the filter, remove the bound probe by washing the filter with 2 mM EDTA, pH 8, plus 0.1% SDS, for 30–60 min at 80 °C. The filter can then be reprobed (no pre-hybridization is necessary).

2.3 Analysis of transcripts by reverse transcription-polymerase chain reaction (RT-PCR)

In common with Northern blot analysis, PCR (as part of a two-stage process) can be used to detect the presence of a specific mRNA species in a mixed RNA population (*Protocol 7*). The initial stage involves the production of a cDNA copy of the RNA under test using reverse transcriptase. This cDNA is then used as the template for PCR, which is carried out using primers designed to detect the RNA of interest. The PCR products can then easily be analysed on an agarose gel for the presence of a correct sized band. The cDNA product can then be further analysed by restriction digest mapping or sequencing, or can be cloned into a vector plasmid.

When designing PCR primers, the following should be taken into consideration:

- Primers should be between 15 and 30 bp long.
- The pair of primers should have similar G + C content.
- There should be minimal secondary structure (i.e. self-complementarity), and low complementarity to each other.

5: Analysis of transcriptional control in DNA virus infections

The reverse transcriptase reaction can either be primed with oligo(dT), or by using a specific primer. The advantage of using oligo(dT) is that the reaction will produce cDNA copies of the mixed RNA population, allowing several mRNA species to be analysed from a single RT reaction. However, if the region to be amplified by PCR is several kb from the poly(A) tail of the RNA, it is more usual to use a specific primer for the RT stage, to ensure that the reverse transcriptase copies the region of interest.

The PCR reaction is extremely sensitive to Mg^{2+} ion concentration, so this will need to be optimized for each set of primers: if the Mg^{2+} concentration is too low, the primers will not form stable hybrids, so no product will be produced. On the other hand, at high Mg^{2+} concentrations the primers will hybridize non-specifically to the template, and a ladder of fragments will be produced. The amount of $MgCl_2$ suggested in *Protocol 7* should therefore be treated as a starting point, and it may be necessary to titre the $MgCl_2$ at 0.2 mM intervals between concentrations of 0.8 and 2 mM.

PCR can be made quantitative by constructing an artificial RNA competitor template which contains the same primer binding sites with a different distance between them compared with the transcript of interest, thus producing a PCR product of a different size from the test RNA. Such a construct can then be used to generate an RNA transcript *in vitro* which can be added in varying amounts to the test RNA population. The extent to which this artificial RNA template competes with the test RNA template gives a measure of the amount of RNA of interest in the sample. Alternatively, a second PCR reaction, using primers directed against an RNA species which would be expected to be present at similar levels in the different samples (e.g. actin or GAPDH), can be carried out. The amount of product from this second PCR can then be used to normalize the amount of test product. This second approach is less accurate, since it involves a different set of primers with different specificity from those used in the test PCR.

Protocol 7. Reverse transcription-PCR amplification of mRNA (RT-PCR)

Equipment and reagents

- 1 mg ml^{-1} oligo(dT) (Pharmacia), or specific 3' primer
- Specific PCR primers
- 10 mM dNTP mixture (all four dNTPs; Pharmacia)
- 10 units µl^{-1} placental RNasin (Promega)
- 200 units µl^{-1} AMV reverse transcriptase (BRL)
- 5 units µl^{-1} *Taq* polymerase (Promega)
- 10 × *Taq* polymerase buffer, supplied by manufacturer

Method

1. Using ~20 µg RNA, set up the following reverse transcriptase reaction in 500 µl microfuge tubes: 50 pmol oligo(dT) or specific 3' primer,

Protocol 7. *Continued*

> 10 mmol dNTP mixture 1 µl, 10 × PCR buffer 1 µl, **water** to 8 µl. Mix and incubate at 65°C for 5 min. Cool to room temperature.
>
> 2. Add AMV reverse transcriptase 1 µl, RNasin 1 µl. Incubate at 42°C for 30 min (up to 1 h), then at 94°C for 5 min.
> 3. Put on ice, then add: 50 pmol specific 5' primer, 50 pmol specific 3' primer[a], 10 × PCR buffer 4 µl, 25 mM MgCl$_2$ 3 µl, water to 50 µl, *Taq* polymerase 0.5 µl. Mix thoroughly, then overlay with mineral oil.
> 4. Heat the reaction mix to 95°C for 5 min in a thermal cycler to denature the DNA.
> 5. Cycle 30–40 times at 94°C for 40 s, 55°C for 45 s, and 72°C for 1 min. Incubate at 72°C for 5 min to ensure complete extension of all amplified molecules.
> 6. Run an aliquot of the reaction product on a 1% agarose gel. If the correct sized bands cannot be seen, then the annealing temperature (50–65°C) and MgCl$_2$ concentrations (final concentration 0.8–2 mM) should be adjusted. The cDNA can now be subcloned by standard procedures (5), or analysed by restriction enzyme digestion or sequencing.
>
> [a] This should be added at this stage if the cDNA was prepared using oligo(dT).

2.4 Analysis of transcripts by *in situ* hybridization (ISH)

Whilst the methods described in the previous sections all give useful information concerning pattern and level of gene expression, more detailed information concerning the spatial pattern of gene expression within tissues can be obtained using *in situ* hybridization (ISH). In this method, labelled probes are used to detect mRNAs within cells in tissue sections, allowing the exact cells producing the transcript of interest to be identified. Generally, RNA probes are used, as they are very sensitive, although DNA oligonucleotides can be useful for distinguishing transcripts with few sequence differences. The other advantages of using RNA probes are that non-specific background can be reduced by treating the sections with RNase during the post-hybridization wash steps (11), and that antisense transcripts can be used as controls for non-specific binding. The DNA sequences required for producing RNA probes should be cloned into a suitable plasmid, such as pBluescript®, which contains bacteriophage T3 and T7 promoters, using standard methods (5).

2.4.1 Probe preparation

Until relatively recently, ^{35}S-labelled radioactive probes have been most commonly used, as they provide the best compromise in terms of the length of autoradiographic exposure required and the resolution of the signal. Auto-

5: Analysis of transcriptional control in DNA virus infections

radiographic detection of the radioactive signal provides a final readout in the form of silver grains which can be counted, permitting an approximation of the number of mRNA transcripts. More recently, non-radioactive probes (prepared using digoxigenin-UTP (DIG-UTP)) have become common, since they offer the advantages of safety and speed. Non-radioactive probes are detected by the formation of a coloured precipitate when the slides are treated with the substrate solution. This type of probe is the one most frequently used in our laboratory. Routinely, these are prepared using DIG RNA Labelling Kit (SP6/T7) according to the manufacturer's instructions (Boehringer Mannheim). Once prepared, an aliquot of the probe reaction mix (2 μl) should be run on a 1% TBE agarose gel at 200 V for 5 min. A faint band should be visible. The probes are then purified using the Nucleon EasiRNA kit (Scotlab). If a large probe (>1.5 kb) is to be used it will need to be hydrolyzed (as described in *Protocol 9*) before RNaid extraction.

2.4.2 Tissue preparation

The following Protocols (*Protocols 8–10*) describe the methods we routinely use in our laboratory to analyse gene expression in neuronal tissues. A detailed description of tissue preparation for ISH are beyond the scope of this chapter, but briefly, the tissues are routinely fixed in 2–4% paraformaldehyde (*Protocol 18*) before being embedded in paraffin wax and sectioned. Sections are then collected on slides which have been precoated with 2% (v/v) 3-aminopropyl triethoxysilane (Sigma) (in acetone). It is important to mount the sections onto coated slides to retain the sections during subsequent steps. Further details of tissue preparation and alternative ISH methods can be found in the Practical Approach series volume entitled 'In situ *hybridization*' (12).

As described in *Protocol 8*, the sections are pre-treated with proteinase K and acetic anhydride before hybridization of the probe. Proteinase K treatment increases the signal, presumably by improving the accessibility of the tissue to the probe. This treatment can be omitted if ISH is carried out on frozen tissues which have been sectioned on a cryostat. Acetic anhydride treatment improves the signal-to-background ratio by acetylating amino residues that might otherwise bind the probe non-specifically.

Protocol 8. Pretreatment of sections

Equipment and reagents

- Tissue sections on coated slides (see above)
- Absolute ethanol, 70% ethanol, and 50% ethanol
- PBS
- Proteinase K: 10 mg ml^{-1} made up in water. Store at −20°C
- Proteinase K buffer: 0.1 M Tris-HCl pH 8.0, 50 mM EDTA pH 8.0
- 20 × SSC: 3 M NaCl, 300 mM sodium citrate
- TEA buffer: 0.1 M triethanolamine-HCl pH 8.0 (Sigma)
- Acetic anhydride

Protocol 8. *Continued*

Method

1. De-wax the sections by incubating the slides in xylene for 10–20 min at room temperature. Transfer the slides to fresh xylene, and incubate for a further 10–20 min. Transfer to absolute ethanol, and incubate for 10 min.
2. Rehydrate the tissues by incubating the slides for 5 min in each of the following solutions: 70% ethanol, 50% ethanol, and PBS (twice). Incubate in water for 1 min.
3. Drain the slides, place them horizontally, and overlay the sections with 1 µg ml^{-1} proteinase K (add to prewarmed proteinase K buffer just before use). Incubate for 30 min at 37 °C.
4. Wash the slides briefly in water, then in freshly prepared 0.1 M TEA buffer.
5. Acetylate the tissues as follows: add 40 ml acetic anhydride to a dry container (**N.B.: acetic anhydride is toxic and volatile**), then simultaneously add 200 ml TEA buffer together with the slides, and mix rapidly by dipping the slides up and down. Incubate the slides in this mixture for 10 min at room temperature.
6. Wash the slides twice in PBS for 5 min at room temperature.
7. Dehydrate the tissues as follows: wash the slides briefly in 2 × SSC, then pass the slides through 50%, 70%, and absolute ethanol as for Steps 1 and 2. Air-dry the sections, and store at 4 °C.

Protocol 9. Alkaline hydrolysis of ISH probes

Equipment and reagents

- Hydrolysis buffer: 80 mM NaHCO$_3$, 120 mM Na$_2$CO$_3$, pH 10.2
- Neutralization buffer: 13.33 µl 3 M sodium acetate pH 5.2, 2 µl glacial acetic acid, 187 µl water
- ISH probe prepared according to the manufacturer's instructions (Boehringer Mannheim)

Method

1. Add 50 µl of the hydrolysis buffer to the probe reaction mix.
2. Incubate at 60 °C for time t, where:

$$t \text{ (min)} = \frac{\text{(original length, kb} - \text{required length, kb)}}{0.11 \text{ (original length, kb} \times \text{required length, kb)}}$$

3. Add 50 µl neutralizing buffer to stop the reaction.
4. Purify the probe using the Nucleon EasiRNA kit (Scotlab).

5: Analysis of transcriptional control in DNA virus infections

Protocol 10. Hybridization and washing of sections

Equipment and reagents

- Denhardt's solution: 1% (w/v) polyvinyl pyrrolidone, 1% (w/v) BSA, 1% (w/v) Ficoll 400
- Pre-treated tissue sections (Protocol 8)
- DIG-labelled probe (prepared according to the manufacturer's instructions)
- Hybridization mixture: 50% formamide, 5 × SSC (Protocol 8), 5 × Denhardt's solution, 250 µg ml-1 yeast tRNA, 500 µg ml-1 herring sperm DNA
- TE buffer: see Protocol 5
- Buffer 1: 0.1 M Tris-HCl pH 7.5, 0.15 M NaCl
- Buffer 2: 1% (w/v) blocking agent (Boehringer Mannheim), 0.5% (w/v) BSA in buffer 1 (this can be dissolved by heating to 60 °C with stirring)
- Buffer 3: 0.1 M Tris-HCl pH 9.5, 0.1 M NaCl, 50 mM $MgCl_2$ (filter sterilize before use)
- Buffer 4: 4.5 µg ml-1 5-bromo-4-chloro-3-indolyl phosphate, 3.5 µg ml-1 Nitroblue tetrazolium, 0.2 mg ml-1 levamisole (all from Boehringer Mannheim) in buffer 3
- Anti-digoxigenin-alkaline phosphatase (DIG-AP) Fab fragments (Boehringer Mannheim)

Method

1. Add the riboprobe (~100 ng per section) to the hybridization mix, and incubate at 75°C for 3 min.
2. Apply the hybridization mixture to each section (just enough to cover the section is sufficient).
3. Gently lower a clean coverslip onto the section.
4. Place all the slides horizontally in a plastic box, together with tissue paper soaked in 50% formamide 5 × SSC, and seal the box to form a humidified chamber.
5. Incubate overnight at 50–65°C.[a]
6. Remove the slides from the hybridization chamber, and immediately place them in a slide rack, immersed in 5 × SSC (pre-warmed to the hybridization temperature) for 30 min at 50°C, to allow the coverslips to fall off.
7. Transfer the slides to 0.2 × SSC, pre-warmed to the hybridization temperature. Incubate for 1 h at the hybridization temperature, changing the solution after 30 min.
8. Transfer to fresh 0.2 × SSC[b], and incubate for 5 min at room temperature.
9. Transfer to buffer 1, and incubate for 10 min at room temperature.
10. Transfer to buffer 2 (block buffer), and incubate for 30 min to 1 h at room temperature.
11. Place all the slides horizontally in a plastic box, together with tissue paper soaked in water. Dilute anti-DIG-AP antibody 1:500 in buffer 2, add sufficient to cover the sections, then seal the box to form a humidified chamber. Incubate for 2–8 h at room temperature, or alternatively, overnight at 4°C.
12. Wash the slides in buffer 1 for 5 min at room temperature.

Protocol 10. *Continued*

13. Incubate the slides in buffer 3 for 2 min at room temperature.
14. Place buffer 4 (developing buffer) on the slides, and incubate in a humidified, dark chamber until colour develops (1–5 h). Stop the reaction by washing the slides in TE buffer, pH 8.
15. Counterstain the cell nuclei in the sections with methyl green. Mount in 50% glycerol-TE, pH 8 (v/v).

[a] The optimal hybridization temperature will need to be determined for each probe. However, 50 °C is normally suitable.
[b] If non-specific background problems occur, this wash stage can be carried out using 0.1 × SSC.

3. Analysis of cloned promoter sequences using reporter gene constructs

One of the most used methods of analysing the pattern of expression from a promoter sequence is to link it to a reporter gene encoding a novel enzymatic activity that can be easily assayed. A number of reporter genes are available, including chloramphenicol acetyl transferase (CAT) (13), b-galactosidase (14), and luciferase (15). More recently, the cloning of the gene for the green fluorescent protein (GFP) (16) has meant that tissue-specific patterns of expression in live cells can be analysed easily using fluorescence microscopy and FACS analysis.

A number of plasmids containing these reporter genes downstream from multiple cloning sites suitable for the insertion of promoter sequences are commercially available (e.g. pCAT®-Vectors from Promega). Once prepared, these reporter constructs can be introduced by transfection into different cell types to analyse their tissue-specific activity (by use of representative tissue-culture cell lines, or primary cells), or to analyse the effect of different transcription factors on the promoter sequence by co-transfecting them into cells, together with expression constructs encoding the proteins of interest. In our laboratory, we have used these types of constructs to demonstrate that the specific forms of Oct-2 which are expressed in neuronal cells can inhibit the basal activity of the HSV IE promoters, and also block its transactivation by the viral protein Vmw65 complex, which is essential for high expression of the IE genes during lytic infection (Section 1).

Although the co-transfection of such plasmids *in vitro* can give a lot of useful information concerning the interaction of viral promoters with viral and cellular factors, a more biologically relevant approach is to introduce these reporter constructs into a non-essential gene in the viral genome, so as to produce recombinant indicator viruses. These viruses can then be used to examine the activity of the promoter in a viral context. This can be achieved either by infection of tissue-culture cell lines engineered to over-express the proteins of interest, so that their effects on the promoter during infection can

5: Analysis of transcriptional control in DNA virus infections

be assessed, or the indicator viruses can be used to infect animals, enabling the promoter's activity during infection *in vivo* to be monitored. A complementary approach to this is to produce transgenic animals carrying the reporter constructs. The production of such animals allows the tissue specificity of the promoter to be assessed with and without virus infection. These approaches will be discussed further in Sections 3.3 and 3.4.

3.1 Methods of transfection

A number of different methods have been developed to enable the introduction of DNA into cultured or primary cells, the chosen method depending to a large extent on the particular cell line to be transfected. The most commonly used methods are calcium phosphate-mediated transfection (13), DEAE-dextran (17), electroporation (18), and lipofection (19). We have found that calcium phosphate-mediated transfection is effective for many cell types, and it is the method presented here in Protocol 11. Electroporation involves the brief application of a high-voltage electric pulse to cells, resulting in the generation of pores in the plasma membrane which allow uptake of DNA by the cells. This method gives high transfection efficiencies, but needs to be optimized to each cell line. Lipofection (which is useful for cell types refractile to calcium phosphate transfection) relies on the use of cationic liposomes which complex with DNA in such a way that the complex can be taken up by cells (19). The mixture of cationic liposomes required for high efficiency gene transfer needs to be optimized for different cell types, and mixtures of liposomes optimized for standard cell lines are commercially available (e.g. Transfectam® from Promega, and Lipofectin™ from Gibco BRL).

Whichever transfection method is chosen, the DNA used should be purified from contaminating RNA using a commercially available kit (e.g. Qiagen tip-100), or by a traditional method such as caesium chloride density-gradient centrifugation. Since transfection efficiency can vary between experiments, it is important that all the plasmids to be compared are transfected at the same time, and that adequate replicates are carried out.

Protocol 11. Transient transfection using calcium phosphate

Equipment and reagents

- Carrier herring sperm DNA: resuspend DNA at 1 mg ml-1 in water overnight at 37°C. On the following day, extract twice with phenol, once with chloroform:IAA (24:1), then precipitate the DNA with ethanol. Following a 70% ethanol wash, air-dry the DNA, then resuspend at 1 mg ml-1 in water. Store 1 ml aliquots at –20°C.

- HEPES-buffered saline (HBS), 1× stock: 140 mM NaCl, 5 mM KCl, 0.7 mM Na_2HPO_4, 5 mM D-glucose, 20 mM HEPES. Adjust pH to 7.05 with 1 M NaOH, filter sterilize, and store at 4°C. The pH is absolutely critical.

The volumes given here are for transfecting cells in 35 mm plates; adjust the volumes accordingly for other plate sizes.

Protocol 11. *Continued*

Method

1. On the day before transfection, cells should be plated out into 35 mm dishes from freshly trypsinized cells, so that on the day of transfection they will be 50–70% confluent and actively growing.

2. To prepare the calcium phosphate precipitate for 35 mm dishes, set up the following solutions:

   Tube A: 400 µl 1 × HBS.

   Tube B: thoroughly mix together 1–10 µg plasmid DNA, 10 µg carrier herring sperm DNA, and 31 µl 2 M $CaCl_2$ (tissue culture grade).

   Carefully add the contents of tube B dropwise to tube A, flicking tube A between additions. Incubate the tube at room temperature for 20–40 min to enable the precipitate to form.

3. Remove the medium from a 35 mm dish, and gently add the DNA mixture to the cells. Incubate the cells at 37 °C for 20–40 min.

4. After the incubation period, carefully overlay with 1 ml complete medium (without removing the DNA mix), and continue incubation at 37 °C for 4–7 h.[a] A fine-grained precipitate will form, and attach to the surface of the cells.

5. Aspirate the mixture from the cells, and wash each dish twice with serum-free medium, and then feed with complete medium and place at 37 °C.

6. Harvest the cells 2–4 days after transfection.

7. Remove DNA mixture, and wash the dishes twice with serum-free medium.

8. Add 1 ml 25% DMSO in 1 × HBS, and incubate at room temperature for 1–4 min. **Do not exceed this time as the DMSO is toxic to cells**.

9. Quickly remove the DMSO mixture, and wash the cells twice with serum-free medium. Feed the cells with complete medium, and incubate as above.

[a] After this stage an optional DMSO boost can be given to increase transfection efficiency, although it is not suitable for all cell types due to the toxicity of DMSO.

3.2 Analysis of promoter activity

Since the analysis of the level of reporter-gene expression is based on the measurement of the enzymatic activity that accumulates during the transfection period, it is important to be able to distinguish transcriptional differences from transfection efficiencies. This is particularly important where a comparison is being made between mutated forms of the promoter

5: Analysis of transcriptional control in DNA virus infections

sequence or transcription factors, or where transactivation is measured in the presence or absence of an inducer. One method is to co-transfect another plasmid expressing a different enzyme from that used in the reporter construct, under the control of a promoter whose activity is unchanged in different cell types. The activity of this second enzyme can then be used to normalize the activity of the reporter gene. Thus, β-galactosidase can be used to normalize CAT activity as follows:

- by comparing the amount of activity of each enzyme in a given amount of protein from the transfected cell lysates (using the Bradford protein assay, ref. 20, or by a commercially available kit).
- by carrying out CAT assays using a volume of lysate corresponding to a given amount of β-galactosidase activity.
- by carrying out both assays on a given volume of transfected cell lysate.

Protocol 12 describes a method of preparing lysates from transfected cells which are suitable for analysis of both CAT and β-galactosidase activity by the methods described in *Protocols 13 and 14*, as well as allowing β-galactosidase activity to be assayed by the Galacto-Light™ kit (Tropix), which we routinely use in our laboratory. It should be noted, however, that if the amount of control plasmid is similar to the amount of reporter plasmid, then competition for limiting transcription factors can occur. An alternative to using a second enzymatic assay to normalize the activity of a reporter gene is to measure the amount of plasmid uptake during transfection by loading the lysate onto a nylon membrane, and hybridising this with a probe prepared from the reporter gene, as described in *Protocol 15*.

Protocol 12. Preparation of transfected cell lysate

Equipment and reagents
- Microcentrifuge (e.g. Eppendorf)
- Sterile plastic microcentrifuge tubes
- Phosphate-buffered saline (PBS)

Method

1. Wash the cells twice with ice-cold PBS, scrape into 1 ml PBS, then transfer them to a 15 ml microcentrifuge tube kept on ice.
2. Pellet the cells by spinning for 20 s at maximum speed in a microcentrifuge.
3. Aspirate the supernatant, and resuspend the cell pellet in 100 μl 0.25 M Tris-HCl pH 7.5.
4. Disrupt the cells by freeze-thawing. To freeze-thaw, immerse the tubes in liquid nitrogen for 1 min, then transfer to a 37°C water bath, and incubate until thawed. Repeat the cycle three times.

Protocol 12. *Continued*

5. Pellet cell debris by spinning the tubes at maximum speed in a microcentrifuge for 5 min. Transfer the supernatant to fresh tubes. The samples can be stored at −20 °C

The bacterial CAT enzyme modifies and inactivates chloramphenicol by mono- and diacetylation. Therefore, determining the percentage of total chloramphenicol converted to the monoacetate form gives an estimate of the transcriptional activity. The CAT activity of the lysates produced in *Protocol 12* can be assayed either by using the direct scintillation method (21), or chromatographically as detailed in *Protocol 13*. Although the direct scintillation method gives a more accurate measure of the level of CAT activity in the transfected cell lysate, the chromatographic method allows a larger number of samples to be handled more readily.

Protocol 13. Chloramphenicol acetyl transferase assay

Equipment and reagents

- [^{14}C]-chloramphenicol (40–50 Ci mmol^{-1}) (Amersham International)
- TLC plates (BDH Merck)
- 4 mM acetyl CoA (Boehringer Mannheim). Acetyl CoA is very unstable, and should be made up fresh each time.

Method

1. Set up the following reaction mixture for each sample: 70 μl 0.25 M Tris-HCl pH 7.5, 35 μl water, 20 μl cell lysate[a], 1 μl [^{14}C]-chloramphenicol (40–50 Ci mmol^{-1}), 20 μl 4 mM acetyl CoA. Incubate the reaction mix for 30–60 min at 37 °C. The incubation period can be extended to 2–3 h if activity is low. In this case, however, it is necessary to substitute 40 mM acetyl CoA for the 4 mM solution, since it is unstable over long incubation times.

2. Extract the chloramphenicol with 1 ml ethyl acetate, by vortexing for 1 min. Spin at maximum speed in a microcentrifuge for 5 min. Transfer the top organic layer (containing all the forms of chloramphenicol) to a fresh tube.

3. Desiccate the ethyl acetate under vacuum. This will take approximately 2 h.

4. Resuspend the chloramphenicol samples in 10 μl ethyl acetate, and spot them onto silica gel TLC plates (along a line approximately 2.5 cm from the bottom). At least 1 cm should be left between samples.

5. Meanwhile, equilibrate a chromatography tank for 1–2 h with fresh 19:1 chloroform:methanol.

5: Analysis of transcriptional control in DNA virus infections

6. Place the TLC plate in the tank, taking care not to submerge the sample spots in the solvent. Leave to run until the solvent front is close to the top.

7. After air-drying, expose the chromatography plates to either X-ray film or to a Phosphor-Imager (Bio-Rad) screen, so that the percentage of converted chloramphenicol can be established.

[a] Depending on the cell type and promoter to be assayed, the amount of cell lysate in the reaction mix may need to be varied. For the assay to be in the linear range, the amount of total chloramphenicol converted to the monoacetate form should not exceed 50%. Therefore, after initially assaying 20 µl of lysate, the assay may need to be repeated with less extract, to ensure that the amount of product is kept in the linear range.

Protocol 14. β-galactosidase assay[a]

Equipment and reagents

- Reaction Buffer: 60 mM Na_2HPO_4, 40 mM $NaH_2PO_4.2H_2O$, 1 mM $MgSO_4.7H_2O$, 10mM KCl, 40 mM 2-mercaptoethanol
- ONPG (o-nitrophenyl-β-D-galactopyranoside) (Sigma)

Method

1. Set up the following reaction mix: 40 µl cell lysate[b], 2 µl 0.1 M DTT, 6 µl glycerol, 500 µl reaction buffer.

    Vortex, and incubate at room temperature for 5 min.

2. Add 100µl (2 mg ml^{-1}) ONPG (o-nitrophenyl-β-D-galactopyranoside), and incubate the reaction at 37°C until a visible yellow colour is achieved (this can take from 5 min to 24 h).

3. Stop the reaction with 250 µl 1 M Na_2CO_3.

4. Measure the change in absorbance in a spectrophotometer at 420 nm.

[a] Based on the method of Herbomel *et al.* (14).
[b] Mammalian cells contain a eukaryotic isozyme for β-galactosidase, therefore it is important to include a blank, non-transfected cell lysate from the same cells in the assay.

Protocol 15. Slot blot analysis of transfected cell lysates

Equipment and reagents

- RNase A: 10 mg ml^{-1} made up in water. Store at –20°C.
- 20 × SSC: 3 M NaCl, 300 mM sodium citrate
- Denaturation solution: 1.5 M NaCl, 0.5 M NaOH
- Proteinase K: 10 mg ml^{-1} made up in water. Store at –20°C. Dilute to 100 µg ml^{-1} before use
- Neutralization solution: 1.5 M NaCl, 0.5 M Tris-HCl pH 7

Protocol 15. *Continued*

Method

1. Add 2 μl 10 mg ml^{-1} RNase A to 30 μl of cell lysate prepared as for *Protocol 12*. Incubate at 37 °C for 30 min.

2. Add 2 μl 100 μg ml^{-1} proteinase K, and incubate for a further 30 min at 37 °C.

3. Heat for 10 min at 95 °C to denature the enzymes.

4. Add 60 μl 20 × SSC, then incubate again at 95 °C for 10 min. Put the samples on ice. Prepare a known amount of the transfected plasmids (to act as standards) in the same way.

5. Meanwhile, assemble the slot-blot apparatus according to the manufacturer's instructions: e.g., place two sheets of Whatman 3MM chromatography paper presoaked in 2 × SSC on the base of the apparatus, and overlay these with a sheet of nylon hybridization membrane (e.g. Hybond N, Amersham) presoaked in 2 × SSC, then close the apparatus.

6. Apply the vacuum to the apparatus, and load the samples into the appropriate wells.

7. When all the samples have passed through the membrane filter, disassemble the apparatus, and place the membrane onto Whatman 3MM paper soaked in denaturation solution. Incubate at room temperature for 5 min.

8. Transfer the membrane onto another piece of Whatman 3MM paper soaked in neutralization buffer, and incubate for a further 5 min.

9. Allow the membrane to air-dry, then cross-link the DNA to the membrane (either by baking at 80 °C or UV cross-linking) according to the manufacturer's instructions.

10. Hybridize the filter overnight using standard methods (15). Expose the membrane to either X-ray film or to a Phosphor-Imager screen (Bio-Rad), so that the amount of plasmid uptake can be calculated from the known standards.

3.3 Preparation of indicator viruses for the analysis of promoter activity during infection

As discussed previously, although the analysis of reporter constructs by transfection into tissue culture cells gives information about the effects of different factors on viral promoters, this method of analysis does not accurately reflect the activity of the promoter during the viral replication cycle. This can be analysed, at least in part, by super-infecting transiently

5: Analysis of transcriptional control in DNA virus infections

transfected cells. However, a more biologically relevant method is to produce recombinant indicator viruses that contain the reporter construct within their genome, by introducing it into a non-essential gene so that normal viral gene expression is not interrupted (such as UL43 or US5 for HSV, ref. 2). Such viruses allow the promoter activity within the viral genome (as opposed to within a plasmid) to be examined in tissue-culture cell lines. Moreover, by dissecting and staining the relevant host tissues after infection, these viruses can be used to determine the pattern of expression from the target promoter during infection of a host (*Protocol 18*).

Such an approach has been used within our laboratory to examine the interaction of Oct-2 isoforms with HSV immediate-early promoters in a viral context. In these experiments, we prepared four viruses in which either a full-length or a mutated form of the IE1 promoter linked to *LacZ* was introduced into the UL43 gene, which were then used to infect BHK cells engineered to over express activating (Oct-2.1) or repressing, neuronal (Oct-2.4 and 2.5) forms of the Oct-2 protein (22). These experiments demonstrated that these transcription factors were indeed able to interact with the IE1 promoter when set in a viral context, thus strengthening the argument that cellular octamer-binding proteins may well be responsible for silencing IE gene expression, leading to latent (as opposed to lytic) infection in neuronal cells.

3.3.1 Production of recombinant viruses

The methods given below are for the preparation of HSV recombinant viruses. Methods for the production of recombinant viruses from other virus families, such as Adenovirus and Vaccinia virus, can be found elsewhere (23, 24). Obviously it is most convenient to use 'coloured' reporter genes (such as GFP or β-galactosidase) so that:

- recombinant progeny producing 'coloured' plaques can be easily separated from white wild-type plaques during purification.
- the activity of the promoter can be easily detected, both in tissue-culture cells and *in vivo*.

Recombinant virus production involves:

- Construction of a recombinant plasmid containing the promoter-reporter gene construct inserted into recombination 'arms' consisting of viral sequences, to allow insertion into the viral genome by homologous recombination. It is important to remember that at least 200 bp of viral sequence is needed either side of the insert for recombination to occur, although 200 bp is very minimal, and 1–1.5 kb is recommended for optimal efficiency.
- Transfection of the plasmid and viral DNA (which is infectious) into tissue culture cells. It is important for good recombination efficiency to linearize the shuttle plasmid, using sites within the plasmid vector sequence.

1) Construction of a "shuttle" plasmid contaning the promoter/reporter gene construct inserted into recombination "arms".

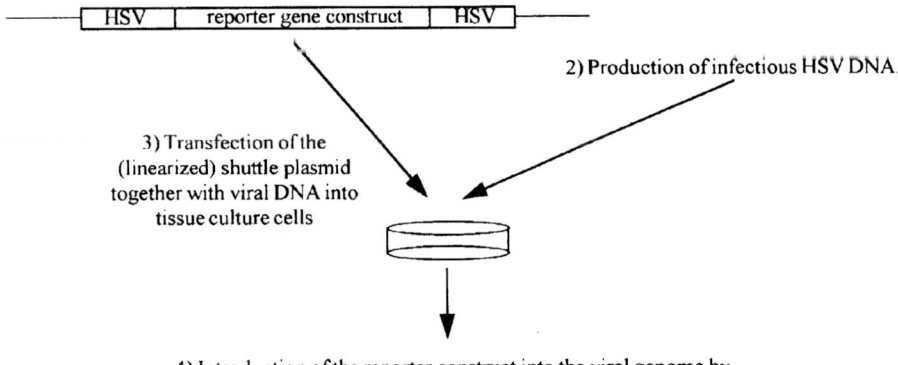

2) Production of infectious HSV DNA.

3) Transfection of the (linearized) shuttle plasmid together with viral DNA into tissue culture cells

4) Introduction of the reporter construct into the viral genome by homologous recombination. Transfection allowed to proceed until complete cytopathic effect is observed.

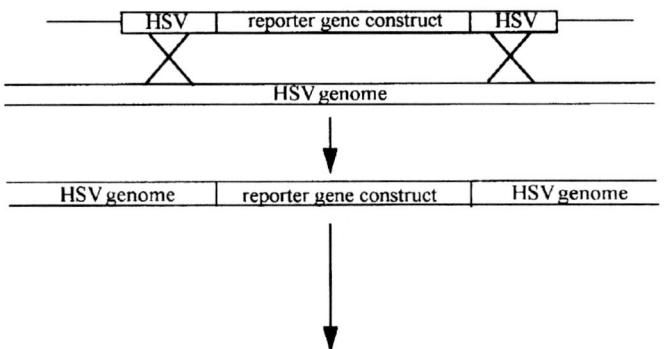

5) Isolation of recombinant plaques by plating out transfected cell lysate onto permissive cell monolayers and identifying plaques produced by recombinant virus. Multiple rounds of isolation carried out until no wild type virus can be detected.

6) Confirmation of the correct insertion of the reporter gene construct by southern hybridisation.

Figure 1. Production of recombinant HSV indicator viruses. A recombinant plasmid, containing the promoter/reporter gene construct inserted into recombinant 'arms' consisting of viral sequences, is co-transfected into permissive cells together with infectious HSV DNA. Recombinant virus (resulting from homologous recombination between the viral and plasmid DNA) is then purified away from the wild-type virus by plaque purification.

5: Analysis of transcriptional control in DNA virus infections

- Plating out the virus produced following such a transfection (which will contain wild-type and recombinant virus) onto permissive cells. Depending on the choice of reporter gene, recombinant plaques can either be selected by fluorescence microscopy (for GFP), or by use of a chromogenic substrate (*LacZ*)
- Screening for, and plaque purification of, the recombinants.

The process of transfecting and screening for recombinant viruses is illustrated in *Figure 1*.

The following methods outline the preparation of infectious HSV DNA, the transfection of HSV and plasmid DNA, and the subsequent purification of recombinant progeny virions. More detailed methods for the general culture of HSV are given elsewhere (25). Once prepared, the recombinant indicator viruses can be used either to infect different tissue-culture cell lines, or cell lines engineered to over-express the cellular transcription factor of interest, or they can be used to infect a host so that the pattern of gene expression during replication *in vivo* can be examined. By using two different marker genes (e.g. *Lac Z* and GFP), viruses can be constructed so that the activity of more than one viral promoter can be examined in parallel. Tissue-culture cells infected with recombinant indicator, which contain *Lac Z* as the reporter gene, can be analysed either by staining with an X-gal based stain (*Protocol 17*), or alternatively cell lysates can be prepared and assayed as described in *Protocols 12 and 14*. *Protocol 18* gives the methods we use in our laboratory to examine β-galactosidase activity in neuronal tissues following intra-cranial or footpad inoculation in mice. Analysis of the expression of GFP can easily be determined by sectioning the tissue samples (if large), and observing by fluorescence microscopy.

We prepare HSV DNA from 80–90% confluent BHK 21/c13 cells grown in roller bottles or $600 cm^2$ plates, which have been infected with HSV at a multiplicity of infection (m.o.i.) of 0.001 plaque-forming units (PFU) per cell, and incubated until complete cytopathic effect (CPE) is seen. Routinely we use 5–10 roller bottles, but this can be scaled up or down as necessary (*Protocol 16*).

Protocol 16. Preparation of infectious HSV DNA

Equipment and reagents

- Proteinase K buffer: 0.01 M Tris-HCl pH 8, 5 mM EDTA pH 8, 0.5% SDS
- Proteinase K: 10 mg ml^{-1} stock made up in water. Store at –70°C

Method

1. Harvest cells into the medium, and transfer to 200 ml centrifuge pots. Pellet the cells and virus by spinning in a Beckman JA14 (or equivalent) for 2 h at 12000 r.p.m., 4°C.

Protocol 16. *Continued*

2. Resuspend pellet(s) in a total of 10 ml of proteinase K buffer, and add proteinase K to a final concentration of 50 μg ml^{-1}.

3. Incubate at 37 °C with shaking for 5-6 h.

4. Add 10 ml water, then extract with an equal volume of phenol:chloroform:IAA (24:1) by gently inverting the tube for 10 min, followed by centrifugation at 10000 r.p.m. for 10 min in an SW40 (or equivalent). Remove the top aqueous layer, and re-extract 2-3 times more with phenol:chloroform:IAA until there is no visible interface.

5. Extract once with chloroform:IAA (24:1) as for Step 4, but this time mix and spin for 5 min.

6. Gently precipitate the DNA with 2 volumes of ethanol. Pellet the DNA by spinning in a bench-top centrifuge at 3500 r.p.m. for 10 min.

7. Wash pellet with 5 ml 70% ethanol, then air-dry until no ethanol is visible (it is important not to overdry the DNA, as this will make it difficult to resuspend).

8. Resuspend the DNA in sterile distilled water by incubating at 37 °C. Routinely we would resuspend the DNA from 10 roller bottles in 3-5 ml.

9. Run 5-10 μl of the DNA on a 0.7% agarose gel to check that the DNA is intact.

10. Transfect various volumes of the DNA (10-50 μl) into BHK cells to determine the optimal amount for transfection (the lowest volume that gives complete CPE in 3-5 days).

3.3.2. Transfection of HSV DNA

This is essentially carried out in the same way as transient transfection (*Protocol 11*); the optimal amount of HSV DNA for transfection (as determined in *Protocol 16*) is co-transfected together with various amounts of the linearized shuttle plasmid (1- to 50-fold molar excess). Maximum recombination efficiency is usually reached at a tenfold molar excess. As the excess of plasmid DNA, and hence the total amount of DNA present increases, the overall transfection efficiency will decrease. The infection is then allowed to proceed until all of the cells are rounded, i.e. 100% cytopathic effect (CPE) is observed (usually 3-5 days). The resulting primary virus stock (containing wild-type and recombinant progeny) is then harvested, and the viral progeny are then screened by plating out onto permissive cells and selecting 'coloured' plaques, as detailed in *Protocol 17*.

5: Analysis of transcriptional control in DNA virus infections

Protocol 17. Screening for recombinant virus

Equipment and reagents
- X-gal stain: 50 ml PBS, 0.07 g potassium ferrocyanide, 0.05 g potassium ferricyanide, 50 µl 1 M $MgCl_2$. Warm to 37 °C, then add 100 µl 150 mg ml^{-1} X-gal (5-bromo-4-chloro-3 indolyl β-D-galactopyranoside) dissolved in DMSO (it is important to warm the stain to 37 °C to prevent precipitation of the X-gal)
- CMC: 1.6% carboxymethyl cellulose in water. Autoclave to sterilize

Method

1. Once 100% CPE is observed, harvest the transfected cells by scraping them into the medium using a rubber policeman. Transfer the resulting stock to a sterile 15 ml tube, then disrupt the cells by three freeze–thaw cycles (*Protocol 11*). This will produce 2 ml of cell lysate.

2. Plate out tenfold serial dilutions of the resulting cell lysate from 10^{-1} to 10^{-6} onto permissive cells plated out onto 35 mm dishes. Overlay the infected monolayers with CMC/FGM (a 1:2 mix of 1.6% CMC and full growth medium). Incubate the plates at 37 °C for 2 days.

3. Select for recombinant progeny by:

    (a) GFP: identifying fluorescent plaques using an inverted fluorescence microscope with a fluorescence optical filter. Plaques can be picked directly through the CMC mix.

    (b) β-galactosidase: remove the CMC mixture, and wash the cells twice with PBS. Add 2 ml per well of X-gal stain. Incubate the plates at 37 °C until blue plaques are observed.

    In both cases, using a Gilson P20 (or equivalent) and a fine pipette tip, pick the plaque into 100 µl serum-free medium, taking care to minimize the amount of any adjacent wild-type virus taken up with the recombinant.

4. Disrupt the cells by three freeze–thaw cycles. Plate out 50, 10, and 1 µl of the lysed plaque onto permissive cells, overlay, and incubate for 2 days as before. Repeat the screen until no wild-type 'white' virus is observed. Once purified, recombinant viruses can easily be checked by Southern blot analysis (see methods elsewhere in this book).

Protocol 18. Staining tissues to identify β-galactosidase activity

Equipment and reagents

- 0.2 M phosphate buffer: Solution A: 6.24 g Na$_2$PO$_4$ in 100 ml water; Solution B: 22.64 g NaH$_2$PO$_4$ in 500 ml water. Mix 9.5 ml of solution A, 40.5 ml of solution B, and 50 ml of water. Adjust to pH 7.4
- 4% paraformaldehyde fixative. Mix 1 g paraformaldehyde with 10 ml water. Heat to 60 °C and add 2 drops 1 N NaOH; the solution should clear (**care: do not exceed 65 °C**). Cool, then add 12.4 ml 0.2 M phosphate buffer, and 2.6 ml water. Adjust to pH 7.3. Store aliquots at –20 °C.
- X-gal stain: 11 ml PBS, 32 mg potassium ferricyanide, 42 mg potassium ferrocyanide, 4 ml 0.1% (v/v) NP-40, 2 ml 0.2% (w/v) sodium deoxycholate, 400 μl 100 mM MgCl$_2$. Warm to 37 °C, then add 500 μl 40 mg ml^{-1} X-gal (5-bromo-4-chloro-3 indolyl β-D-galactopyranoside) in DMSO (it is important to warm the stain to 37 °C to prevent precipitation of the X-gal)

Method

1. Dissect out the tissues of interest (e.g. dorsal root ganglion (DRGs), or brain), transfer to ice-cold PBS, and keep on ice until all the dissection has been carried out.

2. Transfer the tissue to the fixative solution, and incubate on ice for 1 h (larger pieces of tissue should be perfusion-fixed).

3. Remove the fixative, and wash the samples three times with PBS.

4. Add sufficient stain solution to cover the tissue, and incubate at 37 °C until blue cells are observed (up to 18 h).

5. Wash the tissues once with PBS, then transfer to 80% glycerol (v/v in water). Store at 4 °C until required.

3.4 Using transgenic animals to analyse the tissue-specific expression of viral promoters

The production of transgenic animals carrying the viral promoter of interest linked to a reporter gene allows the tissue specificity of the promoter to be assessed with and without virus infection. This enables the tissue-specific pattern of expression from a promoter to be compared directly with the tissue specificity of the virus, allowing an assessment to be made of its role in viral tropism. Thus, Baskar *et al.*, by producing transgenic mice containing *LacZ* linked to the human cytomegalovirus (HCMV) major immediate-early promoter (MIEP), were able to demonstrate that the tissue-specific pattern of expression from this promoter overlapped with the tissue-specific pattern of HCMV expression (26). This indicates that the absence or presence of the cellular factors involved in the control of this promoter play a major role in determining the cellular tropism of HCMV.

4. Analysis of transcriptional control by mutagenesis

Once the pattern of expression from a viral promoter has been defined by use of a reporter gene construct, as described in Section 3, the promoter sequence can be modified by mutagenesis, and alterations assessed for phenotypic consequences *in vitro* or *in vivo,* thus allowing important regions to be defined. Mutated promoters can then either be assessed using a reporter gene, or introduced into the viral genome in order to change the promoter sequence by homologous recombination (Section 3), enabling the role of specific DNA sequences in patterns of gene expression *in vivo* to be analysed. Conversely, mutagenesis can also be used to modify functional domains within viral or cellular transcription factors, allowing important residues to be identified.

4.1 Types of mutation

The simplest type of mutagenesis employed in defining the region within a promoter sequence that confers a specific activity is to use naturally occurring restriction sites to construct a set of deletion mutants. This type of approach can be used to analyse the individual role of different motifs within a promoter, by removing the sequences containing the motifs and analysing the effect. A variation of this method is to use the 'filling in' or 'chewing back' of naturally occurring restriction sites to produce more subtle changes to the sequence. In both cases, T4 DNA polymerase (which has both polymerase and exonuclease activity) can be used.

Once a region that has a functionally important effect has been identified, it can be cloned into a vector in which either a minimal form of the homologous promoter or a heterologous promoter drives the reporter gene, to determine if it can confer a specific pattern of regulation on these promoters (27). The use of a minimal homologous or heterologous promoter can also be used to analyse the effect of manipulating the number, spacing, and context of response elements, to determine the effect on transactivation. For example, in our laboratory we used this method to examine the role of the spacing and context of the two TAATGARAT elements in the HSV IE3 promoter in its interaction with the different isoforms of the octamer-binding proteins Oct-2 and Brn-3 (28). This study showed that the interaction of the promoter with Brn-3 isoforms was dependent on the correct spacing of the binding sites, whereas the interaction with Oct-2 was dependent on both the spacing and the context of the elements (with the requirements being different for different isoforms).

However, it is often not convenient to use naturally occurring restriction sites within the target sequence to create deletion mutants. In this case, nested sets of deletions can be generated by progressively removing sequences from one end of the target DNA, using specific exonucleases such as *Bal*31 or exonuclease III, thus producing a random range of different sized clones. Methods for these procedures can be readily found elsewhere (29).

Finer analysis can be carried out using the insertion, deletion, or alteration of shorter sequences by methods such as linker scanning (30) (which can map to within approximately 10 bp), and site-directed mutagenesis (to map specific nucleotides within the promoter sequence). All of these methods require the viral promoter sequence to be cloned into a plasmid or phagemid vector, which should be carried out using standard methods (5).

4.2 Linker scanning mutagenesis

Although deletion analysis of promoter regions (either using naturally occurring restriction enzymes sites, or by the creation of nested deletions by exonuclease digestion) allows the outer borders of regulatory regions to be defined, these regions are often too large to allow further analysis by the substitution of individual bases. In this case it is necessary to use another form of analysis, such as linker scanning mutagenesis (30), to define important residues further. The aim of linker scanning mutagenesis is to produce a series of mutants in which small overlapping sequences along the sequence of interest have been individually replaced by a common linker sequence. This allows the replacement of important nucleotide or amino acid residues without changing the spacing between promoter elements or protein domains. The method has three stages:

- The generation of randomly spaced breaks in the sequence of interest.
- The deletion of a short sequence on each side of the breaks, and the insertion of linkers.
- Identification and characterization of the mutants generated.

A number of protocols for producing randomly spaced breaks within the DNA sequence by chemical means have been devised (30, 31). These methods, however, are extremely time consuming. A simpler means of producing breaks in the DNA is to linearize the DNA by partial digestion with one (or two) restriction enzymes that cut frequently in the target sequence. The linkers can then easily be inserted. However, the major disadvantages of this method is that using restriction sites to linearize the DNA means that the sequence is not analysed as fully as chemical means allow, and also that the linkers are introduced into the sequence without the concurrent deletion of the same number of nucleotides, so that the precise spacing between sequences is not maintained.

4.3 Oligonucleotide insertion

The simplest approach to introducing large changes within a target sequence is to introduce an oligonucleotide cassette into the DNA using naturally occurring restriction sites, thus allowing the analysis of the role of a specific

5: Analysis of transcriptional control in DNA virus infections

amino acid sequence or nucleic acid motif. By introducing a cassette encoding a particular nucleic acid motif upstream of a minimal homologous promoter, this approach can also be used to demonstrate the role of the motif in the pattern of expression of the promoter of interest. We have used this type of approach to examine the interaction of cellular octamer-binding proteins with the TAATGARAT motif in HSV immediate-early (IE) gene promoters (22). In these experiments we constructed four different reporter constructs:

- full length IEI promoter linked to β-galactosidase.
- minimal IE1 promoter (which had the upstream region containing the TAATGARAT motifs abolished) linked to β-galactosidase.
- minimal IE1 promoter which had a simple TAATGARAT motif inserted upstream, linked to β-galactosidase.
- minimal IE1 promoter which had an overlapping octamer-TAATGARAT motif inserted upstream, linked to β-galactosidase.

These reporter constructs were then inserted into the HSV genome to produce four recombinant indicator viruses, as discussed in Section 3.3.1. We showed that deletion of the upstream IE promoter region containing the TAATGARAT motifs abolishes the inhibitory effect of Oct-2.4 and Oct-2.5 on the viral IE promoter. However, this inhibitory effect can be restored by the addition of a single TAATGARAT motif to the minimal promoter, thus demonstrating that this particular motif can mediate the effects of transcription factors when it is located within the viral genome (22).

When designing oligonucleotides for insertion into the target DNA sequence, they should ideally have sequences for two different restriction sites at either end (so that only the desired combination can be generated), and they should also contain a unique restriction site to aid identification of the recombinants. Although it is not usually necessary, phosphorylation of the oligonucleotides will increase the efficiency of ligation (*Protocol 19*).

Protocol 19. Oligonucleotide insertion

Equipment and reagents

- Paired specific oligonucleotides
- T4 polynucleotide kinase (Promega)
- 10 mM ATP (Sigma)
- T4 DNA ligase (Promega)
- Vector plasmid

Method

1. Oligonucleotide phosphorylation (optional). Add the following components to a microfuge tube: 100 pmol[a] oligonucleotide, 2.5 μl 10 × kinase buffer, 5 units T4 polynucleotide kinase, 2.5 μl 10 mM ATP, water to 25 μl.

Protocol 19. *Continued*

> Incubate the reaction at 37 °C for 30 min, followed by 70 °C for 10 min to inactivate the kinase. Store at −20 °C until needed.

2. Digest the target DNA with appropriate enzyme(s), and gel purify by standard methods (5).
3. Set up two ligation reactions (without ligase), one containing ~100 ng of each oligonucleotide, and one without oligonucleotides (control) (5). Overlay with 50 μl mineral oil.
4. To anneal the oligonucleotides, heat the reaction mixture to 95 °C, then cool slowly to room temperature in a large beaker of water.
5. Add ligase below the mineral oil, mix, and incubate at room temperature for 1 h (or overnight at 16 °C), before transforming the mixture (below the mineral oil) into competent cells (by standard methods, ref. 5).
6. Pick colonies, miniprep, and identify recombinants by the unique restriction site(s) which they contain, or if necessary, by a colony lift using an end-labelled oligonucleotide as a probe (5).

[a] In general, ng of oligonucleotide = pmol of oligonucleotide × 0.33 × N, where N = length of oligonucleotide in bases.

4.4 Site-directed mutagenesis

Once putative regulatory sequences have been identified using the methods described above, small specific changes to the target DNA sequence can be made using site-directed mutagenesis (SDM). SDM can also be used as a powerful tool for the analysis of transcription factors, either by inserting stop codons to produce a series of truncated proteins, or for analysing the role of individual amino acids during transactivation. The original methods devised by Hutchison *et al.* (32) and Kunkel (33), although reliable, were laborious and time consuming. However, advances in molecular biology have enabled simpler systems to be devised, and the two main approaches currently available are based on the use of 'repair oligonucleotide' systems, and PCR. These will be discussed in general below, and are described in detail in the Practical Approach series volume entitled *Directed mutagenesis* (34), and in Chapter 3 of this volume. The design of the mutagenic oligonucleotides, and the general considerations for all site-directed mutagenesis procedures, are fundamentally the same:

- The larger and more complex a mutation, the lower the efficiency with which it will be generated. Larger changes are generally more efficiently carried out using oligonucleotide insertion (Section 4.3).

- The replacement of nucleotides is more efficient than the deletion or insertion of sequences.
- Larger oligonucleotides are generally more efficient than shorter ones. The optimal size is dependent on the number of mismatches, but in general 20–25 mers should be used to replace 1–3 bases in the centre of a sequence, and for larger changes there should be ~15 unchanged bases on either side of the mutation.
- When designing the mutagenic oligonucleotide, palindromic sequences should be avoided, as this could lead to the formation of secondary structures, leading to reduced efficiency. The sequence should also optimally have a G + C content of ~60%.
- Designing the mutagenic oligonucleotide in such as way as to introduce a unique restriction site into the target sequence will aid the identification of mutants.

4.4.1 Unique site elimination-based mutagenesis

In common with the early SDM methods, single-stranded DNA (preferably generated by superinfection with bacteriophage, but alkali-denatured dsDNA can be used with lower efficiency) is hybridized to a mutagenic oligonucleotide which is complementary to the ssDNA, except for a region of mismatch near the centre. Following hybridization, the oligonucleotide is extended with T4 DNA polymerase to create a double-stranded structure. The nick is then sealed with T4 DNA ligase, and the duplex structure is transformed into an *E. coli* host. However, in these methods a second oligonucleotide, which either deletes a unique restriction site within the vector, or alternatively repairs an inactivated antibiotic resistance gene, is included in the reaction, thus allowing mutated plasmids to be selected (these methods are all based on the original system of Deng and Nickeloff, ref. 35). In the first method, selection of mutants is carried out by digesting the mutagenesis reaction mixture with the appropriate restriction enzyme prior to transformation, resulting in the growth of colonies which contain plasmid with the deleted site. A number of commercially available kits (e.g. U.S.E from Pharmacia, and Chameleon from Stratagene) are based on this method of selection.

In the second method, mutagenesis is carried out on sequences cloned into a vector containing two antibiotic-resistance genes, one of which has been inactivated by a small change in its sequence. In this case, selection of mutants is carried out using resistance to the second antibiotic, which will only occur in those colonies containing the repaired site. The addition of a third oligonucleotide to the mutagenesis reaction which inactivates the first antibiotic resistance gene ensures that multiple rounds of mutagenesis can be carried out to make sequential changes to the target sequence without further cloning. This method is the basis of the Altered Sites® kit available from Promega.

Once selected, the colonies should be checked for the mutation by miniprep and restriction-digested, and then sequenced to ensure that the required change(s) has been made.

4.4.2 PCR mutagenesis

Point mutations, insertions, deletions, or even short regions of complete degeneracy can be readily engineered, using mismatched primers followed by cloning of the PCR product as a restriction fragment. This can be carried out on DNA cloned in any plasmid vector, offering the advantage of speed and efficiency over conventional mutagenesis methods. However, the dis-

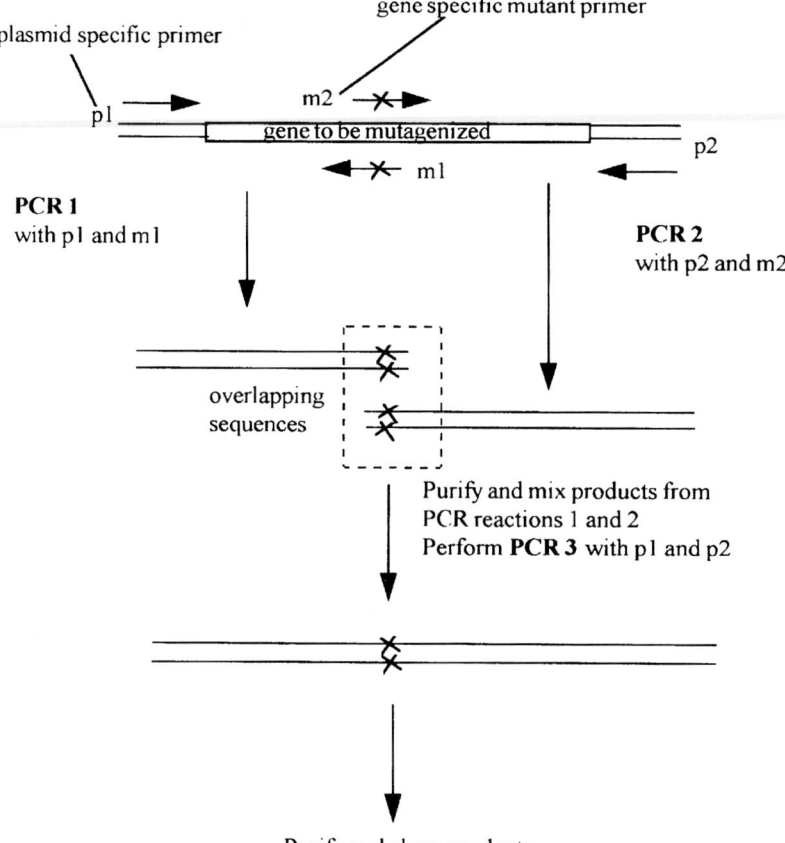

Figure 2. PCR mutagenesis. Two initial PCR reactions (1 and 2) are performed using a plasmid-specific primer (p1 or p2) and a gene-specific mutagenic primer (m1 or m2) to generate two mutant PCR products which overlap at one end. A second PCR reaction is then performed using the two products from PCR reactions 1 and 2, together with the primers p1 and p2, resulting in a full-length, mutated sequence.

advantage of this method is that PCR itself can introduce mutations, which means that the entire target DNA sequence needs to be checked for unwanted changes. The method routinely used in our laboratory (*Protocol 20*), based on that described by Ho *et al.* (36), requires two plasmid-specific primers, allowing amplification of the entire cloned sequence, and two complementary mutagenic primers (*Figure 2*). Two initial PCR reactions are performed using a plasmid-specific primer (p1 or p2) and a gene-specific mutagenic primer (m1 or m2), to generate two mutated PCR products which overlap at one end. This overlap allows annealing of the two products during a second round of PCR using the plasmid-specific primers alone, resulting in the production of a full-length sequence containing the changed nucleotides. The plasmid-specific primers used can in many cases be the M13 forward and reverse universal sequencing primers, which themselves contain several restriction sites aiding the cloning of the mutant sequence. If these are unsuitable, primers containing suitable restriction sites (placed at least 5 bases from the 5' end to allow efficient digestion) can easily be designed, or the product can be cloned into one of the commercially available plasmids designed to include an overhanging 3'-T (e.g. pGEM®-T vector systems from Promega) which anneals with the overhanging 3'-A which *Taq* polymerase adds to all PCR products. The design of the gene-specific mutagenic primers is discussed in Section 4.4.

Protocol 20. PCR mutagenesis

Equipment and reagents

- Specific PCR 3' and 5' primers
- 10 mM dNTPs (all four dNTPs, mixed; Pharmacia)
- 5 units μl^{-1} *Taq* polymerase (Promega)
- 10 × *Taq* polymerase buffer, supplied by the manufacturer

Method

1. Set up two PCR reactions, containing template DNA and either primers p1 and m1 or p2 and m2. We routinely use *Taq* polymerase and buffer from Promega: 5 μl 10 × buffer, 3 μl 25 mM MgCl$_2$, 2 μl 10 mM dNTPs, 1 pmol each primer p1 + m1 or p2 + m2, 100 ng template, water to 50 μl. Overlay with 100 μl mineral oil.

2. Heat the reaction mixture to 98°C for 5 min in a thermal cycler to denature the DNA. Remove from the cycler, then add 50 units of *Taq* polymerase below the mineral oil.[a]

3. Cycle 10–15 times[b] at 95°C for 30 s, 65°C for 30 s, and 72°C for 1 min. Incubate at 72°C for 10 min to ensure complete extension of all amplified molecules.

4. Run an aliquot of the PCR reactions on a 1% agarose gel to separate the product and the primers. Excise the band, and elute the DNA into

Protocol 20. *Continued*

10 µl (e.g. using Hybaid Recovery™ DNA purification kit II (Hybaid)). If the correct-sized bands cannot be seen, then the annealing temperature (50–65 °C) and MgCl$_2$ concentrations (final concentration 0.5–5 mM) should be optimized.

5. Set up a second PCR reaction mixture containing 100 ng each of p1 and p2, together with 2.5 µl of each of the primary PCR products, with optimization if necessary.

6. Make up the reaction mixture to 400 µl with water, and extract once with chloroform:IAA (24:1) to remove traces of mineral oil, then phenol-extract and precipitate the DNA. Cut the product with the appropriate restriction enzyme(s), purify, and ligate into the plasmid vector as usual.

[a] Adding the enzyme at 98 °C prevents non-specific extension occurring during the heating of the reaction mix.
[b] In general, the number of cycles should be kept as low as possible to minimize the incorporation of errors.

5. Identification of cellular transcription factors involved in the control of viral transcription

The previous two sections of this chapter have described methods that can be employed to examine the role of previously identified viral and cellular transcription factors in the control of viral gene expression. However, viral gene expression may also be controlled by as yet unidentified cellular transcription factors. The aim of this section is to introduce briefly methods (such as the DNA mobility-shift assay, and South-western blotting) that can be used to identify cellular transcription factors that may be involved in the control of viral transcription. Further details of these and other relevant methods can be found in the Practical Approach series volume entitled *Transcription factors* (37).

5.1 The DNA mobility-shift assay

Since it provides a powerful assay for the presence of DNA-binding proteins capable of binding to a promoter sequence, the DNA mobility-shift assay (also known as a band shift assay) can be used to determine the potential for a gene to be transcribed in a particular cell type. The method relies on the ability of a protein to bind to a radiolabelled DNA fragment (probe) *in vitro*, followed by electrophoretic separation of the DNA–protein complexes from the unbound DNA on non-denaturing polyacrylamide gels, allowing one or more bound proteins to be identified (38, 39). In general, the larger the

5: Analysis of transcriptional control in DNA virus infections

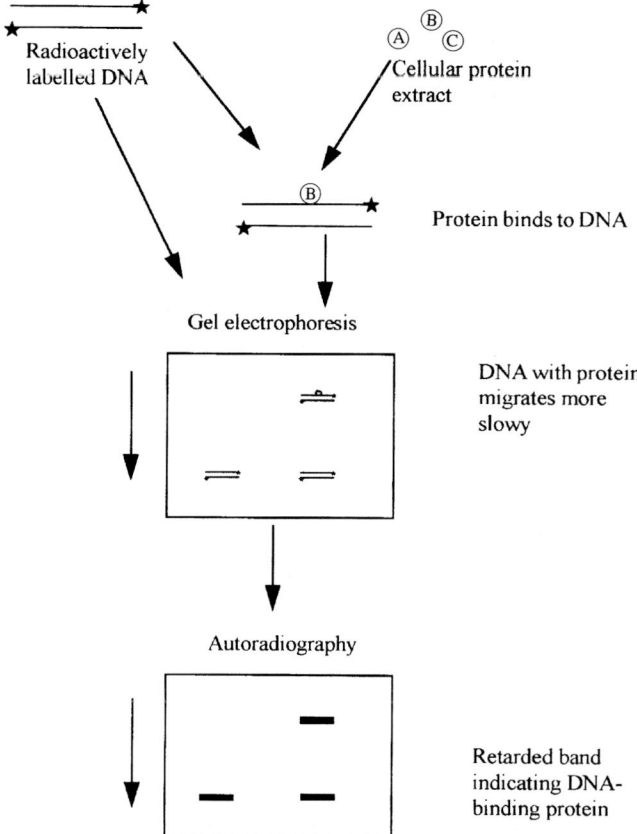

Figure 3. DNA mobility-shift assay. Binding of a cellular protein (B) to the radioactively labelled DNA causes it to move more slowly upon gel electrophoresis, and hence results in the appearance of a retarded band upon autoradiography to detect the radioactive label.

DNA–protein complex that is formed, the greater the extent of mobility retardation within the gel. The principle of the DNA mobility-shift assay is illustrated in *Figure 3*.

In our laboratory we have used the DNA mobility-shift assay to examine protein extracts from cell types that are permissive (BHK cells) and non-permissive (neuronally derived ND7 cells) to lytic infection by HSV1. In these assays, we used the TAATGARAT sequence demonstrated to be important in immediate-early gene transcription (see Section 1 and ref. 40). We showed that both permissive fibroblasts and non-permissive neuronal cells expressed a TAATGARAT-binding protein identified as Oct-1 (mol. wt. 100 kDa), which is known to bind to the TAATGARAT sequence as part of a complex with the virion protein Vmw65 and the cellular protein HCF (Section 1). An additional complex which is present only in neuronal cells has been demonstrated to be Oct-2, which is smaller than Oct-1 (mol. wt. 60 kDa). This finding implicated Oct-2 as a repressor of HSV immediate-early gene expression in neuronal cells (41, 42).

The following methods (*Protocols 21–23*) are intended as an introduction to the DNA mobility-shift assay. Alternative methods and further details can be found in the Practical Approach series volume entitled *Transcription factors* (37).

5.1.1 Preparation of protein extracts

Protein extracts may be prepared from whole cells or isolated nuclei. Whole-cell extracts require less steps in their preparation, and so this may be the method of choice for small tissue samples, since there are fewer stages at which protein will be lost or damaged. Whole-cell extracts should also be prepared from frozen samples, since freezing damages the nuclear membrane, preventing preparation of intact nuclei.

Protocol 21. Preparation of nuclear protein extracts[a]

Equipment and reagents

- Phosphate-buffered saline (PBS)
- Buffer A: 10 mM HEPES pH 7.9, 1.5 mM MgCl$_2$, 10 mM KCl, 0.5 mM DTT, 0.5 mM phenylmethylsulfonyl fluoride (PMSF)
- Nonidet P-40
- Buffer C: 5 mM HEPES pH 7.9, 26% (v/v) glycerol, 1.5 mM MgCl$_2$, 0.2 mM EDTA, 0.5 mM DTT, 0.5 mM PMSF

Method

Samples should be kept on ice at all times; centrifugation should be carried out at 4°C.

1. Prepare cells as follows:
   (a) Adherent and suspension cells: harvest cells (5×10^7 to 1×10^8), and centrifuge at 250 *g* for 10 min.
   (b) Tissues: grind the tissue to a fine powder in liquid nitrogen before resuspending in buffer A.

2. Wash with PBS. Centrifuge at 250 *g* for 10 min.

3. Resuspend the pellet in 5 volumes of buffer A.

4. Incubate on ice for 10 min. Centrifuge at 250 *g* for 10 min.

5. Resuspend the pellet in 3 volumes of buffer A. Add NP-40 to 0.05%, and homogenize with 20 strokes of a tight-fitting Dounce homogenizer to release the nuclei.

6. Successful release of the nuclei may be checked by phase-contrast microscopy (keep a sample from before the addition of NP-40 for comparison).

5: Analysis of transcriptional control in DNA virus infections

> 7. Centrifuge at 250 g for 10 min to pellet the nuclei.
>
> 8. Resuspend the pellet in 1 ml buffer C. Measure the total volume, and add NaCl to a final concentration of 300 mM. Mix well by inversion.
>
> 9. Incubate on ice for 30 min.
>
> 10. Centrifuge at 24 000 g for 20 min at 4°C.
>
> 11. Aliquot the supernatant, and snap-freeze in dry ice-ethanol. Store at −70°C.
>
> ^a Modified from that described by Dignam et al (43).

^aModified from that described by Dignam et al (43).

Whole-cell extracts are made by a modification of the method given in *Protocol 21*. Harvest the cells, and wash with PBS. Resuspend in 1 ml of buffer C, and homogenize with 20 strokes of a tight-fitting Dounce homogenizer. Add NaCl to produce a final concentration of 300 mM, and continue from Step 9 of the nuclear extract preparation method.

5.1.2 Preparation of probe DNA

Either a restriction fragment or a synthetic oligonucleotide probe may be used in this assay. However, if a restriction fragment is used, the size of the fragment is normally kept below 250 bp to enable clear distinction of the probe from any complexes. The type and size of probe used depends on the nature of the investigation (i.e. whether binding to a putative promoter region or to a specific DNA sequence is being analysed). *Protocol 22* gives the method we routinely use to label oligonucleotide sequences. Methods for labelling fragment probes are described in Sambrook et al (5).

> **Protocol 22.** End-labelling oligonucleotide probes for DNA mobility-shift assay
>
> *Equipment and reagents*
> - Specific paired probe oligonucleotides
> - [γ-^{32}P]ATP at 10 mCi ml^{-1} (Amersham Life Sciences)
> - STE buffer: 10 mM Tris-HCl pH 8.0, 100 mM NaCl, 1 mM EDTA
>
> *Method*
> 1. Anneal the separate oligonucleotides as in *Protocol 19*, Step 4.
> 2. Mix 2 pmol of annealed oligonucleotide with 20 µCi [γ-^{32}P]ATP in 50 mM Tris-HCl pH 7.6, 10 mM MgCl$_2$, 5 mM DTT, 0.1 mM EDTA, and 4 units of T4 DNA kinase.
> 3. Incubate at 37°C for 30 min.

Protocol 22. *Continued*

4. Add 200 µl STE to the reaction, and separate the labelled oligonucleotide from unincorporated label by centrifugation through a 1 ml Sephadex G25 column (prepared as in *Protocol 5*). The probe should be recovered in ~200 µl, of which 1 µl (10 fmol of DNA) is used per binding reaction.

Protocol 23. DNA mobility-shift assay

Equipment and reagents
- Poly dI-dC (Pharmacia)
- ^{32}P-labelled probe (*Protocol 22*)
- Protein extract (*Protocol 23*)

Method

1. Set up a 20 µl binding reaction containing 4% Ficoll, 20 mM HEPES pH 7.9, 1 mM MgCl$_2$, 0.5 mM DTT, 50 mM KCl, 2 µg poly dI-dC, 10 fmol ^{32}P-labelled probe, and approximately 2 µg of whole-cell or nuclear protein extract (20).
2. Incubate on ice for 40 min.
3. Run samples[a] on a 4% polyacrylamide gel (0.25 × TBE). The gel should be pre-run at 150 V for about 2 h before electrophoresis (the current will drop from 20–30 mA to approximately 10 mA during this time). The samples are then run for 2.5 h (or until the Bromophenol blue marker dye has run about two-thirds of the way down the gel).
4. Dry the gel onto filter paper (1 h, 80 °C, with vacuum), and expose the gel to either X-ray film or to a Phosphor-Imager screen (Bio-Rad).

[a] Do not add any loading dye to the samples, as the Ficoll in the binding buffer provides the density required for loading. Bromophenol blue in glycerol may be added to a spare track as a marker.

Once cell extracts have been demonstrated to contain proteins capable of binding to the sequence of interest, the specificity of this interaction can be measured by including a large excess (onefold, tenfold, and a hundredfold) of an unlabelled oligonucleotide competitor in the assay. The dilutions of unlabelled oligonucleotide are added to the binding reaction prior to the addition of the cell extract (*Protocol 23*, Step 1). If the protein can also bind to this unlabelled oligonucleotide, it will compete for the protein, leaving less available for binding to the labelled oligonucleotide. This will lead to a reduction (or elimination) of the band corresponding to the complex formed by that protein. Both specific and non-specific competitor sequences should be used, to demonstrate that the complex is DNA sequence-specific. Identification of proteins within a complex may be made by including an antibody to a previously characterized transcription factor in the binding reaction. If this antibody reacts with the protein of interest, it will either prevent its binding to DNA (so eliminating the complex), or cause further electrophoretic retardation and 'super-shift' the complex. This type of analysis is carried out by pre-incubating the cell extracts with the antisera for ~30 min on ice, before they are added to the binding reaction (*Protocol 23*, Step 1). A control binding reaction containing the antiserum in the absence of cellular extract should also be included, since serum may contain proteins capable of binding the DNA probe. The mobility-shift assay can also be used to examine the interaction of a viral protein with the cellular protein, either by comparing the band patterns obtained from infected or uninfected cells, or by adding a small amount of purified viral protein and investigating whether a super-shifted complex is formed (1).

The DNA mobility-shift assay can also be used to define the interaction of DNA-binding proteins with individual motifs or residues within the promoter sequence, following mutation of these sites (as described in Section 4). DNA probes can also be designed so that the co-operative binding of two or more transcription factors to adjacent binding sites can be examined. The assay can also be used to analyse the binding of purified transcription factors, thus allowing the role of specific regions to be examined following mutagenesis of the cloned gene (Section 4).

It should be remembered that, although the DNA mobility-shift assay is a powerful, but simple technique for the identification and study of DNA-protein binding, the demonstration of DNA binding *in vitro* does not necessarily mean that a particular gene is under the control of a particular transcription factor *in vivo*. It is therefore necessary to examine the effect that these factors have on promoter activity when linked to a reporter gene construct (as described in Section 3).

5.2 South-western blotting

In this method, a cell-free extract containing the DNA-binding protein of interest is first electrophoresed on an SDS-polyacrylamide gel, and then the separated proteins are transferred to a nitrocellulose membrane, and subsequently probed with a radioactively labelled nucleotide. After washing off non-specifically bound DNA, the relative molecular mass of the protein can be determined by comparison with marker proteins of known size, following exposure of the membrane to X-ray film. Briefly, this technique involves the following steps:

- The proteins from the cellular extract are separated on a standard SDS-polyacrylamide gel, then transferred to a nitrocellulose membrane.
- The proteins are denatured using guanidine hydrochloride.
- The proteins are then renatured.

- The filter is then incubated in blocking buffer to reduce non-specific hybridization.
- The filter is then hybridized overnight with a [^{32}P]-labelled concatamerized oligonucleotide probe.
- The blot is washed to remove non-specifically bound probe, and then exposed to X-ray film.

Further details of this method can be found in Sambrook *et al* (5).

5.3 Methods for isolating cloned transcription factors

Although the DNA mobility-shift assay and South-western blotting can detect the presence of proteins within cell extracts that are capable of binding to specific DNA sequences, unless the transcription factor has already been identified these techniques do not provide information on the protein itself. Ultimately, such information must be obtained by cloning the cDNA encoding the transcription factor. Details of methods for such cloning procedures are beyond the scope of this chapter, but in brief, this can be done in one of two ways:

- A suitable oligonucleotide probe containing the DNA sequence of interest can be used to screen an appropriate cDNA expression library (see ref. 44 for further details).
- The protein can be purified by sequence-specific DNA affinity chromatography, using a bait DNA that has a high specificity for the protein of interest (this sequence can be determined using the DNA mobility-shift assay). Once purified, the protein can be used to produce antibodies for screening an appropriate cDNA expression library (see ref. 44 for further details). Alternatively, peptide sequence data can be obtained, and this information used to design oligonucleotide probes to screen a cDNA library (see ref. 45 for further details).

Once a suitable full-length cDNA clone of the cellular transcription factor has been obtained, detailed analysis of its interaction with the viral promoter sequence of interest can be carried out as described in Sections 3 and 4.

Acknowledgements

The authors would like to thank David Thomas, Robert Coffin, Liz Ensor, and Lisa Melton for providing advice and protocols during the preparation of this chapter.

References

1. Preston, C. M., Frame, M. C., and Campbell, M. E. M. (1988). *Cell,* **52,** 425.
2. Roizman, B., and Sears, A. E. (1996). in *Field's virology* (3rd edn) (ed. B. N. Fields, D. M. Knipe, P. M. Howley), p. 2231. Lippincott-Raven Publishers, Philadelphia.

3. Roizman, B., and Sears, A. E. (1987). *Annu. Rev. Microbiol.*, **41**, 543.
4. Dix, I., and Leppard, K. N. (1993). *J. Virol.*, **67**, 3226.
5. Sambrook, J., Fritsch, E. F., and Maniatis, T. (1989). *Molecular cloning: a laboratory manual* (2nd edn). Cold Spring Harbor Laboratory Press, NY.
6. Chomczynski, P., and Saachi, N. (1987). *Anal. Biochem.*, **162**, 156.
7. McMaster, G. K., and Carmichael, G. G. (1977). *Proc. Natl. Acad. Sci. USA*, **74**, 4835.
8. Goldberg, D. A. (1980). *Proc. Natl. Acad. Sci., USA* **77**, 5794.
9. Rigby, P. W. J., Dieckmann, M., Rhodes, C., and Berg, P. (1977). *J. Mol. Biol.*, **113**, 237.
10. Feinberg, A. P., and Vogelstein, B. (1983). *Anal. Biochem.*, **132**, 6.
11. Cox, K. H., Deleon, D. V., Angerer, L. M., and Angerer, R. C. (1984). *Devel. Biol.*, **101**, 485.
12. Wilkinson, D. G. (ed.) (1992). In situ *hybridization: a practical approach.* IRL Press, Oxford.
13. Gorman, C. M. (1985). In *DNA cloning: a practical approach* (ed. Glover, D. M.), Vol. 2, p. 143. IRL Press, Oxford.
14. Herbomel, P., Bourachit, B., and Yaniv, M. (1984). *Cell*, **39**, 653.
15. De Wet, J. R., Wood, K. V., Pehuea, M., Helinski, P.R., and Subromani, S. (1987). *Mol. Cell. Biol.*, **7**, 725.
16. Chalfie, M., Tu, Y., Euskirchen, G., Ward, W. W., and Prasher, D. C. (1994). *Science*, **263**, 802.
17. Queen, C., and Baltimore, D. (1983). *Cell*, **33**, 741.
18. Potter, H., Weir, L., and Leder, P. (1984). *Proc. Natl. Acad. Sci. USA*, **81**, 7161.
19. Felgner, P. L., and Ringold, G. M. (1989). *Nature*, **337**, 387.
20. Bradford, M. (1976). *Anal. Biochem.*, **72**, 248.
21. Eastman, A. (1987). *BioTechniques*, **5**, 730.
22. Thomas, S. K., Coffin, R. S., Watts, G., Gough, G., and Latchman, D. S. (1998). *J. Virol.*, **72**, 3495.
23. Caravokyri, C., Pringle, C. R., and Leppard, L. N. (1993). *J. Gen. Virol.*, **74**, 2819.
24. Ramseyewing, A., and Moss, B. (1995). *Virology*, **206**, 984.
25. Harland, J., and Brown, S. M. (1998). In *Herpes simplex virus protocols* (ed. S. M. Brown, and A. R. MacLean), p. 1. Humana Press, Totowa, NJ.
26. Baskar, J. F., Smith, P. P., Nilaver, G., Jupp, R. A., Hoffmann, S., Peffer, N. J., Tenney, D. J., Colberg-Poley, A. M., Ghazal, P., and Nelson, J. A. (1996). *J. Virol.*, **70**, 3207.
27. Luckow, B., and Schutz, G. (1987). *Nucl. Acids Res.*, **15**, 5490.
28. Dawson, S. J., Yu-Zhen, L., Rodel, B., Möröy, T., and Latchman, D. S. (1996). *Biochem. J.*, **314**, 439.
29. Docherty, K., and Clark, A. R. (1993). In *Gene transcription: a practical approach* (ed. B. D. Hames, and S. J. Higgins), p. 111. IRL Press, Oxford.
30. McKnight, S. L., and Kingsbury, R. (1982). *Science*, **217**, 316.
31. Luckow, B., Renkawitz, R., and Schutz, G. (1987). *Nucl. Acids Res.*, **15**, 417.
32. Hutchison, C. A. III, Phillips, S., Edgell, M. H., Gillam, S., Jahnke, P., and Smith, M. (1978). *J. Biol. Chem.*, **253**, 6551.
33. Kunkel, T. A. (1985). *Proc. Natl. Acad. Sci. USA*, **88**, 488.
34. McPherson, M. J. (ed.) (1991). *Directed mutagenesis: a practical approach.* IRL Press, Oxford.

35. Deng, W. P., and Nickeloff, J. A. (1992). *Anal. Biochem.,* **200,** 81.
36. Ho, S. N., Hunt, H. D., Horton, R. M., Pullen, J. K., and Pease, L. R. (1989). *Gene,* **77,** 51.
37. Latchman, D. S. (ed.) (1999). *Transcription factors: a practical approach.* (2nd edn). Oxford University Press, Oxford.
38. Fried, M., and Crothers, D. M. (1981). *Nucl. Acids Res.,* **9,** 6505.
39. Garner, M. M., and Revzin, A. (1981). *Nucl. Acids Res.,* **9,** 3047.
40. Perry, L. J., Rixon, F. J., Everett, R. D., Frame, M. C., and McGeogh, D. J. (1986). *J. Gen. Virol.,* **67,** 425.
41. Kemp, L. M., Dent, C. L., and Latchman, D. S. (1990). *Neuron,* **4,** 215.
42. Lillycrop, K. A., and Latchman, D. S. (1992). *J. Biol. Chem.,* **267,** 24960.
43. Dignam, J. D., Lebovitz, R. M., and Roeder, R. G. (1983). *Nucl. Acids Res.,* **11,** 1475.
44. Cowell, I. G., and Hurst, H. C. (1999). In *Transcription factors: a practical approach* (2nd edn). (ed. D. S. Latchman), p. 123. Oxford University Press, Oxford.
45. Editors in Chief: Nicolas, R. H., Hynes, G., and Goodwin, G. H. (1999). In *Transcription factors: a practical approach* (2nd edn). (ed. D. S. Latchman), p. 97. Oxford University Press, Oxford.

6

Identification and analysis of *trans*-acting proteins involved in the regulation of DNA virus gene expression

ADRIAN WHITEHOUSE and DAVID M. MEREDITH

1. Introduction

We have been interested in identifying and characterising the genes that regulate viral gene expression in Herpesvirus saimiri (HVS). This is a lymphotrophic *rhadinovirus* (γ-2 herpesvirus) of squirrel monkeys (*Saimiri sciureus*), which persistently infects its natural host without causing any obvious disease. However, HVS infection of other species of New World primates can result in lymphoproliferative diseases (1). The genome of HVS (strain A11) consists of a unique internal low G + C-content DNA segment (L-DNA) of approximately 110 kbp which is flanked by a variable number of 1444 bp high G + C-content tandem repetitions (H-DNA) (2). Analysis indicates that it shares significant homology with other γ herpesviruses: Epstein–Barr virus, bovine herpesvirus 4, Kaposi's sarcoma-associated herpesvirus (or human herpesvirus 8), and murine gammaherpesvirus 68.

We have recently identified and characterized the two major transcriptional activating genes which regulate HVS gene expression. One transcriptional activator, encoded by open reading frame (ORF) 50, produces two transcripts. The first, ORF 50a, is spliced, contains a single intron, and is detected at early stages during the productive cycle, whereas the second, ORF 50b, is expressed later, and is produced from a promoter contained within its second exon (3). In addition, we have shown that both transcripts can transactivate a range of viral promoters by binding to a consensus ORF50-recognition sequence, CCN_9GG, within the promoters of transactivated genes, leading to an increase in mRNA levels (3). The second transcriptional regulator, encoded by ORF 57, is homologous to genes identified in all classes of herpesviruses. We have recently shown that the ORF 57 gene product can transactivate a range of viral genes independent of the target gene promoter sequences, and is mediated at the post-transcriptional level (4).

In this chapter, the methods utilized to study transcriptional control genes will be discussed.

2. Identification of transactivating proteins

Our groups have been studying the two major transcriptional regulatory genes encoded by HVS. The expression of reporter genes in mammalian cells in culture has proved invaluable, and is an ideal system to study the transactivation of viral genes and transcriptional regulatory sequences.

This section describes methods of identifying transcriptional activators in a newly sequenced virus. For example, transactivating genes can be identified by transient transfection reporter assays. A range of specific viral and heterologous promoters can be cloned upstream of a range of reporter genes, such as chloramphenicol acetyl transferase (CAT), β-galactosidase, or luciferase. These constructs can then be used to screen for genes which encode transcriptional activators by cotransfection of the promoter reporter construct in the absence and presence of the transactivator. If you wish to identify transactivating genes encoded by a newly identified virus, portions of the genome can be cotransfected separately. The genes encoding the transactivator can then be pinpointed by using increasingly smaller fragments of the genome. The section describes the transfection of mammalian genes, and the variety of reporter genes which can be utilized for such a genome trawl.

2.1 Transfection of mammalian cells

The choice of transfection method is dependent upon the cells used. The most common methods for transient expression involve mediated DNA uptake by calcium phosphate or cationic lipids. Calcium phosphate transfection protocols have already been described in Chapters 3 and 5. A lipofection method which differs slightly from that given in Chapter 3 is described here.

Protocol 1. Liposome-mediated transfection of adherent mammalian cells

Reagents
- Dulbecco's Modified Eagle's Medium (DMEM) (no FCS)
- Liposome: Lipofectamine (Life Technologies), or DOTAP (Boehringer Mannheim)

Method

1. Grow the cells at 37°C in 60 mm culture dishes until they reach 50–80% confluence.
2. For each transfection, dilute 1–2 µg of DNA into 100 ml serum-free DMEM. Add this solution to 25 ml of liposome reagent (2 mg ml^{-1}) diluted in 100 ml of serum-free medium, and mix gently.

6: Identification and analysis of trans-acting proteins

3. Incubate this mixture for 20 min at room temperature to allow formation of DNA–liposome complexes.
4. Wash the cells once with 2 ml of serum-free medium.
5. Add 200 µl liposome–DNA mixture dropwise to the washed cells. Incubate the cells for 2–24 h at 37°C in a humidified incubator in an atmosphere of 5% CO_2.
6. After 18–24 h replace the medium with fresh medium.
7. Harvest the cells after 24–48 h.

Assays for measurement of CAT and β-galactosidase activity have already been described in Chapter 5. A luciferase assay, modified from the method described by Wet *et al.* (5) is described in *Protocol 2*. This assay is based on the oxidation of beetle luciferin with concomitant production of light photons.

Protocol 2. Luciferase assay

Equipment and reagents

- Dry ice-ethanol bath
- 37°C waterbath
- 1.5 ml microcentrifuge tubes
- 5 ml luminometer vials (e.g. Sarstedt)
- Microcentrifuge (e.g. Eppendorf)
- Luminometer (e.g. Lumat LB 9507 from EG&G Berthold)
- 10 µg ml^{-1} D-luciferase
- 10 mM luciferin in 30 mM glycylglycine, pH 7.8
- 0.1 M potassium phosphate pH 7.5, 1 mM EDTA
- Luciferase reaction buffer: 30 mM glycylglycine pH 7.8, 2 mM ATP, 15 mM $MgSO_4$

Method

1. Resuspend cell extracts in 1.5 ml of 0.1 M potassium phosphate pH 7.5, 1 mM EDTA.
2. Disrupt the cells by three cycles of freeze-thawing in a dry ice-ethanol bath and a 37°C waterbath. Pellet the cell debris by spinning the tubes for 60 s in a microfuge.
3. Aliquot 350 µl of luciferin reaction buffer into 5 ml luminometer vials, and place on ice.
4. Add various volumes (10–100 µl) of cell extract.
5. Add 0.1 volumes of luciferin solution.
6. Place reaction in the tube holder of the luminometer, and add 10 µl of luciferin substrate solution. Record the light emission for 10 s.

3. Analysis of the mechanism of transactivation

There are several ways in which transactivating proteins can regulate viral gene expression. We have shown that the two major transcriptional regulating genes encoded by HVS act by different mechanisms to transactivate genes. The ORF 50 gene products transactivate a range of viral promoters, by binding to a consensus ORF50-recognition sequence, CCN_9GG, within the promoters of transactivated genes, leading to an increase in mRNA levels, whereas ORF 57 transactivates viral genes by a post-transcriptional mechanism.

3.1 Characterization of RNA

To determine which mechanism a transactivator uses, RNA analysis must be performed. This can be achieved by a number of methods. Initially, transfection experiments are performed where the plasmid expressing the transactivated gene is cotransfected in the absence and presence of the transactivator. The RNA from these cells should then be extracted, and the RNA analysed. Comparisons can then be made between RNA levels from the gene alone, and in the presence of the transactivator.

Protocol 3. Extraction of total RNA

Equipment and reagents

- DEPC-treated H_2O: All aqueous solutions should be treated with diethyl pyrocarbonate (DEPC) to destroy RNase. Add the DEPC to the solution to a concentration of 0.2%, incubate overnight at 37°C, then autoclave to break down residual DEPC.
- Refrigerated benchtop centrifuge (e.g. MSE Mistral)
- Trizol reagent (Life Technologies)
- Chloroform
- Isopropanol

Method

N.B.: Since RNA is highly labile, always use disposable gloves and sterile glass and plasticware for RNA work.

1. Lyse the cells (1×10^6 cells are sufficient for 30–40 μg of total RNA) using Trizol reagent (follow the manufacturer's instructions).
2. Add 0.2 ml of chloroform (0.2 ml), vortex the solution for 20 s, and incubate at 20°C for 15 min.
3. Centrifuge the samples for 15 min at 4°C (6000 r.p.m.), and precipitate the aqueous phase containing nucleic acids, using 0.5 ml of isopropanol.
4. Wash the pellet with 70% ethanol, and resuspend in 50 ml of DEPC-treated water.
5. Store at –70°C.

3.1.1 Northern blotting

Northern blotting is used to analyse specific RNA molecules in a population of total RNA. The RNA is separated by electrophoresis in a denaturing

6: Identification and analysis of trans-acting proteins

agarose gel, transferred to nitrocellulose, and specific RNA species are detected by hybridization with a radioactively labelled probe.

Protocol 4. Electrophoresis of RNA in denaturing gels, and Northern blot transfer

Equipment and reagents

- Agarose gel electrophoresis equipment
- UV transilluminator
- Oven at 80°C
- Nitrocellulose membrane (e.g. Hybond-N, Amersham)
- Agarose
- 10 × Mops buffer stock: 0.2 M Mops, 50 mM sodium acetate, 0.01 M EDTA
- 40% (v/v) formaldehyde (Analytical Grade)
- Deionized formamide
- RNA loading buffer: 30% glycerol, 1 mM EDTA, 0.01% Bromophenol blue
- Ethidium bromide (10 mg ml^{-1}). **N.B.: Ethidium bromide is a carcinogen**
- 20 × SSC: 3 M NaCl, 0.3 M tri-sodium citrate
- 3MM paper (Whatman)

Method

1. Prepare a 1% agarose gel in 10 × Mops buffer and formaldehyde to give final concentrations 1 × and 2.2 M, respectively. Cast the gel in a fume hood, and allow the gel to set for at least 30 min.

2. Prepare the samples of RNA by mixing the following: RNA (up to 20 µg) 9 µl, 10 × Mops buffer 2 µl, formaldehyde 7 µl, formamide 20 µl. Mix by vortexing, incubate at 65°C for 5 min, and then chill on ice.

3. Add 5 µl of RNA loading buffer.

4. Also prepare RNA markers of the appropriate size.

5. Pre-run the gel for 10 min, then load the RNA samples. For a gel of 20 cm, perform electrophoresis at 60 V for 6–8 h. The marker dye should have migrated approximately 15 cm.

6. Remove the gel, and rinse in DEPC-treated H$_2$O for 10 min.

7. Cut off the RNA markers, and stain with ethidium bromide for 30 min.

8. Photograph the visualized bands under UV transmission.

9. Soak the gel in 20 × SSC, then lay it on a wick of three sheets of 3MM paper soaked in 20 × SSC, avoiding bubbles. Lay the filter, soaked in 20 × SSC, onto the gel, then apply three layers of 3MM (cut to the size of the gel and soaked in 20 × SSC). Finally apply a wad of paper towels, cut to size, followed by a 500 g weight. Allow the DNA to transfer overnight for 16 h.

10. Dismantle the apparatus, and bake the filter for 2 h at 80°C.

Protocol 5. Production of DNA probe

Reagents

- DNA at a concentration of 2.5–25 ng µl^{-1}
- Megaprime kit (Amersham)
- [α-^{32}P]dCTP (10 mCi ml^{-1}, 5000 Ci mM^{-1})
- DNA polymerase (Klenow fragment)
- ProbeQuant G-50 Sephadex G50 Micro Columns (Pharmacia Biotech)

Method

1. Denature the DNA probe for 3 min at 90°C, then chill on ice for 5 min.
2. Add 10 µl of labelling buffer (10 × labelling buffer including 200 µm dGTP, dATP, dTTP, and random hexanucleotide primers).
3. Add 3 µl [α-^{32}P]dCTP and 1.0 µl of DNA polymerase (Klenow fragment), incubate at 37°C for 30 min.
4. Stop the reaction by the addition of 5 µl of 0.2 M EDTA.
5. Separate unincorporated [α-^{32}P]dCTP from the radiolabelled probe using G-50 Sephadex G50 Micro Columns (follow the manufacturer's instructions).

Protocol 6. Northern blot hybridization

Equipment and reagents

- Deionized formamide
- 10% SDS
- 10 × SSC
- Salmon sperm DNA, 10 mg ml^{-1} (denatured by boiling)
- 50 × Denhardt's solution: dissolve 1% w/v Ficoll, 1% w/v polyvinyl pyrrolidine, 1% w/v BSA in sterile H$_2$0, filter-sterilize, and store at –20°C
- Hybridization oven (e.g. Hybaid)

Method

1. Soak the filter in 1 mM EDTA, and boil for 5 min.
2. Place the filter in a hybridization container, and add the prehybridization mixture: 2.5 ml 10 × SSC, 5 ml formamide, 1 ml 10% SDS, 1 ml 50 × Denhardt's solution
3. Incubate at 60°C for 2 h.
4. Add the denatured radiolabelled probe and 20 µl of denatured salmon sperm DNA. Use approximately 1 × 10^6 c.p.m. probe per ml of hybridization solution.
5. Incubate at 60°C for 16 h.
6. After hybridization wash the filter as follows: four times with 50 ml 2 × SSC, 1% SDS, 65°C, for 5 min; twice with 50 ml 0.1 × SSC, 1% SDS, 50°C, for 15 min

6: Identification and analysis of trans-acting proteins

7. Seal the filter in a heat-sealable plastic bag to keep moist, and expose the filter to X-ray film at −70°C with intensifying screens.

3.1.2 Primer extension

Primer extension can be used to ascertain whether transactivation of a gene is due to an increase in the levels of mRNA, or if the mRNA initiation site is changed in the presence of the transactivator. The strategy of primer extension is to hybridize a radiolabelled oligonucleotide to an RNA transcript. Reverse transcription will then produce a complementary DNA strand, beginning at the primer and ending at the 5′ end of the RNA strand.

Protocol 7. Production of 5′ end-labelled oligonucleotides

Equipment and reagents

- ProbeQuant G-50 Sephadex G50 Micro Columns (Pharmacia Biotech)
- 10 × kinase reaction buffer: 0.25 M Tris-HCl pH 9.0, 50 mM $MgCl_2$, 25% glycerol
- 0.1 M DTT
- [γ-^{32}P]dATP (10 mCi ml^{-1}, 5000 Ci mM^{-1})
- Oligonucleotide (10 ng μl^{-1})
- T4 polynucleotide kinase (Life Technologies)
- 0.5 M EDTA

Method

1. Prepare the following reaction mixture: 5 μl oligonucleotide, 2.5 μl of 10 × kinase reaction buffer, 1 μl DTT, 3 μl [γ-^{32}P]dATP, 13 μl H_2O, 0.5 μl T4 polynucleotide kinase.
2. Incubate the reaction mixture at 37°C for 30 min.
3. Add 2 μl of 0.5 M EDTA to terminate the reaction.
4. Separate unincorporated [γ-^{32}P]dATP from the radiolabelled oligonucleotide using G-50 Sephadex G50 Micro Columns. It is important to note that the oligonucleotide must be longer than 20 bp in order to use the G50 columns.

Protocol 8. Primer extension analysis

Equipment and reagents

- 10 × hybridization buffer: 3 M NaCl, 0.1 M Tris-HCl pH 7.5, 10 mM EDTA
- 10 × extension buffer: 0.6 M NaCl, 0.1 M Tris-HCl pH 8.0, 500 μg ml^{-1} actinomycin D, 80 mM $MgCl_2$; store at −20°C
- ^{32}P-labelled oligonucleotide primer, 0.5 ng μl^{-1} (*Protocol 7*)
- 1.0 M DTT
- Reverse transcriptase (Life Technologies)
- 0.1 M of dATP, dTTP, dCTP, and dGTP stocks
- Total RNA (*Protocol 3*)
- Phenol:chloroform (1:1)
- 10% polyacrylamide–7.0 M urea sequencing gel
- RNA loading buffer (*Protocol 4*)

163

Protocol 8. *Continued*

Method

1. Prepare the following reaction mix for each extension reaction: 8 μl RNA (1–10 μg), 1 μl of 10 × hybridization buffer, 1 μl ^{32}P labelled oligonucleotide primer.
2. Mix the contents, and hybridize for 10 min at 50–60 °C. The hybridization temperature should be approximately 5–10 °C below the estimated T_m of the hybrid.
3. Transfer the tubes to 42 °C, and add 23 μl of prewarmed extension reaction mix. This is made up as follows (volume sufficient for 10 reactions): 33 μl of 10 × extension buffer, 3.3 μl of 1.0 M DTT, 3.3 μl of 0.1 M dATP, 3.3 μl of 0.1 M dCTP, 3.3 μl of 0.1 M dGTP, 3.3 μl of 0.1 M dTTP, 5 units μl^{-1} of reverse transcriptase, DEPC-treated H_2O to 233 μl final volume.
4. Incubate the extension reaction for 20 min at 42 °C.
5. Transfer the tubes to ice, and adjust the volume to 400 μl with H_2O.
6. Extract the reaction with phenol:chloroform, and precipitate the extension products with ethanol.
7. Separate the primer extension products using a 10% polyacrylamide–7.0 M urea denaturing polyacrylamide gel.
8. Dry the gel, and visualize the bands by exposure to X-ray film.

4. Identification of *cis*-acting elements

4.1 Mobility shift assays

We have shown that the ORF 50 gene products transactivate specific viral promoters. We believe that ORF 50 binds to specific sequences, termed ORF 50 response elements, within the promoter of the gene it transactivates. To determine which *cis*-acting sequences within a promoter region interact with ORF 50 or other DNA-binding proteins, mobility shift assays can be performed using oligonucleotides which span the promoter region. DNA-binding proteins are incubated with radiolabelled oligonucleotides or DNA fragments. The reaction mixtures are then analysed by gel electrophoresis. If a DNA-binding protein binds to a specific DNA sequence, a protein-DNA complex will be produced which will reduce the mobility of the DNA. This interaction can be visualized as a retarded complex. In order to perform mobility shift assays, the DNA-binding protein should be purified. This can be achieved by the following method (*Protocol 9*).

A specific DNA-binding protein can also be produced by *in vitro* translation. The DNA sequence must be cloned into a standard *in vitro* translation

6: Identification and analysis of trans-acting proteins

vector under the control of a transcription promoter and translation start site. The most convenient method to perform *in vitro* translation is to use the TNT Coupled Rabbit Reticulocyte Lysate translation systems (Promega). This circumvents the time-consuming and often difficult protein purification processes necessary in other expression systems.

Protocol 9. Extraction of DNA-binding proteins from mammalian cells

Modification of the method described by Andrews and Faller (6)

Equipment and reagents

- Vortex mixer
- Microcentrifuge (e.g. Eppendorf)
- Buffer 1: 10 mM Hepes-KOH pH 8.0, 1.5 mM $MgCl_2$, 10 mM KCl, 0.5 M DTT, 1mM 4-(2-aminoethyl) benzenesulfonyl fluoride (AEBSF, Calbiochem)
- Buffer 2: 20 mM Hepes-KOH pH 8.0, 30% glycerol, 420 mM NaCl, 1.5 mM $MgCl_2$, 10 mM KCl; 0.5 M DTT, 1 mM AEBSF)
- Microcentrifuge tubes
- Transfected cells (1×10^6 cells)
- Phosphate-buffered saline (PBS)

Method

1. Transfect cells (as previously described) with a mammalian expression vector over-expressing the DNA-binding protein.
2. Wash the cells in ice-cold PBS, scrape the cells off, and place in a microfuge tube.
3. Pellet the cells for 10 s, and resuspend the cells in 500 μl of ice-cold Buffer 1.
4. Leave the cells on ice for 10 min, then vortex for 10 s.
5. Pellet the cells for 10 s, and resuspend the cells in 100 μl of ice-cold Buffer 2.
6. Leave the cells on ice for 30 min.
7. Remove the cellular debris by centrifugation for 2 min at 4°C.
8. Store the supernatant containing the DNA-binding proteins in aliquots at –70°C.

Protocol 10. *In vitro* translation of a specific DNA-binding protein

Reagents

- Plasmid containing a DNA-binding protein sequence under the control of a T3, T7, or SP6 promoter
- TNT *in vitro* transcription-translation system (Promega)
- Ribonuclease inhibitor
- Radiolabelled amino acid ^{35}S-methionine (1000 Ci $mmol^{-1}$) at 10 mCi ml^{-1} (Amersham)

Protocol 10. *Continued*

Method

Firstly check that the DNA-binding protein is being correctly translated.

1. Make up the following reaction mixture: 25 µl TNT Rabbit Reticulocyte Lysate, 2 µl Reaction Buffer, 1 µl RNA polymerase (T3, T7, or SP6), 1 µl Amino Acid Mixture minus ^{35}S-Met, 1 µl Ribonuclease inhibitor, 1 µg DNA plasmid, H_2O to 50 µl.
2. Incubate the reaction for 90 min at 30 °C.
3. Analyse the translated product using standard denaturing gel electrophoresis.
4. Dry the gel (60 °C for 2 h), and visualize the bands by exposure to X-ray film.
5. If the correct-sized translation product is observed, repeat the reaction without radiolabelled amino acids. This translation product can then be used in a mobility shift assay.

However, in some cases there may be possible interference from the rabbit reticulocyte endogenous proteins in functional assays. This can be avoided by depleting endogenous DNA-binding proteins from reticulocyte lysate translation systems, as described by Ebel and Sippel (7).

Protocol 11. Depletion of endogenous DNA-binding proteins

Equipment and reagents

- TNT Rabbit Reticulocyte Lysate (Promega)
- Biotin-labelled oligonucleotide containing consensus binding sequence
- Streptavidin-coated magnetic beads (Dynal)
- Phosphate-buffered saline (PBS)
- Magnetic particle concentrator (Dynal)

Method

1. Thaw 200 µl of TNT rabbit reticulocyte lysate quickly.
2. Mix with 25 pmol of biotinylated oligonucleotide, and leave for 10 min.
3. Incubate with 1×10^7 streptavidin-coated magnetic beads.
4. Wash four times in PBS.
5. Remove the beads, together with the DNA–protein complexes, through a magnetic particle concentrator.
6. The supernatant can then be used in an *in vitro* transcription-translation reaction (*Protocol 10*).

6: Identification and analysis of trans-*acting proteins*

Protocol 12. Mobility shift assay

This protocol can be divided into four stages:
(a) preparation of a radiolabelled DNA probe containing a particular protein-binding site
(b) preparation of non denaturing gel
(c) the binding reaction, in which protein is bound to the DNA probe
(d) electrophoresis of protein–DNA complexes

A. *Preparation of radiolabelled DNA probe containing a particular protein-binding site*

DNA fragments of 20–300 bp in length may be used as probes. However, in our experience probes of 30–50 bp are the most appropriate, as longer probes may contain multiple binding sites, and make interpretation of the assay difficult. Small DNA fragments or annealed oligonucleotides containing the putative protein binding site are end-labelled with [γ-^{32}P]dATP using T4 polynucleotide kinase, as previously described (*Protocol 7*).

B. *Preparation of non-denaturing gel*

Equipment and reagents

- Protein gel electrophoresis unit (e.g. Bio-Rad)
- 10 × TBE electrophoresis buffer: 0.9 M Tris base, 0.9 M boric acid, 25 mM EDTA, adjust to pH 8.3 with glacial acetic acid
- 40% acrylamide
- 5 × gel running buffer: 0.24 M Tris-HCl pH 8.0, 1.9 M glycine, 0.01 M EDTA
- 2% bisacrylamide
- 10% ammonium persulfate
- TEMED

Method

1. Assemble glass plates using 0.75 mm spacers.
2. Prepare the following non-denaturing protein gel mixture: 5 ml acrylamide (40%), 3.3 ml bisacrylamide (2%), 31.7 ml distilled H_2O, 10 ml gel running buffer, 300 μl ammonium persulfate, 60 μl TEMED.
3. Pour the gel between the plates and insert the comb; leave to set (for optimal results use a comb with teeth >7 mm wide).
4. Place gel in the tank, and fill with 1 × TBE running buffer.
5. Pre-run the gel for 30 min at 100 V.

C. *Preparation of binding reactions*

Reagents

- [γ-^{32}P]dATP radiolabelled probe
- 10 × binding buffer: 40% glycerol, 10 mM EDTA, 50 mM DTT, 0.1 M Tris-HCl pH 8.0, 1M NaCl, 1mg ml^{-1} BSA
- Non-specific carrier DNA, e.g. poly(dI-dC) (Pharmacia)
- DNA-binding protein (crude extract or *in vitro* translation product)

Protocol 12. *Continued*

Method

1. Prepare the binding reaction as follows: 1 μl probe (5000–20 000 c.p.m.), 1.5 μl buffer, 7.5 μl H$_2$O, 1 μl poly(dI-dC), 5 μl protein.
2. Mix the reaction gently, and incubate for 15 min at 20°C.

D. *Running and analysis of non-denaturing gel*

Reagents
- 10 × loading buffer: 30% glycerol, 1 mM EDTA, 0.01% Bromophenol blue

Method

1. Add 10 μl of 10 × loading buffer to the binding reactions.
2. Load onto the non-denaturing gel (see B).
3. Electrophorese for 2–3 h at 100 V to give good separation of the free probe and protein–DNA complexes.
4. Stop the gel when the Bromophenol blue is approximately 4 cm from the bottom of the gel.
5. Separate the gel equipment and plates.
6. Place a piece of Whatman 3MM paper on the gel, and peel off the gel attached to the paper.
7. Dry the gel (80°C for 2 h), and visualize the bands by exposure to X-ray film.

In order to analyse how specific the protein interaction is with the DNA sequence, competition assays can be used. This is necessary because most protein preparations will contain both specific and non-specific DNA-binding proteins. For a specific competitor the same unlabelled DNA can be used as a probe. In addition, a non-specific competitor can be any fragment of DNA with an unrelated sequence.

Protocol 13. Competition mobility shift assay

Method

1. Assemble the binding reaction as previously described (*Protocol 11*).
2. In addition, add to the labelled probe increasing amounts of unlabelled specific (e.g. oligonucleotide or restriction fragment) and non-specific (e.g. poly(dI-dC)) competitors.
3. Incubate and electrophorese as described in *Protocol 12*.

6: Identification and analysis of trans-*acting proteins*

A further control is to use antibodies in the mobility shift binding assay to identify specific proteins present in the DNA-protein complex. Addition of a specific antibody which interacts with a protein involved in complex formation can form an antibody-protein-DNA ternary complex which can further reduce mobility of the complex in the gel, resulting in a supershift. It is advisable to purify the antibody prior to use in this assay and heat to 56 °C for 30 min, as nuclease activity may degrade the probes.

Protocol 14. Antibody supershift assay

Method

1. Assemble the binding reaction as previously described (*Protocol 12*).
2. In addition, add varying small amounts of a specific antibody against the DNA-binding protein (set up a control reaction with a non-specific control antibody).
3. Incubate and electrophorese as described in *Protocol 12*.

4.2 Purification of transcription factors

At the start of any purification procedure, it is essential to have developed a rapid and sensitive assay for protein function. Nearly all successful protein purification procedures rely on speed, to minimize either protein degradation, denaturation, or modification. It will be evident from the previous sections that, in general, transcription factor assays are not rapid. It is not surprising, therefore that only a very small number of virus transcription factors have been purified.

The starting point in devising a purification strategy is the reason purifying the protein, and what yield and purity are required. This may seem obvious, but these matters need to be considered carefully. To purify the protein to make antibodies, a yield of 20–200 μg may be required. Alternatively, protein–protein interaction may be the main interest, in which case purification to homogeneity is not necessary, and the yield needed will probably only be a few micrograms. On the other hand, crystallography may need a continuing supply of tens of milligrams. Each is best achieved through different purification strategies.

Transcription factors are not usually abundant proteins. Except in rare instances, it is highly likely that the protein will have to be produced using heterologous expression systems. Purification of a protein to homogeneity will require a very large quantity of virus-infected cells. With the exceptions of herpes simplex type 1, adenovirus 2 or 5, or vaccinia, the chances are that it will not be possible to produce sufficient virus-infected cells to obtain a reasonable yields at the end of a multi-step purification. Even the experienced protein purifier may be lucky to retain 10% of the total starting material. A minimum starting point to recover a few micrograms may well be 10^9–10^{10} virus-infected cells.

We do not intend to cover cloning and expression procedures in this chapter, as the methods are well described in other volumes in this series. We would note, that in our experience, it is impossible to predict which expression system will be optimal for trouble-free recombinant protein production. In particular, many transcription factors are modified through phosphorylation and O-glycosylation. These may in some cases have fundamental effects on the activity of the transcription factor. Protein phosphorylation and dephosphorylation is a dynamic process, so it may be advisable to determine whether phosphorylation of the protein is important for activity. We would note that as far as we are aware, phosphorylation of these virus proteins is carried out by host-cell protein kinases, so that if the transcription factor is active in a transfection assay, it is either being phosphorylated correctly, or phosphorylation is not important for its activation

4.2.1 Immunoaffinity purification

The first step in any purification strategy is to understand the protein. In many cases a gene may be identified as encoding a transcriptional activator, but the identity of the protein in the virus-infected cell is not known. A recommended starting point, therefore, is production of an antibody reagent. Most commonly, this will be produced using an immunogen produced via recombinant DNA means, or alternatively peptide. The procedures used for heterologous expression and antibody production are described in many other volumes in this series, so we make no attempt to provide protocols here. We would suggest that for prokaryotic expression, only a small portion, for example 10–20% of the sequence, should be expressed. Larger fusion proteins are often more susceptible to intracellular degradation during fusion protein induction in *E. coli*.

Our experience with different proteins over an extended period is that there are no hard and fast rules over the choice of expression system. The choice of peptide sequence synthesized for immunization is controlled by the potential solubility, and the location of that sequence within the polypeptide. Although there are software packages designed to predict potential antigenic regions, we have not found them useful. As a rule of thumb, we would suggest synthesis of several peptides, preferably based on both amino- and carboxy-termini, as well as some internal sequence. Peptides should be purified to the highest level you can afford, and be 12–15 amino acids in length.

Successful purification of the protein may be achieved in one step through immunoaffinity purification (*Protocol 17*), using an antipeptide or anti-fusion protein antibody, although again there are no hard and fast rules. Conditions used for eluting a protein from such a column are harsh, because the only way to release a protein from an antibody is to cause structural changes, that is,

6: Identification and analysis of trans-acting proteins

denaturation. You may be fortunate to have an antibody from which you can elute the protein using the cognate peptide, but this is most likely to occur with a monoclonal anti-peptide antibody. Polyclonal antibodies are usually of too high an affinity to be able to utilize this procedure.

There are numerous methods for covalently coupling antibody to a chromatography support. We favour two of the more traditional approaches, because of their ease and reliability. Cyanogen bromide (CNBr)-Sepharose links antibody through a covalent attachment via primary amine groups in lysine residues. The disadvantage with this procedure is that the orientation of the antibody with respect to the bead surface is random, so that the antigen binding site may be physically masked. In our experience this does not seem to pose too many problems, although the binding capacity of the immunoaffinity matrix varies with each preparation. The advantage with hydrazide-activated supports is that antibody molecules are all linked through the carbohydrate residues attached to the Fc portion of the antibody. This means that all antigen-binding sites are oriented facing away from the bead surface. Both types of matrix are best purchased commercially. The hydrazide linking can be purchased as a kit from Bio-Rad, the Affi-Gel Hz Immunoaffinity kit. All reagents required are provided, along with full detailed instructions. CNBr-Activated Sepharose may be purchased from several sources. This matrix deteriorates when damp, therefore we recommend purchasing small quantities, and using fresh material each time a column is made.

Protocol 15. Antibody conjugation using CNBr

Reagents

- 1 mM HCl
- CNBr-activated Sepharose (Pharmacia)
- 0.1 M sodium bicarbonate, pH 8.5–9.0
- Phosphate-buffered saline (PBS)
- Purified antibody, 3–5 mg ml^{-1} in PBS per 1 ml hydrated beads
- 1 M Tris-HCl (or any other primary amine) pH 8.5

Method

1. Hydrate the beads in 10 ml of 1 mM HCl, and wash with a further 100 ml of HCl, using a sintered glass funnel.
2. Wash the beads with a further 100 ml of distilled water.
3. Resuspend the beads in 5 ml of sodium bicarbonate solution, and place in a sealable vessel.
4. Add the antibody, and immediately check the absorbance at 280 nm, to measure the initial protein concentration. The sample may then be returned to the vessel.
5. Incubate the beads at room temperature for 1 h with constant agitation.
6. Measure the absorbance again; the protein concentration should have dropped to <5% of the starting level.

Protocol 15. Continued

7. Allow the beads to settle, aspirate the supernatant, add 5 ml of 1 M Tris-HCl, and incubate for 10 min, to block any residual active groups on the matrix.
8. Finally, wash the matrix with PBS. The matrix is now ready for pre-treatment prior to use.

4.2.2 Cell fractionation

The nuclear localization of transcription factors allows an initial simple enrichment procedure. Infected cells may be disrupted and nuclei isolated, as a starting material for purification. Purified nuclei may be then flash-frozen and stored at −70°C, in order to accumulate large quantities of starting material. Cell disruption using a Dounce homogenizer or sonication is not an option, because of the hazard from infection. We favour initial lysis of cells using the detergent Brij-96, which does not disrupt the integrity of the nuclear membrane.

Protocol 16. Preparation of nuclei

Materials
- Lysis buffer: 1% Brij 96 in 50mM Tris-HCl pH 8.3, 140 mM NaCl, 0.2 mM EDTA, 100 mM leupeptin, 1 mM pepstatin, 1 mM PMSF
- Virus-infected cells (freshly pelleted from suspension)

Method

All procedures should be carried out on ice or at 4°C.

1. Resuspend the cell pellet in 3 ml lysis buffer per 5×10^7 virus-infected cells.
2. Disrupt the cells by shearing, using a syringe and three passes through a 26-gauge needle.
3. Centrifuge the lysate at 4000 g for 15 min to pellet the nuclei.
4. Aspirate the supernatant, and dispose using appropriate precautions.
5. Resuspend the nuclei in 5 volumes of the lysis buffer, mix, and repeat the centrifugation step.
6. Either flash-freeze the nuclei using liquid nitrogen, or continue processing.

One step purification of some transcription factors may be achieved through immunoaffinity chromatography. Most may be released in a soluble form in a high ionic strength buffer. For ease, we detail a procedure which has

6: Identification and analysis of trans-acting proteins

worked well in our hands. Obviously it needs to be optimized for each protein. It is difficult to assess the capacity of a column and determine how much affinity matrix will be required. Generally a 1 ml column should have an antigen binding capacity of at least 0.5 mg. This means that a lysate from $>10^9$ nuclei may be applied.

Protocol 17. Immunoaffinity purification

Equipment and reagents

- Antibody-conjugated chromatography matrix
- Nuclear pellet
- Wash buffer: as solubilization buffer, but without detergent
- Elution buffer: 0.2 M glycine-HCl pH 2.2
- Solubilization buffer: 1% Triton X-100 in 20 mM Hepes pH 7.5, 0.5 M NaCl, 1mM DTT, 0.2 mM EDTA, 100 mM leupeptin, 1 mM pepstatin, 1 mM PMSF
- Neutralising buffer: 0.5 M Tris-HCl pH 8.5

Method

All procedures should be carried out on ice or at 4°C

1. Lyse the nuclei with solubilization buffer, using 1 ml per 5×10^7 nuclei.
2. Stand on ice for 15 min, and mix occasionally with gentle vortexing.
3. Centrifuge the lysate at 50 000 g for 1 h at 4°C
4. Apply the lysate to the immunoaffinity column using a peristaltic pump at a flow rate of approximately 0.1 ml min^{-1}. Save the unbound material. Ensure that application of the sample is not hurried, as there will be insufficient time for antibody and antigen to interact.
5. Wash the column with at least 50 ml of wash buffer, at a flow rate of up to 1 ml min^{-1}.
6. Allow the buffer to drain into the chromatography bed.
7. Block the flow out of the column, and apply 1 ml of elution buffer. Mix the beads thoroughly using a sterile Pasteur pipette, and allow to stand for 5 min.
8. Collect the eluate by restoring the flow, and add an equal volume of neutralising buffer.
9. Repeat this process two or three more times.
10. Analyse the eluates using SDS-PAGE. In most cases, nearly all the material will be in the first two fractions collected.

4.3 Conventional purification procedures

As one starts to understand the role of the transcription factor, the association with other proteins becomes important. There are numerous ways of studying this process. During conventional purification procedures, it may be that one or more protein(s) appears to co-purify, consistently. The co-elution of a protein from a chromatography column does not necessarily mean that the proteins physically associate within the cell, and there may be other trivial explanations. If the purification strategy sequentially utilizes different properties, such as charge, hydrophobicity, and size, and stoichiometric amounts of protein(s) continue to co-purify, it is likely that there may be some relevant association.

We recommend that at the start of devising a purification strategy, the ability of a protein to bind to and elute from various chromatography matrices should be tested. This can be carried out using small quantities of starting material. As with all aspects of protein purification, there are many options and no hard and fast protocols can be given. Our advice is to start with a procedure which maximizes release of your protein in a native, soluble form. We recommend that all buffers contain an extensive range of protease inhibitors and a reducing agent. In addition, phosphatase inhibitors, such as sodium fluoride, should be added. A second useful step is concentration of the protein by precipitation with ammonium sulfate. The advantage of this is that a protein precipitate, redissolved, will still contain a high concentration of the salt, and this sample is ideal for binding to a hydrophobic interaction matrix, such as phenyl Sepharose. The protein will typically be eluted from such a matrix by reducing the salt concentration and addition of ethylene glycol. This sample is ideal for loading directly onto ion-exchange matrices. Stepwise elution from a small ion-exchange column allows sample concentration, again, ideal for chromatography by size fractionation. This dilutes the sample, which may be concentrated again via binding to ion-exchange matrices.

In conclusion, we recommend that conventional protein purification from virus-infected cells should be undertaken when absolutely unavoidable. It is an expensive and time-consuming process.

Acknowledgements

Work in the authors' laboratories is supported in part from grants from Yorkshire Cancer Research, The Candlelighters Trust, the West Riding Medical Trust, the Medical Research Council, and the Wellcome Trust.

References

1. Fleckenstein, B., and Desrosiers, R. C. (1982). In *The herpesviruses* (ed. B. Roizman). Vol. 1, p. 253–332. Plenum Press, New York.
2. Albrecht, J. C., Nicholas, J., Biller, D., Cameron, K. R., Biesinger, B., Newman, C., Wittman, S., Craxton, M. A., Coleman, H., Fleckenstein, B., and Honess, R. W. (1992). *J. Virol.*, **66**, 5047.

3. Whitehouse, A., Carr, I. M., Griffiths, J. C., and Meredith, D. M. (1997). *J. Virol.*, **71**, 2550.
4. Whitehouse, A., Stevenson, A. J., Cooper, M., and Meredith, D. M. (1997). *J. Gen. Virol.*, **78**, 1411.
5. Wet, J. R., Wood, K. V., DeLuc, M., Helinski, D. R., and Subramani, S. (1987). *Mol. Cell. Biol.,* **7**, 725.
6. Andrews, N. C., and Faller, D. V. (1991). *Nucl. Acids Res.,* **19**, 2499.
7. Ebel, T. T., and Sippel, A. E. (1995). *Nucl. Acids Res.,* **23**, 2076.

7

Interaction of DNA virus proteins with host cytokines

ALSHAD S. LALANI, PIERS NASH, BRUCE T. SEET, JANINE ROBICHAUD and GRANT MCFADDEN

1. Introduction

It has become evident that large DNA viruses have adapted a variety of strategies to modulate the host immune response to virus infection. Recently, a variety of DNA virus-encoded gene products have been identified with activities targeted towards modulating the cytokine networks of the inflammatory response of their host. Members of the poxvirus family were the first DNA viruses shown to encode secreted versions of cellular cytokine receptors, and currently a dozen virus proteins that interact with cytokines have been identified in both the poxvirus and herpesvirus families (*Table 1*) (1, 2). Viral cytokine receptor mimics (also known as 'viroceptors') are soluble or membrane-bound proteins that may share significant homology to the ligand-binding domains of cellular cytokine receptors, and function by sequestering cytokine ligands from interacting with their cognate cellular receptors. Although many viral cytokine-binding proteins were previously identified by

Table 1. Examples of DNA virus proteins that interact with host cytokines

Virus	Open reading frame	Function
Myxoma virus	M-T1	secreted CC-chemokine binding protein
Myxoma virus	M-T7	secreted IFN-γ receptor homologue, and CC-, CXC-, C-chemokine binding protein
Myxoma virus	M-T2	secreted TNF receptor homologue
Vaccinia virus	B15R	secreted IL-1β receptor homologue
Vaccinia virus	B18R	secreted IFNα/β binding protein
Tanapox virus	n/d	secreted IL-2/IL-5/IFN-γ binding protein
Epstein–Barr virus	BARF-1	Secreted CSF binding protein
Cytomegalovirus	US28	7-TM CC-chemokine receptor (CCR) homologue
Human herpesvirus-8	ORF74	7-TM CXC-chemokine receptor (CXCR) homologue

their sequence relationship to cellular proteins in the database, more recently a number of virus-encoded soluble proteins have been identified which bind to and inhibit cytokines with high affinity, but do not share any sequence identity with known cellular cytokine or immunomodulatory receptors. This chapter describes methods used in the identification, and the biochemical and functional analysis of novel virus-encoded cytokine-binding proteins.

2. Identification of novel virus soluble cytokine-binding proteins

Viral cytokine-binding proteins are generally detected either as secreted glycoproteins or as transmembrane cell-surface proteins. The techniques described in this section are designed to allow the identification of novel cytokine-binding proteins primarily, though not exclusively, from the class of secreted glycoproteins. The production of concentrated supernatants containing secreted viral proteins is described in Section 2.1. This is followed by two crude screening methods for testing the presence of a viral cytokine-binding protein. While both the gel mobility cross-linking shift assay, and the ligand blot overlay (Sections 2.2 and 2.3) may yield a false-positive result, they are quick and easy first-stage screening methods. By comparison, Sections 2.4 and 2.5 detail methods that are less prone to artefactual binding, but are also more costly and difficult to set up.

2.1 Generation of secreted virus proteins from infected cells

Most viruses utilize the pre-existing cellular export pathways, and hence most secreted cytokine-binding proteins have a cleavable N-terminal signal sequence, and are processed through the endoplasmic reticulum and Golgi apparatus prior to secretion. Accordingly, the collection of supernatants from virus-infected cells is the first stage toward cytokine-binding protein identification and purification.

Protocol 1. Generation of serum-free supernatants containing secreted viral proteins

Equipment and reagents
- Serum-free tissue culture medium (appropriate to cells)
- Tissue culture medium supplemented with 10% serum
- PBS
- 3000 mol. wt. cut-off CENTRIPREP-3 spin protein concentrators (AMICON)

Method
1. Infect an 80% confluent monolayer of cells ($>10^7$ cells) with virus at a multiplicity of infection of 10 in a small volume (i.e. 5 ml in a 150 cm^2

flask) of serum-containing tissue culture medium. Note: it is also imperative to prepare mock-infected cultures in parallel for control studies.
2. Incubate the cultures for 1 h at 37°C with frequent rocking.
3. Wash the infected cells three times with generous volumes of PBS to remove any unabsorbed virus or residual serum.
4. Add a small volume (i.e. 10 ml for a 150 cm^2 flask) of serum-free medium to the infected cells, and incubate the cultures in a humidified 37°C, 5% CO_2 incubator for 2–24 h to allow for maximal early or late viral gene expression.
5. Collect the supernatants, and transfer to a centrifuge tube.
6. Centrifuge the supernatant at 4°C for 1 h at 25 000 g to remove virus particles and cellular debris.
7. Concentrate the supernatant tenfold using a CENTRIPREP-3 concentrator in a 4°C centrifuge.
8. Store the concentrated crude supernatant containing the complete spectrum of secreted viral proteins in aliquots at 4°C, or −80°C for long-term storage.

2.2 Use of chemical cross-linking to detect viral cytokine-binding proteins

The cross-linking shift assay to detect non-covalent complexes between a test cytokine and one or more viral proteins in a crude supernatant sample has been previously described (3). This method is a crude screen, and will yield a positive result in cases where the binding affinities may not be physiologically relevant, so long as some weak interaction does take place. While the potential for artefacts encountered with the use of cross-linkers cannot be overlooked, this method still provides a powerful tool for screening large numbers of samples without the need to optimize the conditions.

Protocol 2. Gel mobility cross-linking shift assay

Equipment and reagents
- 200 mM 1-ethyl-3-(3-dimethylaminopropyl)-carbodiimide hydrochloride (EDC) (Sigma) chemical cross-linker in 0.1 M potassium phosphate buffer pH 7.5[a]
- 10 mM sodium phosphate buffer pH 7.5
- 1 M Tris-HCl pH 7.5
- ^{125}I-labelled cytokine (can be purchased from commercial vendors, or cytokines can be radiolabelled by standard radio-iodination methods[b])
- 2 × SDS-gel loading buffer (100 mM Tris pH 6.8, 200 mM dithiothreitol, 4% w/v SDS, 0.2% w/v Bromophenol blue, 20% v/v glycerol)
- SDS-PAGE apparatus
- Vacuum gel dryer
- Autoradiography film and cassettes with intensifying screens

Protocol 2. *Continued*

Method

1. Incubate 10 μl of concentrated serum-free virus-infected or mock-infected cellular supernatants (*Protocol 1*) with 1–5 μl (0.1–2.0 nM) of radiolabelled cytokine for 2 h at room temperature with occasional mixing. If necessary, add 10 mM sodium phosphate buffer to make the final reaction volume equal to 15 μl.
2. Add 2 μl of 200 mM EDC chemical cross-linker, mix well, and incubate the samples for 15 min at room temperature.
3. Repeat Step 2.
4. Quench the reaction mixture by adding 2 μl of 1 M Tris-HCl pH 7.5.
5. Add 10 μl of 2 × SDS-gel loading buffer, and boil the samples for 3 min. Load the samples onto an SDS–12% polyacrylamide gel to resolve the protein complexes by electrophoresis.
6. Vacuum-dry, and expose the gel to autoradiography film at −80 °C overnight to detect the presence of novel shifted cross-linked protein complexes.[c]

[a] A variety of non-cleavable chemical cross-linkers with differing functional group specificity, spacer-arm length, and reactivities exist. Readers are encouraged to consult the Pierce catalogue for further details. The amine- and carboxyl-reactive cross-linker, EDC, is just one of many cross-linkers available from Pierce that works well for this application.
[b] Tyrosine-containing cytokines can be radiolabelled with Na-[^{125}I] by iodination reagents such as IODO-BEADS and IODO-GEN (Pierce).
[c] Novel cytokine–viral protein complexes appear as slower mobility-shifted species than the radiolabelled cytokine alone or when incubated with mock-infected supernatants after SDS-PAGE (*Figure 1*).

2.3 Ligand blot overlays for detection of viral cytokine-binding proteins

The ligand blot assay has been used successfully in many cases, not only identify novel protein–protein interactions, but also to characterize these interactions biochemically. As a versatile diagnostic tool for the biochemical characterization of protein–protein interactions, the ligand blot overlay assay has been adapted for use in mapping specific domains involved in protein interactions, characterizing receptor recognition of folded protein intermediates, and to examine the contribution of post-translational modifications in receptor–ligand binding. This method has also been successfully used to identify viral proteins that interact with cytokines (4, 5). The approach involves using a radiolabelled cytokine probe to bind to viral proteins immobilized on a supported nitrocellulose membrane. Although this technique offers a rapid method for identifying potentially novel viral cytokine-interacting proteins, one must consider that insufficient protein renaturation after immobilizing

7: Interaction of DNA virus proteins with host cytokines

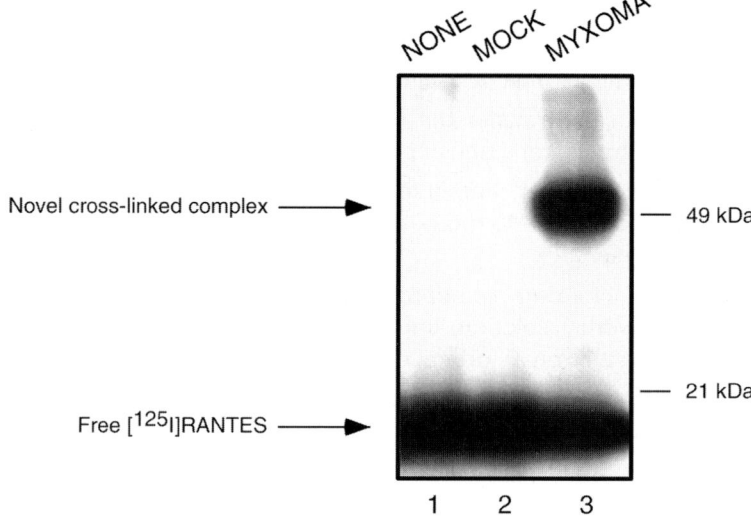

Figure 1. Detection of a secreted myxoma virus chemokine-binding protein by gel mobility cross-linking shift assay. [^{125}I]RANTES chemokine was incubated with PBS (lane 1), concentrated mock-infected (lane 2), or myxoma virus-infected cellular supernatants (lane 3) for 2 h followed by chemical cross-linking, and the resulting protein complexes were resolved by SDS-PAGE (12% acrylamide) and autoradiography. In the absence of any binding partners [^{125}I]RANTES migrates at its predicted 8–10 kDa molecular mass. However, when incubated with myxoma virus-secreted proteins, a shifted protein complex of 50 kDa (see arrow) is observed, suggesting a viral RANTES binding partner of an approximate molecular mass of 40 kDa.

viral proteins onto the fixed membrane support may limit the efficiency of protein–protein interactions. A more extensive description of this technique and its application can be found in another books in the Practical Approach series (6).

Protocol 3. Ligand blot assay

Equipment and reagents

- Blocking solution: 2% (w/v) non-fat skimmed milk powder in 10 mM Tris-HCl pH 7.4, 140 mM NaCl, 0.02% NaN$_3$
- Wash solution: 10 mM Tris-HCl pH 7.4, 140 mM NaCl, 0.05% Tween-20, 0.02% NaN$_3$
- Supported nitrocellulose membrane
- Overlay solution: 10 ng ml^{-1} [^{125}I]cytokine prepared in blocking solution
- SDS-PAGE apparatus
- Electrophoretic transfer apparatus
- Orbital shaker or rocking platform

Method

1. Resolve 10–20 µl of concentrated serum-free virus-infected and mock infected cellular supernatants (*Protocol 1*) under non-reducing

Protocol 3. *Continued*

   conditions by SDS-PAGE. In parallel, stain a duplicate gel either with Coomassie brilliant blue or by silver staining.

2. Electrophoretically transfer the gel onto a nitrocellulose membrane, using an electrophoretic transfer apparatus at 200 mA for 1–2 h.

3. Following the transfer, immerse the transferred membrane with an ample volume (i.e. 10–15 ml) of blocking solution for 4–24 h at 4°C on an orbital shaker.

4. Replace the blocking solution with 10 ml of the [^{125}I]cytokine-containing overlay solution, and incubate the membrane for 4 h at room temperature on an orbital shaker.

5. Decant the radioactive overlay solution, and wash the nitrocellulose membrane for 10 min with ample volumes (10–15 ml) of wash solution at room temperature on an orbital shaker.

6. Repeat Step 5 twice.

7. Air-dry the nitrocellulose membrane on filter paper for 10 min. Place the membrane between Saran-wrap™ or a plastic sheet.

8. Expose the nitrocellulose membrane to autoradiography film for a minimum of 24 h.[a]

9. Develop the film to visualize the presence of membrane-bound [^{125}I]cytokines to viral proteins.

[a] Exposure times may vary depending on the activity of the radiolabelled cytokine or efficiency of protein interactions

2.4 Immunoprecipitation of viral cytokine-binding proteins

There is a wide range of variations of the technique of co-immunoprecipitation to detect novel cytokine-binding proteins. Here we describe one such application.

Protocol 4. Co-immunoprecipitation of viral cytokine-binding proteins

Equipment and reagents

- TRAN[^{35}S]-LABEL (800–1200 Ci mmol^{-1}) (contains both [^{35}S]L-methionine and [^{35}S]L-cysteine) (ICN Pharmaceuticals)
- Cys and Met amino acid-free tissue culture medium
- [^{35}S]Metabolic labelling medium: 100 µCi ml^{-1} TRAN[^{35}S]-LABEL prepared in serum-free Cys/Met-free tissue culture medium
- Cytokine
- Anti-cytokine antibody
- PBS
- Protein A-Sepharose beads[a]
- Screw-cap microfuge tubes
- Nutator or rotator
- SDS-PAGE apparatus

7: Interaction of DNA virus proteins with host cytokines

Method

1. Generate ^{35}S-labelled concentrated virus-infected and mock-infected cellular supernatants as described in Steps 1–3 of *Protocol 1*.

2. Add 10 ml of [^{35}S]Metabolic labelling medium to a 150 cm^2 flask of infected cells, and incubate the cultures in a humidified 37°C, 5% CO_2 incubator for 2–24 h, to allow for maximal early or late viral gene expression. For metabolic labelling greater than 8 h, it is recommended to supplement the medium with 10% serum-free Cys/Met-containing tissue culture medium.

3. Repeat Steps 5–8 of *Protocol 1*.

4. Mix 500 μl of concentrated ^{35}S-labelled virus-infected or mock-infected supernatants with 50–100 ng of desired cytokine in a microcentrifuge tube, and incubate the samples for 2–24 h at 4°C with gentle agitation.

5. Add an appropriately diluted amount of the respective anti-cytokine antibody to the mixture, and incubate the samples for 1–2 h at 4°C on a nutator.[b]

6. Prepare a slurry of protein A beads by mixing a 1:1 (w/v) amount of protein A-Sepharose beads with 500 μl of PBS, and incubate the beads for 1 h at 4°C.

7. Centrifuge the beads for 30 s at 12 000 *g*, gently aspirate the supernatant, and re-suspend the beads with 500 μl of PBS.[c]

8. Add 50 μl of the protein A-bead 'slurry' to the [^{35}S]viral proteins–cytokine–antibody mixture, and incubate the samples for 1 h at 4°C on a nutator.

9. Centrifuge the samples at 12 000 *g* for 30 s at 4°C using a microcentrifuge, to pellet the protein A-beads. Carefully remove the supernatant using a disposable pipette tip. Add 1 ml of PBS, and re-suspend the beads by vortexing.[c]

10. Repeat Step 9 three times, and remove the last wash as completely as possible without disturbing the bead pellet.

11. Add 50 μl of 2 × SDS-gel loading buffer (*Protocol 2*) to the tube, and boil the samples for 5 min.

12. Centrifuge the samples at 12 000 *g* for 30 s, and carefully collect all the supernatant without disturbing the bead pellet, using a fine pipette tip.

13. Load the supernatant onto an SDS-12% polyacrylamide gel, and resolve the protein complexes by electrophoresis.

Protocol 4. *Continued*

14. Vacuum-dry the gel, and expose it to autoradiography film to detect the presence of novel ^{35}S-labelled viral proteins that specifically co-immunoprecipitate with the cytokine immuno complexes.

[a] The use of protein G may be substituted in cases where protein A may have a poor affinity for some species of antibodies.
[b] The amount of anti-cytokine antibody may need to be determined empirically. We suggest pilot experiments using three different dilutions of antibodies, such as 1:100, 1:500, and 1:1000 for initial assays.
[c] Buffers that contain detergents such as 1% NP-40, 1% Triton-X 100, or RIPA may be substituted for PBS, to increase the stringency of non-specific protein complex formations.

2.5 Use of plasmon resonance for analysis of interactions of cytokines with viral proteins

Surface plasmon resonance (SPR) spectrometry is a relatively new and very powerful method for studying receptor–ligand interactions. In this chapter (this section, and also Section 4.3), we will discuss SPR both as a method for identifying novel proteins, and as a means of characterizing in detail the nature of these interactions. An experimental SPR set-up is shown in *Figure 2*. The technique relies upon the generation of a non-radiative evanescent field produced when the angle of incident electromagnetic radiation (θ_i) is greater than the critical angle, and total reflection occurs. The evanescent field decays exponentially, with maximum strength at the surface, and is enhanced by a thin layer (typically 50 nm) of gold on a glass slide mounted onto the hemicylindrical lens. In this set-up, the energy of the incident light (I_0) is transferred to the valence electrons in the metal, a process referred to as excitation of the surface plasmons, from which the technique gets its name. By scanning the angle of incidence, a minimum in reflectivity is observed at the resonance angle ($\Delta\theta_{res}$). This angle depends on the wavelength of the incident light, the thickness of the gold layer, and the medium adjacent to the gold. In SPR, the refractive index of the evanescent field layer is monitored, while the other factors are constant. In this way, SPR can provide real-time data for the interaction between receptor and ligand, if one of these has been immobilized on the surface of the metal layer.

The advantages of using SPR are that no chemical labelling of a reacting substance is required, and that detailed mechanistic and kinetic data can be obtained in a real-time setting. There are, however, a number of drawbacks to this technique. For purposes of screening for novel cytokine-binding viral proteins, SPR requires that the molecule being investigated be immobilized onto a sensor chip. This makes the technique much more useful for detailed study of a known protein than it is for bulk screening for novel binding partners. Secondly, the detection limit for SPR is generally around 1–10 nM for a 20 kDa protein, and even higher for smaller molecules. As the technology

7: Interaction of DNA virus proteins with host cytokines

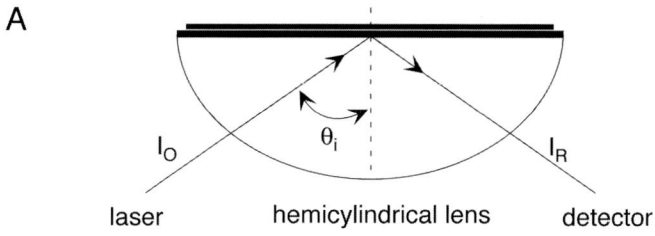

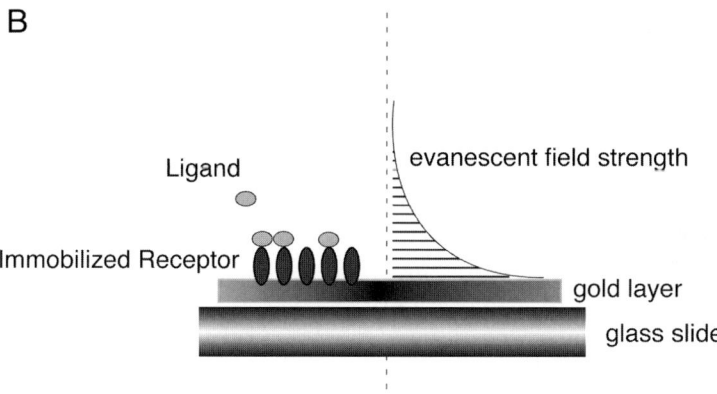

Figure 2. Schematic representation of surface plasmon resonance. (A) The hemicylindrical lens with a gold-coated glass slide, showing the resonance angle θ_i. (B) An expanded view of the surface of an experimental SPR set-up, indicating the immobilized receptor binding to a soluble ligand, with the decaying evanescent field strength away from the surface of the lens.

advances, this limit is being rapidly reduced, and modifications such as the use of liposomes have recently been reported to improve sensitivity to the picomolar range (7). One note of caution: although this method is convenient for acquiring detailed kinetic analysis of protein–protein interactions, it is also subject to a variety of experimental artefacts that may confound analysis and interpretation of the results (8).

Protocol 5. SPR for identifying novel viral cytokine-binding proteins[a]

Equipment and reagents
- Hepes-buffered saline (HBS) (10 mM Hepes, 150 mM NaCl, 3.4 mM EDTA, 0.05% surfactant P20, pH 7.4)
- Activating solution: 0.2 M 1-ethyl-3-(3 dimethylaminopropyl)carbodiimine, 0.05 M N-hydroxy-sulfosuccinimide

Protocol 5. *Continued*

- Concentrated serum-free virus-infected cellular supernatants (*Protocol 1*)
- 50 nM biotinylated cytokine dialysed in HBS
- Streptavidin solution: 0.2 mg ml^{-1} in 10 mM acetate buffer pH 4.2
- 1 M ethanolamine pH 8.5
- 10 mM HCl
- BIAcore biosensor (Pharmacia)
- Research grade CM5 sensor chips (Pharmacia)

Method

1. Set up BIAcore with two flow-cell CM5 sensor chips, running HBS at 5 μl min^{-1} at 25°C.
2. Activate each CM5 flow cell with 50 μl of activating solution.
3. Inject 50 μl streptavidin solution into the cell, and then block with 50 μl 1 M ethanolamine. Approximately 8000 resonance units (RU) of streptavidin should be fixed onto the surface by this procedure.
4. Inject 50 nM biotinylated cytokine onto one of the streptavidin-coated sensor chips for 1 min. The remaining chip is not coated with cytokine, and is used as a negative control.
5. Condition both flow cells with ten 2 min pulses of 10 mM HCl. Record the attachment of cytokine in RU.
6. Dilute the concentrated virus-infected cellular supernatants (*Protocol 1*) in HBS, and inject over the cytokine surface at a flow rate of 50 μl min^{-1}.[b]
7. The appearance of concentration-dependent, saturable binding is evidence of the presence of a viral cytokine-binding protein.

[a] Modified from ref. 9.
[b] The high flow rate will reduce the mass transport effect due to the potential high association rate for binding partners. Several different dilutions of the viral supernatant should be measured, to confirm that any interactions seen are concentration-dependent and saturable.

3. Synthesis and purification of cytokine-binding proteins

While several expression systems are available for the production of viral cytokine-binding proteins, a variety of factors may influence the choice. Presented here are two examples of expression systems routinely used and widely available for the production of heterologous proteins. Purification of expressed proteins, traditionally a time-consuming and costly endeavour has, in recent times, become easier and more cost-efficient, due to the introduction of commercially available standardized columns, as well as the emergence of epitope-tagging technology for use in affinity purification.

3.1 Vaccinia virus expression system

While large-scale production from baculovirus or similar systems has advantages in terms of the amount of material produced, it is possible that certain viral proteins may be more reliably and efficiently produced from a recombinant virus of the same family. A case in point is the myxoma virus protein, SERP-1, for which production in a mammalian system results in low yields. This same protein has not been amenable in either yeast-based expression (*Pichia pastoris*), or *E. coli* expression systems. Therefore, in certain cases, it may be desirable to produce the viral cytokine-binding protein from a closely related recombinant virus system. Vaccinia virus offers a number of advantages for this type of application, and has been extensively used for the production and purification of poxvirus cytokine-binding proteins (10, 11). Vaccinia virus expression has been the topic of an entire chapter in a previous Practical Approach series volume (12), and elsewhere (13). We will therefore restrict our focus to recent advances in this field, and a simple protocol for the production of a recombinant vaccinia virus expression system.

A synthetic early/late vaccinia virus promoter has recently been reported which is capable of driving synthesis of proteins under its control at levels many times that of previously reported vaccinia virus promoters (14). These vectors carry flanking sequences for directed insertion into the thymidine kinase locus of vaccinia virus.

Protocol 6. Recombinant vaccinia virus for the expression of viral cytokine-binding proteins

Equipment and reagents

- cDNA of viral cytokine-binding protein (with intact signal sequence if secretion is desired) cloned into pSC65 (with lacZ marker) (14)
- Lipofectin (Gibco BRL Life Technologies), or other appropriate transfection system
- BGMK or CV-1 (ATCC #CCL26) cells
- H143*tk-* cells (ATCC #CRL8303)
- Bromodeoxyuridine (BUdR) 5mg ml^{-1}; filter-sterilize, aliquot, and store at –20°C

- DMEM supplemented with 10% NBS
- Vaccinia virus strain WR, obtainable from ATCC as VR-119
- Swelling buffer (40 mM Tris-HCl, 8 mM MgCl$_2$, pH 8.5)
- 10 mg ml^{-1} 5-bromo-4-chloro-3-indolyl β-D-galactopyranoside (X-gal) in dimethylformamide
- 2% (w/v) low melting-point agarose; sterilize by autoclaving

A. *Infection/transfection*

1. Plate BGMK cells into six-well plates at 2.5 × 10^5 cells per well, and grow at 37°C to 70% confluence (approximately 24 h).

2. Infect BGMK cells with vaccinia virus WR at a multiplicity of infection (m.o.i.) of 0.05 per cell (2.5 × 10^4 PFU per well) in a volume of 300 μl per well. Adsorb virus by incubating for 2 h at 37°C, occasionally rocking to prevent the monolayer from drying out.

Protocol 6. *Continued*

3. Prepare the transfection mixture according to the manufacturer's directions.
4. Transfect the cells with 1–10 µg of recombinant plasmid DNA by following the transfection protocol.
5. Allow infection/transfection to proceed for another 4 h. Harvest the cells, and re-suspend in swelling buffer.
6. Liberate the virus by freeze-thawing three times, and sonicate in a cup sonicator for three 10 s bursts.

B. *Selecting recombinant vaccinia virus*
1. Plate H143 cells into six-well plates at 2.5×10^5 cells per well, and grow at 37 °C to 70% confluence (approximately 24 h) in medium containing 50 µg ml^{-1} BUdR.
2. Infect with recombinant virus at an m.o.i. of 0.001–0.1, and allow infection to proceed for 24 h in medium containing 50 µg ml^{-1} BUdR.
3. Harvest the virus into swelling buffer as described above, and liberate virus by freeze-thawing three times, and sonicate in a cup sonicator for three 10 s bursts.
4. Repeat steps 1–3 three times to allow drug selection to enhance the percentage of *tk-* virus. These are referred to as blind rounds of selection.
5. Following the three blind rounds of selection, adsorb semi-pure recombinant virus at an m.o.i. of 0.001–0.01 in 300 µl.
6. After 1 h, add back medium containing 50 µg ml^{-1} BUdR.
7. Allow the infection to proceed for 24–48 h.
8. Prepare a mixture of 1 part 2% liquid LMP agarose, and one part 2 × DMEM containing 10% FCS and 100 µg ml^{-1} X-gal at 37 °C.
9. Overlay 1 ml per well of this mixture on top of the infected monolayer. Allow the agarose to solidify, and the place the plates at 37 °C for 3–6 h to allow colour development of blue plaques.
10. Using a Pasteur pipette with bulb, carefully remove a single blue plaque. (The Pasteur pipette must form a seal with the bottom of the plate in order to aspirate the virus-infected cells). Deposit the plaque into 200 µl of swelling buffer. Harvest several dozen plaques in this way.
11. Use the recovered virus to repeat Steps 4–12 for at least three rounds of purification, in order to achieve a clonal line of virus, which should be 100% β-galactosidase positive by titering, and also produce the viral cytokine-binding protein of interest.

3.2 Synthesis of cytokine binding proteins by Baculovirus expression systems

Recombinant Baculoviruses have become useful tools for producing large quantities of protein for biochemical and functional characterization of viral cytokine protein. The system has traditionally exploited the prodigious ability of *Autographa californica* nuclear polyhedrosis virus (AcNPV) to produce the structural coat protein, polyhedrin. Transfer vectors that possess sequences that flank the polyhedrin gene in the wild-type genome are positioned 5' and 3' of the expression cassette on the transfer vectors. Co-transfection with AcNPV transfer vector results in a homologous recombination event within the flanking sequences. The foreign gene, now under the transcriptional control of the strong polyhedrin promoter, is carried by the recombinant virus. Upon infection of cultured insect cells (*Spodoptera frugiperda*), the protein of interest is expressed abundantly and can then be harvested, purified, and subjected to study.

Several secreted poxvirus cytokine-binding proteins have been successfully expressed using Baculovirus expression systems, and appear to retain their ability to interact with their respective ligand counterparts (15–17). There are now many choices in the types of transfer vectors available, types of promoters, varieties of insect cells, and types of media to support them. Kits are available that aid in hastening the process to production (*Table 2*). Since insect cells grow well in suspension and as monolayer cultures, scaling-up is possible by employing spinner flasks for the large-scale production of the protein of interest. Baculovirus transfer vectors that introduce various affinity tags that aid in the process of purification are now available; for example, cleavable histidine and glutathione S-transferase (GST) affinity tags are now commonly available among the various Baculovirus kits, and can allow a relatively simple batch or column purification procedure. Further benefits of Baculovirus expression systems are that proteins produced are similar in structure, biological activity, and immunological activity, since proteins produced are very often properly folded and possess the potential for post-translational modifications such as glycosylation, phosphorylation, acylation, or amidation.

3.3 Fc fusion protein production

Creating a fusion of a viral cytokine-binding protein and the Fc region of an immunoglobulin (Ig) to create an immunoadhesin (18) provides a powerful tool for elucidating the molecular interactions involved in cytokine binding, as well as for identifying novel binding targets. The Fc-fusion protein created in this section can be used in surface plasmon resonance (Section 4.3) and the scintillation proximity assay (Section 4.2), and is particularly useful for obtaining reliable solid-phase binding data. When studying a viral cytokine-

Table 2. Recombinant *Baculovirus* expression kits[a]

Company	Product name	Vectors	General features
Gibco BRL	Bac-to-Bac™	pFastBac™	Blue-white screening in *E. coli* for bacmid DNA
		pFastBac™ HT	Features pFastBac™ plus 6X histidine tag, affinity resin supplied, cleavage site
		pFastBac™ DUAL	Expresses two recombinant proteins simultaneously
Invitrogen	MaxBac 2.0	pBlueBac	Blue-white screening
		pBlueBac His2	Blue-white screening, 6X histidine tag
Clontech	BacPAK™	pBacPAK-His 1-3	6X histidine tag
Pharmingen	BaculoGold™ Starter Package	pVL1392/1393	Contains critical components for Baculovirus expression
	BaculoGold™ Transfection Kit	pVL1392/1393	Contains basic components for Baculovirus expression
	Baculovirus GST Expression Purification Kit	pAcHLT-A,B,C	Provides components to express and affinity- and purify Baculovirus-expressed proteins.
	Baculovirus 6X His Expression Purification Kit	pAcHLT-A,B,C	Provides components to express and affinity- and purify Baculovirus-expressed proteins

[a] This is not an exhaustive, nor a complete description of all available kits or vectors available.

binding protein, it is necessary to test purified protein in solid phase assays to obtain the second-order dissociation rate constant for the interaction of binding protein and ligand. To obtain accurate values from solid-phase binding, it is usually necessary to bind the protein being studied to the solid support. This is done through a linker that allows consistent protein orientation and removes it spatially from the immediate surface of the solid support, where micro-environmental factors may effect the experimental results. Creating an Fc-fusion protein allows both ease of binding to the solid matrix and improves the quality of the solid-phase binding data. Finally, an Fc-fusion protein is easily purified with protein A or protein G beads, allowing a single-step purification scheme. *Protocol 7* details the expression from Cos-7 cells, although any cell line compatible with the promoter on the vector can be used.

7: Interaction of DNA virus proteins with host cytokines

Protocol 7. Production and purification of an Fc-fusion protein

Equipment and reagents
- pIg-Tail vector (R & D Systems)
- Cos-7 cells (ATCC CRL 1651)
- AIMV medium (Gibco BRL Life Technologies)
- Lipofectin (Gibco BRL Life Technologies), or other appropriate transfection system
- 1 M Tris base
- HiTrap protein A column (Pharmacia)
- Wash buffer: 10 mM sodium phosphate, 150 mM NaCl, pH 7.2
- Elution buffer: 100 mM citric acid pH 3
- 1 M Tris-HCl pH 9.0

Method

1. Clone the viral cytokine-binding protein into the pIg-Tail vector, and produce sufficient quantities of high quality DNA.
2. Seed Cos-7 cells in a 175 cm^3 flask at 10^6 cells per flask in AIMV medium.
3. The following day, transfect the Cos-7 monolayer with 80 μg of recombinant DNA per flask according to the transfection protocol.
4. After the transfection recovery period, replace the AIMV medium.
5. Four days after transfection, collect the tissue culture medium.
6. Centrifuge the recovered medium for 5 min at 3500 *g*, and recover the cell-free supernatant, and adjust the supernatant to pH 8.2 by the addition of 1 min Tris base.
7. Equilibrate a HiTrap protein A column with wash buffer.
8. Apply the supernatant to the HiTrap protein A column.
9. Wash the column with 25 ml of wash buffer to remove non-specifically bound proteins.
10. Elute the Fc-fusion protein with elution buffer, and immediately neutralize the eluant with 0.1 volume of 1 M Tris-HCl pH 9.0.
11. Dialyse the eluant into 0.1 × wash buffer, or other appropriate buffer, for future use.
12. The Fc-fusion protein can be stored as aliquots at −80 °C, or lyophilized and stored.

3.4 Purification of secreted viral cytokine-binding proteins by fast protein liquid chromatography (FPLC)

The extensive biochemical characterization of a biologically active protein often depends on its isolation and purification away from crude extract. The use of column chromatography has proved to be a useful and efficient method to separate secreted viral proteins from supernatants obtained from cells infected with virus. The basic principle behind chromatography is the differential

Table 3. The physical properties exploited by various chromatographic methods

Method	Property exploited
Ion-exchange chromatography	Net charge/distribution of charged groups
Gel filtration/size exclusion chromatography	Size and shape
Hydrophobicity-interaction chromatography	Hydrophobicity
Affinity chromatography	Affinity for ligands, antibodies, etc.

Table 4. Examples of columns used in chromatographic purification of proteins[a]

Column	Application	Reference
Anion exchange (AEX)		
HiTrap Q (1 ml/5 ml)	Strong AEX. Useful for methods screening, sample capture step, or sample concentration	
Mono Q HR 5/5	Strong AEX. Useful for semi-preparative high resolution IEX chromatography	11, 20–22
Cation exchange (CEX)		
HiTrap SP (1 ml/5 ml)	Strong CEX. Useful for methods screening, sample capture step or sample concentration	30
Mono S HR 5/5	Strong CEX that adsorbs cationic proteins. Useful for semi-preparative high resolution IEX chromatography	
Size exclusion (SEC)		
Superdex 75 HR 10/30	High resolution SEC. Sample volume <250 µl	
Superdex 200 HR 10/30	Fractionates 3–70 kDa (Superdex 75), or 10–600 kDa (Superdex 200)	22
Hiload Superdex 75 16/60	High resolution SEC. Sample volume <5 ml	11
Hiload Superdex 200 16/60	See Superdex HR 10/30 for M_r range	20
Affinity		
HiTrap Protein A (1 ml or 5 ml) Protein A Sepharose 4 Fast Flow	Purification of different forms of IgG	28
HiTrap Protein G (1 ml or 5 ml)	Purification of different forms of IgG[b]	
HiTrap Heparin (1 ml or 5 ml)	Isolation of heparin-binding proteins[c]	30
HiTrap Blue (1 ml or 5 ml)	Isolation of albumin, interferon, coagulation factors, and nucleic acid-binding proteins	
HiTrap Chelating (1ml or 5ml)	Isolation of proteins and peptides containing exposed histidine residues	
Hydrophobicity interaction		
Alkyl Superose HR 5/5	Separates highly hydrophobic proteins	
Phenyl Sepharose HR 5/5	Separates weakly hydrophobic proteins	
HiTrap HIC Test Kit	Contains five different 1 ml HIC test columns for screening methods[d]	

[a] All columns described are available from Amersham Pharmacia.
[b] Differs from HiTrap Protein A in that Protein G binds polyclonal IgG from cow, rat, sheep, and horse with higher affinity.
[c] Although the HiTrap Heparin has not been used to purify viral cytokine-binding proteins, it has been used to isolate the poxvirus chemokine homologue, MC148 (30).
[d] Kit contains Phenyl Sepharose High Performance, Phenyl Sepharose 6 Fast Flow (low sub), Phenyl Sepharose 6 Fast Flow (high sub), Butyl Sepharose 4 Fast Flow, and Octyl Sepharose 4 Fast Flow.

7: Interaction of DNA virus proteins with host cytokines

partitioning between a stationary phase and a mobile phase. In general, by exploiting the known physical properties of the protein of interest, various strategies of separation can be employed (*Table 3*). Commercially available columns are available from several companies (*Table 4*). When isolating a viral protein suspected of binding a particular cytokine, the protein of interest can be monitored either by Western blot, provided a suitable antibody is available, or by its activity through a cross-linking assay (Section 2.2). Due to constraints of space, a thorough discussion on the topic of protein purification is impossible. The techniques described below are based on methods of protein purification which we have routinely used for the purification of secreted poxvirus proteins. These and other purification methods have been more extensively discussed in other books in this series (19).

3.4.1 Ion-exchange chromatography

Ion-exchange chromatography (IEC) is one of the most frequently used methods of purifying protein, because it offers the advantages of high resolving power, high capacity, reproducibility, and relatively low cost. Resolving proteins using IEC relies on the reversible adsorption of charged molecules to an insoluble matrix coated with groups of opposite charge. Elution of proteins involves increasing the ionic strength or changing the pH of the buffer. Resolving a particular protein from a crude extract depends on the unique properties of charge density, charge distribution, and surface hydrophobicity.

The choice of an ion-exchange column depends on the charge characteristics of the protein of interest. Secreted cytokine-binding proteins that have recently been identified in our laboratory have been negatively charged (20–22). This property was exploited for purification by employing a strong anion-exchange column. The anion-exchange column, MonoQ (Pharmacia), has routinely been used in our laboratory during the early stages of purification. The cheaper, disposable HiTrapQ (Pharmacia) shares many of the properties of the MonoQ column, and has been useful for testing purification schemes, as well as for quickly isolating and concentrating the protein of interest from a large batch of crude supernatant. Perhaps one of the greatest benefits of using this particular column is its economy, since it is less than a tenth of the price of its higher-priced counterpart, the MonoQ column. While there are several inexpensive alternative columns available within Pharmacia's HiTrap series, low-cost alternative columns with similar characteristics to the high-priced versions are also available from other companies.

Since the adsorption and desorption of protein depends on the ionic strength and the pH of the buffer, each condition should be optimized to allow the highest binding of the protein of interest. Further, the final charge of a fully processed glycoprotein can be affected by post-translational modifications, such as sialylation or sulfation. When using an anion-exchange column, initial buffer conditions should be adjusted to a pH that is at least 1 pH unit above the isoelectric point of the protein of interest, while the pH should be at least 1 pH unit below the isoelectric point when using a cation exchanger. The ionic strength of the starting buffer should be such that it allows the maximum binding of the protein of interest while allowing of the elution of contaminating proteins. This can usually be determined during trial runs by starting with little or no salt during sample application. Gradually increasing the salt concentrations and monitoring the concentrations at which the protein elutes will allow subsequent runs to begin with an ionic strength that allows maximum binding without elution. To increase resolution, sample sizes and flow rates can be reduced, while the gradient volume can be increased.

Protocol 8. Ion-exchange chromatography: anion-exchange chromatography

Equipment and reagents

- Dialysis tubing, 10 kDa cut-off (Spectrapore)
- Syringe, syringe filter (Millipore)
- 3000 mol. wt. cut off spin concentrator (Millipore)
- MonoQ HR 5/5 (Pharmacia)
- FPLC System (e.g. ÄKTApurifier, Pharmacia)[a]
- Buffer A: start buffer, 25 mM Tris pH 7.5[b,c]
- Buffer B: elution buffer, 25 mM Tris pH 7.5, 1 M NaCl[c]
- SDS-PAGE apparatus

Method

1. Generate viral proteins by infecting cells, and collecting supernatants according to Steps 1–6 in *Protocol 1*.

2. Sample preparation: dialyse the concentrated viral supernatants against buffer A[b] using dialysis tubing (follow the manufacturer's instructions). Filter the sample through a 0.45 μm syringe filter.

3. Equilibration: wash the column with 5 column volumes (5 ml) of buffer A. Wash with the counterion using 5 column volumes of buffer B. Equilibrate with 5 column volumes of buffer A.

4. Sample loading: load the sample using a 10 ml (50 ml for larger samples) superloop at a flow rate of 1 ml min^{-1}. Wash off unbound protein with 5 ml of buffer A. Collect the flow-through in case the protein of interest fails to bind to the column.

5. Sample elution: during initial runs, create a linear gradient over 20 column volumes by gradually increasing the salt concentration (buffer B) from 0–1 M NaCl at a flow rate of 0.5–1.0 ml min^{-1}. Later runs can use step gradients[d] to economize on time and reagents. Collect 0.5–1.0 ml fractions during the length of the elution gradient.

6. Re-equilibration: wash the column with 5 column volumes of buffer B to elute bound proteins. Equilibrate by washing the column with 5 column volumes of buffer A.

7. Detection of protein: analyse the fractions by applying 10–20 µl of each fraction[e] to SDS-PAGE, and visualizing by Coomassie Blue staining and/or Western blotting. If an antibody is unavailable, cytokine-binding proteins can be detected by repeating *Protocol 2* (cross-linking assay) by testing each fraction for activity. Pool fractions that contain the cytokine-binding protein. Apply the sample to further purification steps (*Protocol 9*, for instance) if sufficient purity is not attained.

[a] The system includes superloops, pumps, UV monitor, and fractionator.
[b] Start buffer conditions are typically of low ionic strength. Samples can be dialysed with higher salt concentrations to eliminate the binding of weakly binding anionic proteins, provided the concentrations do not exceed the conditions needed for maximum binding of the protein of interest.
[c] Filter (0.22–0.45 µm filter) and degas all solutions prior to use on column.
[d] Step gradients can be created to maximize the volume over which the elution gradient occurs and minimize the pre- and post-elution gradient volumes.
[e] To detect protein, fractions may need concentrating before application to SDS-PAGE, depending on the concentration of the protein of interest within each fraction.

3.4.2 Size-exclusion chromatography

Size-exclusion chromatography (SEC) is a method that separates proteins or other molecules according to their size. The principle of SEC is based on the partitioning of molecules between a liquid phase surrounding a solid stationary phase composed of a porous matrix. While large molecules are excluded from the pores of the matrix and elute earliest from the column, smaller molecules enter the pores of the stationary phase, and their migration through the column is thereby impeded. Molecules, therefore, elute from an SEC column in descending order of size.

Apart from having applications in the processes of purification, this method has been useful for estimating the molecular weight of a protein, separating monomers from dimers and oligomers, and as a desalting step, and it has also been extended to determining equilibrium constants. In our laboratory, SEC was used to identify monomer and dimer forms of the myxoma virus secreted TNF receptor homologue, M-T2 (20). SEC has, however, primarily been used during the purification of the gamma-interferon receptor homologue, M-T7, as a second step in a two-step purification process applied following IEC (22).

Several factors must be considered when using SEC as a method of purification. First, care must be taken when selecting a column, since such factors as the length, width, type of matrix, and stability of the matrix under various conditions will affect the efficiency of purification of the protein of interest. Fortunately, a wide variety of columns are commercially available, each with defined characteristics that must be matched to the purpose at hand. The length of the column is a significant factor, since greater resolution is achieved from longer columns. The volume of the column is also a factor when considering the sample size involved: a wider column can accommodate a larger sample. The type of matrix will also affect resolution and selectivity, due to variations in pore size, composition, and stability. The fractionation range must also take into account the possible presence of homo- or heterodimers or higher molecular weight oligomers that may be formed from the protein of interest. For instance, a column will exclude a protein oligomer if the column's fractionation range is below that of the size of the complex formed by the protein of interest, and the oligomer will elute just after the void volume. The pH and ionic strength of the sample buffer and eluant are not critical factors in determining resolving power, but should be chosen to preserve the size and stability of the protein of interest. The following protocol describes the method typically used after IEC, in what could be considered a polishing step.

Protocol 9. Size-exclusion chromatography

Equipment and reagents
- Syringe, 0.45 μm syringe filter (Millipore)
- 3000 mol. wt. spin concentrator (Millipore)
- HiLoad™ 16/60 Superdex 200 (Pharmacia)
- FPLC System (ÄKTApurifier - Pharmacia)
- PBS[a]

Method

1. Sample preparation: concentrate the sample to <5 ml (or 0.5–5% of the bed volume of the column) for application to Superdex 200 HiLoad 16/60. Filter the sample through a 0.45 μm syringe filter.

2. Equilibration: the column should be stored in 20% ethanol. Wash the column with 1.5 column volumes of PBS.

3. Sample loading: load the sample at a flow rate of 0.5 ml min^{-1}. Very large proteins or protein aggregates may be immediately excluded during loading or shortly after the loading period. It is advisable to collect the flow-through if the native size of the protein of interest is unknown.

4. Isocratic elution: elute the protein with 1.5 column volumes of PBS at a flow rate of 0.5 ml min^{-1}. Collect 1.0 ml fractions, commencing after the void volume elutes.

5. Detection of protein: detect protein as described in Step 7 of *Protocol 8*. Pool fractions containing the protein of interest. Concentrate fractions if necessary.

[a] Filter (0.22–0.45 μm filter) and degas all solutions prior to use on column.

3.5 Purification by affinity chromatography

As one final approach to purifying a protein, we should give some consideration to affinity chromatography. Affinity chromatography relies on a highly specific interaction between a protein and a ligand. Common examples of this include purification through receptor–ligand interaction, or the use of a monoclonal antibody to purify its target protein. Tagged fusion proteins are often purified in this manner, as is detailed for the Fc-fusion protein detailed in *Protocol 7*. Extensive descriptions are available from the volume in this series dealing specifically with affinity chromatography.

4. Analysis of cytokine-binding partners

Once a viral cytokine-binding protein has been identified, it is of fundamental importance to be able to characterize the nature and biological relevance of the interaction between the viral protein and its cognate ligand. A number of factors are key to dissecting the properties of the binding interaction. The first consideration is whether or not the dissociation constant is sufficient to predict that the viral protein will effectively compete for binding with cellular receptors. Determination of the K_d value is therefore a useful parameter to establish a protein as a bona fide binding partner. There are a number of ways of determining the K_d for an interaction of cytokine and viral protein, and a selection of these are detailed in Sections 4.1, 4.2, and 4.3. It is important to note that accurate values require the use of purified protein, or better still, the use of a purified Fc-fusion protein.

A second line of investigation seeks to prove that the viral cytokine-binding protein is capable of ablating the appropriate cellular response to the host cytokine. Three examples of such methods are detailed in this section. Section 4.4 outlines the procedure used to show that the viral cytokine-binding protein is indeed capable of competitively inhibiting binding of a cytokine to its cognate cellular receptor. The final three sections (4.5, 4.6, and 4.7) describe examples of biological inhibition assays that can be modified for testing specific cytokines. These assays rely on the capacity of cytokines to induce specific cellular responses such as proliferation or cytolysis, which can in turn be inhibited by the viral cytokine-binding protein.

4.1 Solid phase binding

The classical method for quantitative examination of a binding interaction is the solid-phase binding assay. The method detailed in *Protocol 10* is general in nature, and can be adapted to any cytokine-binding protein. The simplicity of this assay, which requires equipment available to most investigators, and its ability to generate data for Scatchard analysis, makes it an ideal starting point for the quantitative analysis of a novel viral cytokine-binding protein. *Figure 3* shows an example of a solid-phase binding experiment conducted for the interaction of interferon γ with the myxoma virus soluble interferon receptor, T7.

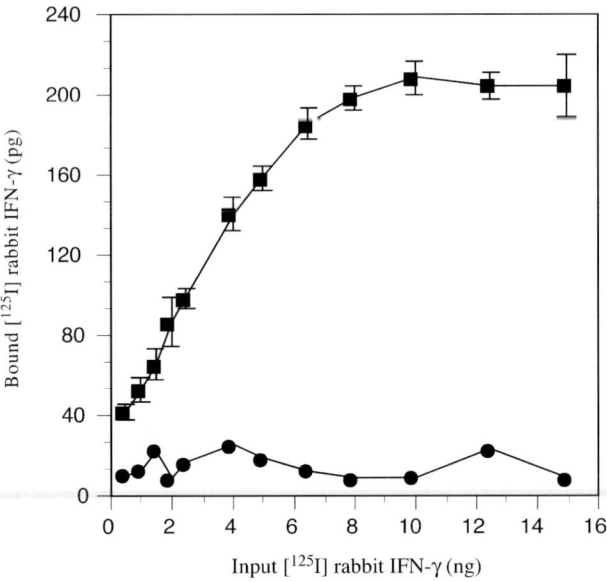

Figure 3. Solid-phase binding analysis of [^{125}I]IFN-γ to the myxoma virus soluble IFN-γ receptor homologue, M-T7. Increasing amounts of [^{125}I] rabbit IFN-γ were incubated with 50 ng M-T7 protein (squares) or BSA (circles) immobilized onto 96-well immunoplates, in the presence or absence of a hundredfold excess of unlabelled rabbit IFN-γ competitor. Specific saturable binding of [^{125}I] rabbit IFN-γ to M-T7 was observed with a K_d of approximately 1.2 nM by Scatchard analysis.

Protocol 10. Binding analysis using a solid phase binding assay

Equipment and reagents

- Maxi-Sorp immuno-microtiter plates (96 flat-bottom wells) (Nunc)
- Purified viral cytokine-binding protein, diluted to 500 ng ml^{-1} in PBS
- Blocking buffer: 5% skimmed milk powder, 0.1% Tween 20 in PBS[a]
- PBS
- ^{125}I-labelled cytokine (2200 Ci mmol^{-1}) (Dupont NEN)
- Unlabelled cytokine
- Gamma counter

Method

1. Dilute purified viral cytokine-binding protein (see Sections 3.4 and 3.5 for purification strategies) to 500 ng ml^{-1} in PBS.

2. Add 100 µl of purified viral cytokine-binding protein to each well of a 96-well Maxi-Sorp immunoplate. Wrap the plate in Parafilm, and incubate at 4°C overnight to allow the soluble viral proteins to coat the bottom of the wells. Prepare at least 48 wells per experiment, to allow

7: Interaction of DNA virus proteins with host cytokines

for triplicate samples. As a control, also prepare parallel wells coated with an irrelevant protein (i.e. BSA).

3. Gently aspirate the wells, add 200 µl of blocking buffer to each well, and incubate the plate on a rocking platform for 1 h.

4. Prepare increasing concentrations (i.e. 0.5–100 ng ml^{-1}) of [^{125}I]cytokine in blocking buffer. In separate tubes, also prepare increasing dilutions of [^{125}I]cytokines containing a hundredfold excess of unlabelled (cold competitor) cytokine in blocking buffer.

5. Gently aspirate the wells, and add 100 µl of increasing concentrations of [^{125}I]cytokine to consecutive wells. In parallel wells, add 100 µl of [^{125}I]cytokine containing a hundredfold excess of unlabelled cytokine. Incubate the plate for 2 h on a rocking platform.

6. Gently aspirate the wells to remove any unbound cytokine, and wash each well three to five times with 200 µl of PBS.

7. Remove individual wells from the immunoplate with a sharp pair of scissors or with a microplate cutter, and measure the bound counts of each well using a gamma counter.

8. Quantify the specific binding of [^{125}I]cytokine to the viral protein by subtracting the radioactive counts bound in the presence of a hundredfold excess of unlabelled cytokine competitor from the total binding observed for each dilution point.[b]

9. Saturable binding of the viral protein to the cytokine can be determined by plotting the measured concentration of bound cytokine against the amount of input cytokine added per dilution point (*Figure 2*). Commercial software packages such as LIGAND (Cambridge Biosoft) are ideal for performing Scatchard-type analysis.

[a] Skimmed milk may be replaced by 1% bovine serum albumin or 1% ovalbumin as an alternative blocking reagent.
[b] Non-specific binding should routinely represent less than 10% of total binding.

4.2 Scintillation proximity assay

The recently developed, and now commercially available, technique known as a scintillation proximity assay (SPA), is a powerful addition to the area of quantitative binding methods. SPA relies on the binding properties of the receptor to bring its radiolabelled ligand into the proximity of a microsphere embedded with scintillant, and thus set up a cascade of emitted light (*Figure 4*). This method is both extremely sensitive, and relatively easy to perform. *Protocol 11* outlines a general method for the analysis of any viral cytokine-binding protein/cytokine pair, with the only requirements being Fc-tagged viral protein (Section 3.3) and radioiodinated cytokine.

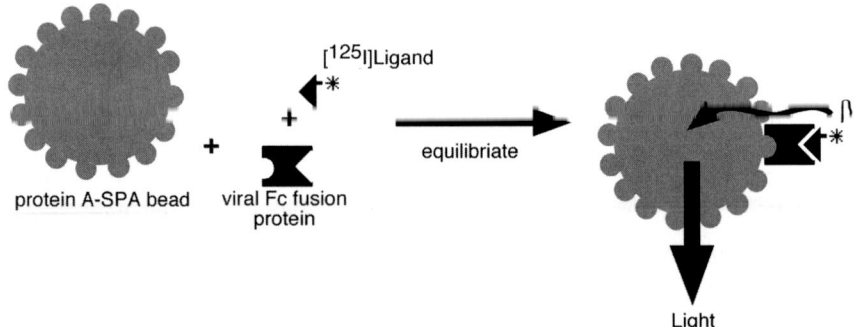

Figure 4. Schematic representation of the Scintillation Proximity Assay (SPA) format to measure binding of [^{125}I]cytokines to viral-Fc fusion proteins. Radioiodinated cytokine ligands which bind to virus cytokine-binding–Fc fusion proteins are brought into close proximity by the protein A-microfluorosphere beads, and stimulate the bead-embedded scintillant to emit light.

Protocol 11. Binding analysis using a scintillation proximity assay (SPA)

Equipment and reagents

- Protein A-fluoromicrosphere scintillant embedded beads (SPA beads, Amersham)
- Binding buffer: PBS, 25 mM Hepes pH 7.4, 1% BSA, 0.05% NaN$_3$
- [^{125}I]cytokine (2200 Ci mmol^{-1}) (can be purchased commercially)
- β-scintillation counter
- Unlabelled cytokine (prepare as a hundredfold excess per dilution point in binding buffer)
- Purified viral cytokine-binding–Fc fusion protein (prepare as 100 ng ml^{-1} in binding buffer)

Method

1. Wash 5 mg of protein A-SPA beads, and resuspend in 1 ml of binding buffer.
2. Aliquot 50 μl of increasing concentrations (i.e. 0.1–2 nM) of [^{125}I]cytokine to two sets of ten microcentrifuge tubes.
3. To the first set of tubes, add 50 μl of binding buffer alone per tube.
4. To the second set of tubes, add 50 μl of a hundredfold excess of unlabelled cytokine per tube. These will be used to determine non-specific binding.
5. To a third set of microcentrifuge tubes, add 150 μl of binding buffer alone. These will be used to determine non-specific binding of the viral-Fc fusion protein to the beads.
6. Add 50 μl of purified viral protein-Fc fusion protein to each of three

7: Interaction of DNA virus proteins with host cytokines

sets of tubes above, and incubate for 2 h at room temperature with gentle agitation.

7. Add 50 μl of protein A-SPA beads to each of the three sets of tubes above. Incubate the mixtures for 2 h at room temperature with gentle agitation to immobilize the viral-Fc protein onto the beads.

8. Carefully transfer the contents of the tubes to scintillation vials, and measure the bead-bound radioactivity in a β-scintillation counter.

9. Determine the specific binding of the [^{125}I]cytokine to the viral fusion protein by subtracting the non-specific binding observed in the presence of excess unlabelled cytokine, or by subtracting the viral-Fc fusion protein binding in the absence of cytokine.

4.3 The use of surface plasmon resonance (SPR) for detailed kinetic studies

Protocol 12. Use of SPR for the analysis of Fc-linked viral cytokine receptors[a]

Equipment and reagents

- BIAcore biosensor
- Fc-tagged viral cytokine-binding protein (Section 3.3)
- Solution of purified cytokine, or concentrated conditioned medium from cytokine-secreting cells
- Research grade CM5 sensor chip
- Hepes-buffered saline (HBS): 10 mM Hepes, 150 mM NaCl, 3.4 mM EDTA, 0.05% surfactant P20, pH 7.4
- 100 mM phosphoric acid

Method

1. Couple goat anti-human IgG, γ chain-specific (GHFC), to the CM5 sensor chip by covalent amine linkage, according to the manufacturer's directions (see also *Protocol 5*).

2. Inject Fc-tagged viral cytokine-binding protein (30–50 μg ml^{-1} in HBS) over the immobilized GHFC at 3–5 μl min^{-1}. A negative control is generated by passing conditioned medium lacking the Fc-tagged viral cytokine-binding protein over the GHFC-CM5 chip.

3. Inject purified cytokine at various concentrations (typically 0.05–20 μg ml^{-1}) over both surfaces at 5 μl min^{-1}. Data from this are analysed as described below.

4. Regenerate the chip with one 10 μl pulse of 100 mM phosphoric acid at 10 μl min^{-1}.

[a] Modified from ref. 23.

4.3.1 Analysis of SPR spectrometry data

Pharmacia supplies kinetic evaluation software (Bia-evaluation software) with the BIAcore instrument that can be used to determine rate constants by linear transformation of the primary data and non-linear fitting of the sensorgrams. The analysis of the data is complicated by several factors (8). The most common error is in the assignment of the number of binding sites. So long as a single binding site is assumed to be present (i.e. A + B = AB), it is possible to analyse binding data under pseudo-first order assumptions. This is clearly not the case for the interaction between a monoclonal antibody and an immobilized antigen, but many cytokine receptors do interact in a simple 1:1 stoichiometry, and can therefore be described by pseudo-first order kinetics (24). It is also important to note that although the ligand and receptor may interact with a 1:1 stoichiometry, certain SPR artefacts may cause the system to fail to be adequately described by a single exponential dependence of response upon time (8). These factors should be considered before assuming a more complex model.

The association stage of the sensorgram can be used to obtain the second order association rate constant by an exponential fit to the integrated rate equation

$$R_t = R_{eq}(1 - \exp(-k_s(t - t_0))) \qquad [1]$$

Where R_t is the amount of ligand remaining bound (in RU) at time t, R_{eq} is the amount of ligand bound in RU at equilibrium, t_0 is the time that injection started, and $k_s = k_{on}C + k_{off}$, where C is the concentration of protein ligand injected over the surface of the sensor chip. The association rate constant, k_{on}, is then determined from the slope of a plot of k_s versus C. The dissociation rate constant is obtained by an exponential fit of desorption data in terms of the integrated rate equation,

$$R_t = R_0\exp(-k_{off}(t - t_0)) \qquad [2]$$

In this way, the basic kinetic constants of on-rate and off-rate can be readily determined in a short period of time with minimal use of material. For more detailed treatment of kinetic and thermodynamic analysis of SPR data, see refs 25 and 26.

4.4 Inhibition of cell-surface binding

From a biological standpoint, a viral cytokine-binding protein is most commonly aimed at blocking the interaction between a host cytokine and its cognate cellular receptor. Using radiolabelled cytokine, the ability of a novel viral cytokine-binding protein to inhibit competitively the binding of cytokine to the surface of target cells can be tested *in vitro* using *Protocol 13*. A sample data set showing inhibition of the binding of the chemokines MCP-1 and MIP-

7: Interaction of DNA virus proteins with host cytokines

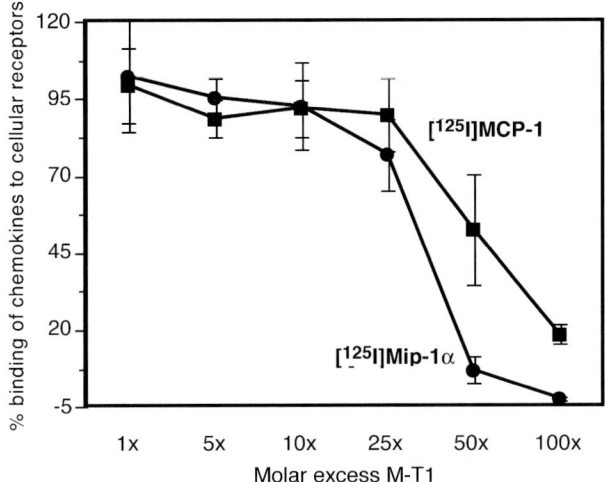

Figure 5. Competitive inhibition of CC-chemokine binding to monocyte receptors by a myxoma virus chemokine-binding protein. Pre-incubation of molar excess amounts of the myxoma virus glycoprotein, M-T1, with [^{125}I]MCP-1 (squares) or [^{125}I]MIP-1α (circles) effectively inhibits CC-chemokine binding to cell surface chemokine receptors on primary monocytes. Non-specific binding was determined in the presence of a hundredfold excess of unlabelled chemokine competitor, and subtracted from the total binding measured. The data are represented as percentage binding (± SD) averaged from triplicate samples.

1α to primary monocytes by the myxoma virus protein T1 is shown in *Figure 5*. Analysis of this data yields a value for the biological inhibitory activity of the viral cytokine-binding protein expressed as an IC_{50}.

Protocol 13. Competitive inhibition of cytokine binding to cell surface receptors

Equipment and reagents

- Binding medium: complete tissue culture medium (i.e. RPMI-1640) supplemented with 1 mg ml^{-1} BSA
- Target cells expressing cognate cytokine receptor
- [^{125}I]cytokine (50 000 c.p.m. per reaction)
- Unlabelled cytokine
- Purified viral cytokine-binding protein
- 10% sucrose prepared in PBS
- Gamma counter

Method

1. Wash the target cells twice with ice-cold binding medium, and re-suspend the cells at 10^7 cells ml^{-1} on ice.
2. To determine the total amount of radiolabelled ligand capable of binding cells in the absence of competitor, prepare a sample reaction

Protocol 13. *Continued*

containing 50 000 c.p.m. of [^{125}I] cytokine per tube, diluted to a final volume of 100 μl. Each sample reaction should be performed in triplicate, and prepared on ice.

3. To assay the viral cytokine-binding protein, prepare a range of concentrations from 0.1–100-fold excess of viral cytokine-binding protein, based on the specific activity of the radiolabelled cytokine, added to separate tubes of [^{125}I]cytokine, each to a final volume of 100 μl.

4. To assay cold competition, prepare a sample reaction containing a hundredfold excess of unlabelled cytokine to a final volume of 100 μl. This set will be used to determine the level of non-specific binding of the cytokine to cell-surface receptors.

5. Incubate all sample reactions at 37 °C for 30 min.

6. Add 200 μl of the cell suspension from Step 1 to each of the sample reactions. Incubate the cells at 4 °C for 1 h with constant gentle agitation.

7. Spin the tubes at 15 000 *g* for 10 min at 4 °C, using a microcentrifuge. Aspirate the supernatant, and re-suspend the cell pellet in 800 μl of 10% sucrose in PBS.

8. Centrifuge the samples at 15 000 *g* for 10 min at 4 °C, using a microcentrifuge to pellet the cells through the sucrose cushion.

9. Aspirate the supernatant, and carefully cut the tip of the tube containing the cell pellet with a razor blade or a sharp pair of needle-nosed pliers. Place the tip in a vial, and measure the radioactivity bound to the cell pellet in a gamma counter.

10. Determine the percentage inhibition of binding of the cytokine to its cognate receptor by the viral protein:

% inhibition of binding = [c.p.m. bound in the presence of viral competitor − c.p.m. of the cold competitor (non-specific binding)] / [total c.p.m. bound in the absence of competitor − c.p.m. of the cold competitor (non-specific binding)] × 100.

4.5 Inhibition of cytokine-induced cytolysis

Biological activity assays are of critical importance for the testing of viral cytokine-binding proteins. These are often exquisitely sensitive assays that can readily detect low levels of a viral cytokine agonist. Furthermore, the demonstration of biological activity is critical for determining the function of a binding protein. The mere demonstration of physical association by methods such as cross-linking or ligand blots is not necessarily evidence of a biologic-

7: Interaction of DNA virus proteins with host cytokines

ally relevant association in the absence of some form of biological inhibition data. *Protocol 14* describes a biological assay for virus protein protection against TNF-induced cytolysis. TNFα from different species can be used in this assay, but viral proteins are often highly species-specific. For instance, myxoma virus M-T2 protein is a potent inhibitor of rabbit TNFα, but has virtually no activity against human or murine TNFα (10).

Protocol 14. Viral protein protection against TNF-induced cytolysis

Equipment and reagents
- 96-well tissue culture plates
- L929-8 cells
- DMEM supplemented with 10% FCS
- Actinomycin D
- TNFα
- PBS
- 50% ethanol, 0.05M NaH$_2$PO$_4$
- Neutral red vital stain

Method
1. L929-8 cells are plated out in 96-well dishes at a density of 5×10^4 cells per well, in the presence of 2 μg ml^{-1} actinomycin D.
2. Incubate the cells for 2 h at 37°C.
3. Simultaneously add TNF and varying molar ratios of purified viral TNF-binding protein, or concentrated viral supernatant, for 18 h at 37°C. TNF should be used at a concentration for which L929-8 cell killing is >95%. Antibody against TNF can be used as a positive control for inhibition of TNF activity.
4. Determine Neutral red vital dye uptake (27). Add 100 μl of 0.05% neutral red in PBS, and incubate at 37°C for 2 h. Add 100 μl of 50% ethanol, 0.05 M NaH$_2$PO$_4$, and determine absorbance at 570 nm with a reference wavelength of 655 nm.
5. Maximum cytolysis is determined for a sample of TNF in the absence of any inhibitor, while maximum viability is determined by L929-8 samples incubated in medium alone. Percentage viability for experimental samples can then be determined according to the formula:

$$\{(OD_{experimental} - OD_{maximum\ killing})/(OD_{maximum\ viability} - OD_{maximum\ killing})\} \times 100\%$$

4.6 Growth inhibition/proliferation assay

Many cytokines are capable of inducing or blocking the proliferation of certain target cells. A viral cytokine receptor that is capable of binding to and inhibiting a host cytokine can be monitored by its ability to prevent enhanced cell growth, or to block growth inhibition. This assay examines the ability of a soluble viral cytokine receptor to block cytokine-induced modulation of proliferation. Cell proliferation can be easily assayed by the uptake of tritiated thymidine, a measure of DNA synthesis.

Protocol 15. Proliferation inhibition assay

Equipment and reagents
- [^{3}H]thymidine
- Appropriate cell line responsive to cytokine (i.e. 3T3 cells for growth factors such as FGF or PDGF, Mv1Lu/CCL-64 cells for TGFβ, SIRC or RL-5 cells for TNFαII
- 96-well tissue culture plates
- Purified cytokine
- Cell harvester (e.g. Packard Filtermate 196 Harvester)

Method
1. Plate cells at a density of 1–2 × 10^4 cells per well, and incubate for 4–24 h at 37°C.
2. Simultaneously add cytokine and purified viral cytokine-binding protein or concentrated viral supernatants at various molar ratios.
3. Add [^{3}H]thymidine (1 μCi per well), and incubate cells for 24–48 h (time will depend on the proliferation activity of the cytokine) at 37°C.
4. Harvest the cells using a cell harvester, and collect incorporated counts onto filter paper using methanol/trichloroacetic acid, or other appropriate precipitation method compatible with the cell harvester being used.
5. Determine the incorporated counts using a scintillation counter. Controls should include wells incubated in medium alone, with cytokine but no inhibitor, and with cytokine in the presence of neutralizing antibody.

4.7 Measuring the effect of viral chemokine-binding proteins on chemokine-induced calcium flux

Chemokines are chemotactic cytokines that play an important role during the inflammatory response to viral infection by regulating the trafficking of leukocytes to sites of infection. Recently, several viral proteins have been identified that disrupt chemokine function; these include soluble binding proteins (21–23, 28), membrane-bound chemokine receptor homologues (29), and chemokine homologues (30–32). One of the earliest consequences of chemokine signalling through its receptor is the increase in intracellular calcium from cytosolic stores. By measuring chemokine-induced calcium fluxes in susceptible cells, viral proteins suspected of disrupting normal chemokine signal transduction can be analysed. The use of fluorescent calcium indicators has become immensely popular in examining this signal transduction event. Below is described a method by which the myxoma virus

7: Interaction of DNA virus proteins with host cytokines

soluble chemokine-binding protein, M-T1, was analysed for its ability to prevent calcium release, by intercepting the chemokine before it has the opportunity to signal through its receptor (33). Readers interested in an extensive description of additional methods and the theory of measuring cellular calcium are encouraged to consult the volume in this series that deals specifically with these issues (34).

Protocol 16. Inhibition of chemokine-induced calcium flux

Equipment and reagents
- Chemokine (MCP-1, R & D Systems)
- Human monocyte THP-1 (ATCC cell lines and hybridomas)
- Fura-2 AM (Molecular Probes) in DMSO
- HBSS (Gibco BRL)
- BSA (Sigma)
- Calibrated RatioMaster Fluorescence Spectrometer (Photon Technology International)

Method

1. Suspend cells in 2 ml serum-free medium with 5 μM Fura-2 AM, for 30 min in the dark at 37°C. Wash 3 times with HBS containing 1 mg ml^{-1} BSA.
2. Re-suspend cells at a final concentration of 3×10^5 cells ml^{-1}.
3. Transfer 6×10^5 cells to a continuously stirred quartz cuvette in the spectrophotometer.
4. Monitor emission of Fura-2 fluorescence once every second at 510 nm at a dual excitation of 340 nm and 380 nm. Calculate calcium concentration according to Grynkiewicz *et al.* (35).
5. Stimulate cells with chemokines or appropriate mock protein in the absence or presence of increasing viral chemokine-binding protein.

References

1. Spriggs, M. K. (1996). *Annu. Rev. Immunol.*, **14**, 101.
2. Barry, M., and McFadden, G. (1998). *Parasitology*, **115**, S89.
3. Upton, C., Mossman, K., and McFadden, G. (1992). *Science*, **258**, 1369.
4. Spriggs, M. K., Hruby, D. E., Maliszewski, C. R., Pickup, D. J., Sims, J. E., Buller, R. M., and VanSlyke, J. (1992). *Cell*, **71**, 145.
5. Symons, J. A., Alcamí, A., and Smith, G. L. (1995). *Cell*, **81**, 551.
6. Soutar, A. K., and Wade, D. P. (1989). In *Protein function: a practical approach* (ed. T. E. Creighton). p. 55. IRL Press, Oxford.
7. Wink, T., van Zuilen, S. J., Bult, A., and van Bennekom, W. P. (1998). *Anal. Chem.*, **70**, 827.
8. O'Shannessy, D. J., and Winzor, D. J. (1996). *Anal. Biochem.*, **236**, 275.
9. Sadir, R., Forest, E., and Lortat-Jacob, H. (1998). *J. Biol. Chem.*, **273**, 10919.
10. Schreiber, M., and McFadden, G. (1994). *Virology*, **204**, 692.

11. Nash, P., Whitty, A., Handwerker, J., Macen, J., and McFadden, G. (1998). *J. Biol. Chem.*, **273**, 20982.
12. Smith, G. L., (1993). In *Molecular Virology* (ed. A. J. Davison, and R. M. Elliott). p. 257. Oxford University Press, New York.
13. Moss, B., and Earl, P. L. (1992). In *Short Protocols in Molecular Biology* (2nd edn) (ed. F. M. Ausubel, J. R. Brent, R. E. Kingston, D. D. Moore, J. G. Seidman, J. A. Smith, and K. Struhl), p. 16.63, John Wiley and Sons, New York.
14. Chakrabarti, S., Sisler, J. R., and Moss, B. (1997). *BioTechniques*, **23**, 1094.
15. Alcamí, A., and Smith, G. L. (1992). *Cell*, **71**, 153.
16. Alcamí, A., and Smith, G. L. (1995). *J. Virol.*, **69**, 4633.
17. Alcamí, A., Symons, J. A., Collins, P. D., Williams, T. J., and Smith, G. L. (1998). *J. Immunol.*, **160**, 624.
18. Ashkenazi, A., and Chamow, S. M. (1997). *Curr. Opin. Immunol.*, **9**, 195.
19. Harris, E. L. V., and Angal, S. (ed.) (1989) *Protein Purification Methods: A Practical Approach*. Oxford University Press, New York.
20. Schreiber, M., Rajarathnam, K., and McFadden, G. (1996). *J. Biol. Chem.*, **271**, 13333.
21. Graham, K. A., Lalani, A. S., Macen, J. L., Ness, T. L., Barry, M., Liu, L-Y., Lucas, A., Clark-Lewis, I., Moyer, R. W., and McFadden, G. (1997). *Virology*, **229**, 12.
22. Lalani, A. S., Graham, K., Mossman, K., Rajarathnam, K., Clark-Lewis, I., Kelvin, D., and McFadden, G. (1997). *J. Virol.*, **71**, 4356.
23. Smith, C. A., Smith, T. D., Smolak, P. J., Friend, D., Hagen, H., Gerhart, M., Park, L., Pickup, D. J., Torrance, D., Mohler, K., Schooley, K., and Goodwin, R. G. (1997). *Virology*, **236**, 316.
24. Wu, Z., Johnson, K. W., Choi, Y., and Ciardelli, T. L. (1995). *J. Biol. Chem.*, **270**, 16045.
25. O'Shannessy, D. J., Brigham-Burke, M., Soneson, K. K., Hensley, P., and Brooks, I. (1994). In *Methods in enzymology* (ed. Johnson, M. L. and Brand, L.). Vol. 240, p. 323. Academic Press, London.
26. O'Shannessy, D. J., Brigham-Burke, M., Soneson, K. K., and Hensley, P. (1993). *Anal. Biochem.*, **212**, 457.
27. Branch, D. R., Shah, A., and Guilbert, L. (1991). *J. Immunol. Meth.*, **143**, 251.
28. Smith, G. L., Symons, J. A., and Alcamí, A. (1998). *Semin. Virol.*, **8**, 409–418.
29. Cao, J. X., Gershon, P. D., and Black, D. N. (1995). *Virology*, **209**, 207.
30. Damon, I., Murphy, P. M., and Moss, B. (1998). *Proc. Natl. Acad. Sci. USA*, **95**, 6403.
31. Kledal, T. N., Rosenkilde, M. M., Coulin, F., Simmons, G., Johnsen, A. H., Alouani, S., Power, C. A., Luttchau, H. R., Gerstoft, J., Clapham, P. R., Clark-Lewis, I., Wells, T. N. C., and Schwartz, T. W. (1997). *Science*, **277**, 1656.
32. MacDonald, M. R., Li, X-Y., and Virgin, H. W. IV. (1997). *J. Virol.*, **71**, 1671.
33. Lalani, A. S., Ness, T. L., Singh, R., Harrison, J. K., Seet, B. T., Kelvin, D. J., McFadden, G., and Moyer, R. W. (1998). *Virology*, **250**, 173–184.
34. McCormack, J. G., and Cobbold, P. H. (ed.) (1991) *Cellular Calcium: A Practical Approach*. Oxford University Press.
35. Grynkiewicz, G., Poenie, M., and Tsien, R. Y. (1985). *J. Biol. Chem.*, **260**, 3440.

8

Analysis of DNA virus proteins involved in neoplastic transformation

KERSTEN T. HALL, MARIA E. BLAIR ZAJDEL and
G. ERIC BLAIR

1. Introduction

The small DNA tumour viruses, simian virus 40 (SV40), polyoma viruses, human adenoviruses (e.g. Ad2, Ad5, and Ad12), and human papillomaviruses (e.g. HPV16 and HPV18) encode at least one oncogene product responsible for transformation of cells in culture and formation of tumours in animals. These oncogene products modulate signal transduction and other cellular pathways to abrogate normal cell cycle control mechanisms, and this results in rapid cell proliferation. To achieve this, virus oncogene products frequently interact with cellular proteins such as pRb, p53, or c-Src.

Here we give a broad overview of the properties of the major transforming proteins encoded by the small DNA tumour viruses, and their cellular targets. A more detailed treatment of these topics is given in articles in Fanning (1). In addition, a number of reviews give thorough background information on the topics of transforming proteins of human adenoviruses (2–5), SV40 (6), polyoma viruses (7), and human papillomaviruses (8, 9). A good general source book for information on oncogenes, including those of DNA viruses, is Hesketh (10).

1.1 Human adenovirus oncoproteins

The human adenovirus genome consists of a linear double-stranded DNA molecule of approximately 36 kb. Two genes, E1A and E1B, which are required for transformation of cells by human adenoviruses, are located in the left 11% of the genome. There are at least 47 human adenovirus serotypes, which have been classified into six subgenera (A to F) according to various properties, including the oncogenicity of viruses in newborn rodents. The serotypic origin of the E1A gene determines the oncogenic phenotype of adenovirus-transformed cells. Viruses belonging to subgenus A (such as adenovirus 12, Ad12) induce tumours with high frequency and short latency, while viruses from subgenus B (such as Ad3 and Ad7) are weakly oncogenic. Adenoviruses from subgenus C (which includes the well studied serotypes Ad2 and Ad5), D, E, and F are non-oncogenic. All human adenoviruses studied so far can transform primary rodent cells in culture; however, only cells transformed by viruses of subgenus A and B are oncogenic in newborn rodents, paralleling the oncogenic properties of the parental viruses.

The E1A gene encodes two major mRNAs, generated by differential splicing, that direct synthesis of two proteins (289 and 243 amino acids in Ad2 and Ad5, and 266 and 235 amino acids in Ad12) which are identical except for an internal peptide segment. Both proteins are located mainly in the nucleus. They are highly phosphorylated, which is at least in part responsible for the discrepancy between their electrophoretic mobility in SDS-PAGE and their theoretical molecular weights. Three highly conserved domains or regions, CR1, CR2, and CR3, have been identified in all the human Ad E1A genes, and their functions defined by mutational analysis in the 289-residue E1A protein. The first conserved domain, CR1, is located between residues 40 and 80, CR2 between 121 and 139, and CR3 between 140 and 185 in Ad2. The CR3 domain is unique to the 289-amino acid species. Mutants in CR1 complement mutations in CR2, demonstrating discrete functional domains that act in *trans* to alter the host cell. Monoclonal antibody competition experiments and mutational analyses have identified two distinct protein binding sites in the CR1 domain. The segment between residues 60 and 80 is involved, together with the amino-terminal amino acids 1–39, in the binding of a large, cellular phosphoprotein, p300, which is a transcriptional co-activator with intrinsic histone acetyltransferase activity (11). E1A mutants that fail to bind p300 also fail to stimulate cellular DNA synthesis in resting cells, and do not immortalize cells in culture. The CR1 segment between residues 40 and 60 forms part of the binding site for the retinoblastoma susceptibility protein pRb and related proteins (p107 and p130). Another region necessary for binding and functional inactivation of these proteins is CR2, with residues 121–126 (DLXCXE) being the binding motif. This explains why CR2 is absolutely required for the mitogenic activity of E1A.

The retinoblastoma susceptibility gene product, pRb, and related proteins (p107 and p130) form complexes with a transcription factor, E2F, that are dissociated by E1A to release transcriptionally active E2F. Expression of E2F can stimulate DNA synthesis in quiescent cells. It has also been established that other host cell-cycle control proteins, such as cyclin A, bind to CR2. E1A proteins can bind p300 and the pRb-related proteins simultaneously, thus potentially facilitating their interaction. Loss of Rb function also results in the induction of p53, by increasing the expression of the tumour suppressor protein p14^{ARF} in human cells (p19^{ARF} in mouse cells), which prevents MDM2-controlled degradation of p53 (12). This is one of the mechanisms by which E1A promotes apoptosis in primary cells.

Conserved region 3, which is dispensable for E1A activities specifically associated with cell-cycle regulation, is responsible for transcriptional activation of other adenovirus promoters. It is also involved, through its zinc finger domain, in activation of certain cellular promoters. It is also interesting to note that the CR2 and CR3 are separated by a 20-amino acid alanine-rich spacer region in oncogenic Ad12 and in the oncogenic Simian adenovirus 7 (SA7) E1A, but not in non-oncogenic Ad5 E1A. The precise function of the alanine-rich domain is not yet known, although it appears to mediate, in part, the oncogenicity of Ad12.

The E1B region encodes two proteins, of 19 kDa and 58 kDa, that are unrelated in amino-acid sequence, but both have anti-apoptotic function (4). These two proteins act independently, and by different mechanisms. The E1B-58 kDa protein directly abolishes the activity of the tumour suppressor, p53, the most commonly mutated protein in human cancer. The p53 protein is a transcription factor that negatively regulates cell proliferation. It is thought that p53 is not constitutively involved in the cell cycle, but is activated in response to DNA damage. A stable complex of p53 and E1B-58 kDa of human subgenus C adenoviruses can be readily detected by immunoprecipitation followed by SDS-PAGE. When p53 is in a complex with E1B-58 kDa protein, it is unable to transactivate transcription. Mutants of the E1B-58 kDa gene that are unable to form a complex with p53 are also unable to transform cells in culture in co-operation with E1A. The E1B-19 kDa protein inhibits apoptosis by a mechanism similar to that of human bcl-2 protein.

1.2 Simian virus 40 and polyomavirus oncoproteins

Simian virus 40 (SV40) and polyomaviruses belong to the papovavirus group of DNA tumour viruses. These viruses possess double-stranded, circular DNA genomes, which are much smaller than those of adenoviruses: for example, SV40 has a genome of 5227 bp. Two out of six proteins encoded by SV40, termed large T and small t antigens, are involved in immortalization and transformation of primary rodent cells. Although SV40 large T antigen is sufficient to cause immortalization and transformation of primary rodent cells, small t antigen enhances large T-induced transformation. SV40 large T and small t antigens are products of the same gene, generated by alternative splicing, and they share an 82 amino acid N-terminus. The SV40 large T antigen is a 708 amino acid nuclear phosphoprotein with multiple biochemical activities. It is composed of several domains, three of which contribute to transformation of cells in culture or formation of tumours in animals. These three domains have been localized to amino acids 1–82 (Region 1), 101–118 (Region 2), and amino acids 351–450 and 533–626 (Region 3).

Region 1 is homologous to the J domain of the DnaJ family of molecular chaperones (13). This region plays an essential role in virion assembly, virus DNA replication, transcriptional control, and oncogenic transformation. It has, therefore, been proposed that the J domain is involved in the rearrangement of multiprotein complexes. Further studies have suggested that the J domain is required for functional inactivation of Rb family proteins.

Region 2 of T antigen, which has been shown to be important for host cell transformation, contains a consensus amino acid sequence, LXCXE, that is required for binding to the tumour suppressor protein, pRb, and to two structurally related proteins, p107 and p130. Large T antigens with mutations in this region are defective for pRb/p107/p130 binding, and for transformation of REF52 and NIH 3T3 cell lines. However, as described above, further results suggest that the J domain of SV40 large T antigen plays a role in the functional inactivation of Rb family proteins. A model has therefore been proposed in which Rb family proteins bind to Region 2, but the J domain directs their inactivation by perturbing the phosphorylation of p107 and p130, and promoting the degradation of p130.

Region 3 of SV40 large T antigen contains a p53 tumour suppressor-binding domain, which is located between amino acids 351 and 626. A spacer segment redundant for binding to p53 divides this region into two parts between amino acids 351 and 450, and between 533 and 626. In one model, it is proposed that amino acids 351–450 are important for direct contact with p53, and amino acids 533–626 are important for regulating this interaction.

A novel functional domain involved in preventing apoptosis independently of inactivation of p53 has recently been mapped to amino acids 525–541 (14). The carboxy-terminal region of the large T antigen also binds p300, pCBP, and p400 transcriptional coactivators, and this activity has been implicated in the transforming mechanism of this protein (15).

Small t antigen of SV40 comprises 174 amino acids. The region between residues 97 and 103 interacts with the protein phosphatase 2A (PP2A) heterodimer, consisting of 36 kDa and 65 kDa subunits. This interaction reduces the ability of PP2A to inactivate ERK1 and MEK1 protein kinases, resulting in stimulation of proliferation of quiescent monkey kidney cells. Small t antigen-dependent assays also identified other regions which had the ability to enhance cellular transformation. These regions are located in the N-terminal part (residues 1–82), which is shared by the small and large T antigens of SV40, and can potentially function as a Dna J domain. Small t antigen can also associate with tubulin, and it has been suggested that this plays a role in its biological function.

The polyomaviruses encode three proteins involved in cellular transformation, termed large tumour antigen (LT), middle T antigen (mT), and small tumour antigen (sT). These three proteins result from the differential splicing of the early region transcript, and contain homologous sequences. The large T antigen of polyoma interacts with the tumour suppressor protein, pRb, and is able to immortalize primary fibroblasts in culture. The Dna J domain located at its N-terminus, particularly the HPDKYG sequence found between residues 42 and 47, is critical for functional inactivation of Rb family proteins, as is also the case with SV40 large T antigen (13). The expression of LT is not, however, sufficient to produce a fully transformed cell phenotype; this requires mT, which is the major transforming protein of the polyomavirus. Mouse polyoma middle T consists of 421 amino acids, and can be divided into at least three domains, some of which are shared with LT and sT. The amino-terminal domain is composed of the first 79 amino acids, and is also present in LT and sT. Adjacent to it, between residues 80 and 192, is a domain that is also present in the polyoma sT, and which contains two cysteine-rich regions, Cys-X-Cys-X-X-Cys, which have also been identified in small t of SV40. Mutation of these cysteines abolishes the ability of mT to transform cells. The remaining 229 amino acids are unique to mT, and contain the major tyrosine phosphorylation site of mouse mT (tyr 315), and a hydrophobic region (approximately 20 amino acids at the carboxy terminus) involved in membrane localization of this protein which is necessary for its transforming activity.

Polyoma mT associates with several cellular proteins. These complexes include protein phosphatase 2A (PP2A), Src family tyrosine kinases, phosphatidylinositol 3-kinase (PI-3K), and an adaptor protein, Shc, and they are necessary for cellular transformation. The existence of large complexes containing mT and all the cellular targets described above has been reported. Most of the mT in a transformed cell is complexed with the serine/threonine-specific protein phosphatase PP2A, which also associates with the small T antigens of polyoma and SV40. PP2A usually consists of the core dimer of a 36 kDa catalytic subunit (C subunit) and a regulatory subunit of 65 kDa (A subunit), which associate with variable regulatory subunits (B subunit) to give a heterotrimer. Polyoma mT and both polyoma and SV40 small T antigens displace the variable B subunit in this enzyme, although there is no sequence homology between them and this subunit. The interaction of PP2A with these tumour antigens has an inhibitory effect on dephosphorylation of *in vitro* substrates such as myosin light chain, myelin basic protein, large T antigen, and p53, and also reduces its ability to inactivate ERK1 and MEK1 protein kinases, which results in cell-cycle progression.

About 10% of mT in a transformed cell is bound to $pp60^{c-src}$ (c-Src). This cytoplasmic protein kinase has several non-catalytic domains, which regulate its enzymatic activity in both a positive and a negative fashion. In normal cells, the activity of c-Src is repressed by phosphorylation of Tyr^{527} in the non-catalytic C-terminal region, which has been mapped as the mT binding site. Molecules of c-Src complexed with mT are not phosphorylated at Tyr^{527}, and this is accompanied by a tenfold increase in kinase activity. However, genetic analysis has shown that association of mT and c-Src is necessary, but not sufficient, for transformation.

Phosphatidylinositol 3-kinase (PI 3-K) is another enzyme with which mT can form a complex. PI 3-K consists of two subunits: a catalytic subunit of

110 kDa (p110), and a regulatory subunit of 85 kDa (p85). The regulatory subunit, p85, contains two SH2 (*Src Homology*) and one SH3 domain. Binding of mT to PI 3-K occurs through the interaction of the SH2 domains of p85 with the PTyr^{315}XXMet sequence of mT when Tyr^{315} is phosphorylated by the associated Src family kinases. Mutation of this tyrosine results in an mT mutant that is able to activate c-Src, but is defective in association with PI 3-K and in transformation. It has been recently reported that mT activates the Ser/Thr kinase Akt in a PI3-kinase-dependent manner (16)

The mouse polyoma middle T also interacts with a family of adaptor proteins (66, 52, and 46 kDa) termed Shc for *Src Homology* 2 (SH2) and *Collagen* (α1) related. A single mutation in middle T antigen, which replaces tyrosine at position 250 by serine, disrupts the association of this protein with Shc, and results in much reduced transformation of cells in culture, and alters the spectrum of tumours and their morphology in inoculated animals (17). Phosphorylated Shc associates with another adaptor protein, Grb2, which is able to bind to SOS, the 150 kDa guanylnucleotide exchange factor for Ras. This targets SOS to the plasma membrane location of Ras, allowing the rapid conversion of Ras from the inactive GDP-bound state to the active GTP-bound state. Transformation by the mouse middle T requires functional Ras. However, the hamster middle T does not have a region homologous to the segment containing tyrosine 250 of mouse middle T, suggesting that the Shc link may not be involved in its transforming mechanism.

Polyoma middle T antigen can also form complexes with phospholipase C-γ-1 and 14-3-3 proteins. Tyrosine 322 of mT, when phosphorylated, becomes the docking site for the SH2 domain of phospholipase C-gamma-1 (PLC-γ-1). Mutation of tyrosine 322 to phenylalanine renders mT defective for interaction with PLC-γ-1, and in transformation (18). Phosphorylation of serine 257 of mT regulates its association with 14-3-3 proteins, which have been linked to cell-cycle control and signalling (19).

1.3 Human papillomavirus oncoproteins

The papillomaviruses also belong to the papovavirus group. The high risk human papillomaviruses 16 and 18 are the best documented examples of viruses as a causative agent in human cancer, with type 16 being particularly prevalent. HPV DNA is found in approximately 90% of cervical carcinomas, and E6 and E7 genes are regularly expressed in these carcinomas. Immortalising or transforming properties of high risk HPV E6 and E7 have been demonstrated in *in vitro* transformation assays, and they are the only virus genes required for these processes (8).

The HPV E7 proteins are small (HPV16 E7 comprising 98 amino acids), zinc-binding phosphoproteins which are localized in the nucleus (9). They are structurally and functionally similar to the E1A protein of subgenus C adenoviruses. The first 16 amino-terminal amino acids of HPV16 E7 contain a

8: Analysis of DNA virus proteins involved in neoplastic transformation

region homologous to a segment of the conserved region 1 (CR1) of the E1A protein of subgenus C adenoviruses. The next domain, up to amino acid 37, is homologous to the entire region 2 (CR2) of E1A. Genetic studies have established that these domains are required for cell transformation *in vitro*, suggesting similarities in the mechanism of transformation by these viruses. The CR2 homology region contains the LXCXE motif (residues 22-26) involved in binding to the tumour suppressor protein pRb. This sequence is also present in SV40 and polyoma large T antigens. The high risk HPV E7 proteins (of, for example, types 16 and 18) have an approximately tenfold higher affinity for pRb protein than the low risk HPV E7 proteins (of, for example, type 6). Association of the E7 protein with pRb promotes cell proliferation by the same mechanism as the E1A proteins of adenoviruses and SV40 large T antigen. Recent studies have shown that E7 promotes degradation of Rb family proteins, rather than simply inhibiting their function by complex formation. The CR2 region also contains the casein kinase II phosphorylation site (residues 31 and 32). The remaining 61 amino acids of E7 protein have very little similarity to E1A; however, a sequence CXXC involved in zinc binding is present in both proteins. The E7 protein contains two of these motifs, which mediate dimerization of the protein. Mutation in one of the two zinc-binding motifs destroys transforming activity, although this mutant is able to associate with Rb protein. Therefore dimerization may be important for the transforming activity of E7.

The HPV E6 are small basic proteins (HPV16 E6 comprising 151 amino acids) which are localized to the nuclear matrix and non-nuclear membrane fraction. They contain four cysteine motifs, which are thought to be involved in zinc binding. E6 encoded by high-risk HPVs associates with the wild type p53 tumour suppressor protein. For association with p53, the E6 protein requires a cellular protein of 100 kDa, termed E6-associated protein (E6-AP). Like SV40 large T antigen and Ad5 E1B-58 kDa, E6 proteins of high risk HPVs abrogate the ability of wild type p53 to activate transcription. However, the mechanism of E6 action is different from that of SV40 large T antigen and the E1B protein, since it involves degradation of p53. It has been shown that E6-dependent degradation of p53 occurs through the cellular ubiquitin proteolysis pathway.

It has recently been reported that invasive cervical carcinomas very frequently harbour HPV16 E6 variants, rather than the prototype (20). This suggests that the risk of developing cervical intraepithelial neoplasia is not the same with all variants of HPV16. Mutations in HPV16 E6 have been identified in regions known to be important for the transforming activity of this protein, and in other regions which are likely to be significant for host immune recognition. In comparison, mutations in HPV16 E7 are rare, and they are located in regions likely to be significant for host immune recognition. No mutations have yet been discovered in regions known to be important for HPV16 E7 transforming activity.

2. Cell systems utilized for study of DNA tumour viruses

Virus oncogenes can be introduced into cells by transfection, in order to study their effects on transformation and immortalization. Both established cell lines and primary cell cultures can be used to generate virus-transformed cells. However, since established cell lines are often aneuploid, or contain mutations in tumour-suppressor genes and proto-oncogenes, it is important to use primary cells to study the effects of virus oncoproteins on immortalization and certain aspects of transformation. In our laboratory, primary cell cultures of rat embryonic fibroblasts or baby rat kidney epithelial cells are used for this purpose.

Protocol 1. Culture of baby rat kidney and rat embryo fibroblasts

Equipment and reagents

- Seven- to 10-day-old baby Wistar rats, or 14- to 16-day-pregnant Wistar rats[a]
- Several pairs of sterile forceps and scissors
- Sterile phosphate-buffered saline (PBS)
- Dulbecco's Modified Eagle's Medium containing 10% fetal calf serum (FCS), and penicillin and streptomycin (100 units ml^{-1} each)
- Tissue culture (laminar flow) hood
- 75 cm^2 plastic tissue-culture flasks
- Trypsin-EDTA solution
- Sterile 100ml bottle containing a stirring bar
- Sterile gauze
- Sterile, disposable plastic 50 ml centrifuge tubes
- Wash bottle containing 70% ethanol
- Haemocytometer
- CO_2 incubator at 37°C, with a humidified atmosphere of 95% air, 5% CO_2
- Temperature-controlled warm room or incubator, with a heat-free magnetic stirrer

Method

1. Outside the tissue-culture laboratory, kill the baby rats or the pregnant rats by an approved method (such as asphyxiation by CO_2). Thoroughly soak the dead animals using a wash bottle of 70% ethanol, to prevent contamination of the primary cell cultures by surface microbes.

2. Remove the embryos from the uterus of 14–16 day pregnant Wistar rats by first cutting through the abdomen with a sterile pair of scissors. Using a second pair of sterile scissors, cut open the uterus, and remove the embryos using a sterile pair of forceps. Collect the embryos in a sterile bacterial Petri dish.

3. Remove the kidneys from the baby rats by first cutting away the skin from the back. Using a second pair of sterile scissors and forceps, carefully cut through the body wall, in a line parallel to the spine. Locate and remove the kidneys, collecting them into a sterile bacterial Petri dish.

8: Analysis of DNA virus proteins involved in neoplastic transformation

4. Working now in a laminar flow hood in the tissue-culture laboratory, rinse the embryos or kidneys several times with PBS, and mince into fine pieces using a sterile pair of scissors.
5. Transfer the minced tissue into a sterile 100 ml bottle containing a stirring bar. Add 20 ml of trypsin-EDTA, and stir the suspension at 37 °C for 10 min on a heat-free magnetic stirrer.
6. After allowing the larger clumps of tissue to settle, remove the upper cell suspension and filter through sterile gauze into 20 ml DMEM/FCS/penicillin/streptomycin (protease inhibitors in FCS inactivate the trypsin).
7. Add a further 20 ml trypsin-EDTA to the remaining tissues, and repeat Steps 5–6. Pool the cell suspensions in sterile disposable plastic 50 ml centrifuge tubes, and centrifuge for 10 min at 1500 g in a bench-top centrifuge at room temperature.
8. Gently resuspend the cell pellet in 10 ml DMEM/FCS/penicillin/streptomycin medium. Remove an aliquot of 100 µl to count the cell density in a haemocytometer.
9. Pipette approximately 4×10^5 cells (in 5 ml) into 75 cm^2 flasks.
10. Incubate approximately 2–3 days at 37 °C in a CO$_2$ incubator.

[a] Experimental use of animals in research is covered by animal welfare legislation. Investigators must ensure that they have met all local and national legal requirements before any experimental protocol is begun.

2.1 DNA transfection of virus oncogenes into mammalian cells

There are several methods of introducing foreign DNA into cells in culture, including calcium phosphate precipitation, electroporation, and a number of commercially available liposome complexes. In our laboratory, the most commonly used technique is the calcium phosphate precipitation method, in which the DNA to be transfected is mixed with a solution of calcium chloride and Hepes-buffered saline (HBS). This results in the formation of a DNA-calcium phosphate precipitate, which is then engulfed by the cells. Calcium phosphate transfection protocols have already been described in Chapters 3 and 5. Alternatively, commercial reagents such as Lipofectamine™ can be used. These reagents contain cationic lipids that form a complex with the DNA. Lipofection protocols are described in Chapters 3 and 6.

2.2 Selection of virus-transformed cells

When primary cells are transfected with a virus oncogene, transformed cells proliferate as foci of rapidly growing cells on a monolayer of normal cells. Over a period of about two weeks, normal cells will die, and the foci can be fixed and counted (if the biological activity of the oncogene is to be quantified), or foci can be picked and expanded into cell lines. The selection of cells transformed by a virus oncogene can also be achieved by co-transfecting with a plasmid containing a selectable marker, such as the neomycin resistance gene. This is often used when established cell lines are transfected with virus oncogenes. Inclusion of neomycin 2–3 days after transfection (e.g. G418 at 500 μg ml^{-1}) in the growth medium will enrich for transformed cells. *Protocol 2* below describes the transformation of NIH3T3 cells with the neo resistance marker as an example (21, 23).

Protocol 2. Focus formation assay[a]

Equipment and reagents
- DMEM containing 10% FCS, and strepto-mycin and penicillin (100 units ml^{-1} each)
- Sterile PBS
- Giemsa Modified solution (BDH)
- Distilled water
- Light microscope
- Sterile, finely drawn Pasteur pipettes
- 24-well tissue culture dishes

Method

1. Grow primary cells to 40% confluence, and transfect with DNA encoding a virus oncogene (see above).

2. Allow the transfected cells to grow for 48–72 h in culture medium containing 10% FCS. After 48–72 h, reduce the concentration of FCS in the medium to 5%. The period of time for which the cells are incubated after transfection will vary according to the cell type.

3. Incubate the cells in medium containing 5% FCS for another 12–14 days, replenishing with fresh medium every 3 days.

4. Examine the cell sheet for the presence of transformed foci, and stain the cells with Giemsa Modified solution. To each dish of cells, add 2 ml of Giemsa Modified solution, and leave for 1 min. Add an equal volume of PBS, and leave for a further 2 min with gentle agitation.

5. Rinse the dish of cells with 2 ml distilled water, and leave to dry in an inverted position. Record the number of colonies.

6. To derive cell lines from foci, take some unfixed plates, remove the medium, and pick macroscopic foci using finely drawn Pasteur pipettes. Transfer the cells into 1 ml of DMEM/10% FCS in a well of a 24-well culture dish. When cells are confluent, they can be trypsinized and expanded.

[a] Adapted from ref. 22.

8: *Analysis of DNA virus proteins involved in neoplastic transformation*

Protocol 3. Selection of cells using neomycin selection

Equipment and reagents
- Lipofectamine™ reagent (2 mg ml^{-1}, Gibco-BRL)
- DMEM/10% FCS
- 100 mm tissue-culture dishes
- Geneticin (G418) at 10 mg ml^{-1} in PBS
- DMEM/20% FCS
- Neomycin-resistance expression plasmid, e.g. pSV2neo (Clontech), and the plasmid containing the virus oncogene

Method
1. Seed NIH3T3 cells at 5×10^5 cells per 100 mm dish.
2. Co-transfect the cells with the pSV2^{neo} plasmid and the plasmid containing the virus oncogene, at a ratio of 1:10 respectively, with Lipofectamine (see Chapters 3 and 6).
3. After 6 h of incubation at 37°C in a CO_2 incubator, add an equal volume of DMEM/20% FCS, and continue the transfection overnight.
4. The following day, change the medium to DMEM/10% FCS.
5. At 48 h post-transfection, change the medium, and replace it with DMEM containing 10% FCS and G418 at 500 µg ml^{-1}.
6. After one week, change the medium, replacing it with fresh DMEM containing 10% FCS and 500 µg ml^{-1} G418.
7. Approximately two weeks after the addition of G418 selection medium, pool any drug-resistant colonies by trypsinization, and seed in triplicate into soft agar to produce well separated colonies (*Protocol 5*).

2.3 Cloning of transformed cells, and assay of cell growth

Once the cells have been transfected with a transforming virus oncogene, the efficiency of transformation can be assessed by counting the number of transformed colonies which arise. In addition, well separated colonies can be picked, and expanded into cell lines. The protocols below describe the measurement of colony formation on either tissue culture dishes or in soft agar. The growth properties of transformed cells can also be quantitatively assessed.

Protocol 4. Determination of colony formation in tissue-culture dishes

Equipment and reagents
- 75 cm^2 tissue-culture flasks
- DMEM/10% FCS
- 0.1% Crystal violet in 2% ethanol
- 10% (w/v) trichloroacetic acid (TCA)

Protocol 3. *Continued*

Method

1. Trypsinize the cells, and seed in duplicate 75cm^2 flasks at 200 cells per flask in DMEM/10% FCS. In addition, lower concentrations of serum can be tested, since many lines of virus-transformed cells can be readily grown in media containing 5% or 2.5% FCS.

2. Change the medium weekly, and after 3 weeks of incubation, fix the colonies in 10% TCA for 20 min, and stain with Crystal violet for 15 min, both at room temperature. The flasks are then washed extensively with distilled water. The colonies can now be counted.

Protocol 5. Measurement of colony formation in soft agar

Equipment and reagents

- 2% Noble agar (Difco) in deionized, distilled water
- DMEM, without serum
- 2 × DMEM, without serum
- Fetal calf serum
- Sterile, plugged, finely drawn Pasteur pipettes
- Water bath at 45°C

Method

1. Prepare the soft agar cloning medium by first autoclaving the Noble agar, cooling, and maintaining at 45°C.

2. In a laminar flow hood, mix 25 ml of 2 × DMEM, 25 ml 2% agar, 30 ml DMEM, and 20 ml FCS in a sterile 100 ml bottle. Pipette 15 ml aliquots into the bases of six 100 mm bacterial Petri dishes. Cover, and allow to set at room temperature. Maintain the rest of the soft agar cloning medium in the 45°C water bath.

3. Trypsinize the cells, and disperse to obtain single-cell suspensions.

4. Prepare a series of tenfold dilutions of cells in DMEM at 10^2, 10^3, and 10^4 cells ml^{-1}, and mix 0.8 ml aliquots of these dilutions with 1.2 ml of the agar mixture. Spread over the set bases, and allow to set. Incubate in a humidified CO$_2$ incubator at 37°C. Ensure that the incubator is thoroughly humidified throughout this period, as the agar can dry and the cells will die. A beaker of water can be left close to the dishes.

5. Colony growth can be observed by periodic observation of the plates with a light microscope.

6. After 7–14 days, macroscopic colonies should be evident in some plates.

7. The colonies can now be counted.

8: Analysis of DNA virus proteins involved in neoplastic transformation

8. Well separated colonies can be picked, and aspirated into 1 ml aliquots of DMEM/10% FCS in 24-well tissue culture plates.
9. These cells can now be expanded (after trypsinization) into clonal cell lines.

Protocol 6. Assay of transformed cell growth

Equipment and reagents
- 25 cm² tissue-culture flasks, or 60 mm tissue-culture dishes
- DMEM/10% FCS
- Inverted light microscope

Method
1. Trypsinize a monolayer of cells that has just become confluent, disperse well, and seed at a density of 2×10^3 cells cm^{-2} per 25 cm² flask, or 5×10^4 cells cm^{-2} per 60 mm dish, in replicate.
2. Change the medium, and replace with fresh DMEM/10% FCS every 2–3 days. Count the number of cells each day for a six-day period. The doubling time of the culture can be determined from the growth slope in the initial logarithmic phase.
3. The life-span of cloned cell lines in culture can be estimated by growing cells in DMEM containing 10% fetal calf serum, and routinely subculturing at a 1:5 ratio when confluent. If cell growth is inhibited, the culture can be considered to have become senescent. It should be possible to subculture fully transformed cells indefinitely.

3. Characterization of virus-transformed cells

Once a primary cell culture has been transformed by the introduction of a virus oncogene, the transformed cells can be characterized by assaying for the expression of virus oncoproteins. Antibodies are an invaluable tool in these studies, and can be used to detect virus proteins by immunofluorescence, Western blotting, or radio-immunoprecipitation. Polyclonal antibodies can be raised by inoculation of an animal, usually a rabbit, with a sample of the virus protein. Usually the rabbit then receives a course of three boosts of the virus immunogen at one-week intervals. A test bleed may be taken ten days after a boost, and the sera obtained tested for the presence of antibody by performing immunofluorescence against virus-transformed cells. The use of expression vectors, such as the pGEX series of vectors, has proved useful in generating antibodies against virus proteins, since the open reading frame encoding the virus protein can be cloned into the expression vector, and over-expression of the antigen induced in *Escherichia coli* by treatment with IPTG.

Antibodies against the spectrum of virus transforming proteins present in a transformed cell can also be obtained from the serum of tumour-bearing animals. The procedure described below yields useful sera containing antibodies against the adenovirus transforming proteins (principally the E1B-58 kDa and E1B-19 kDa proteins). Further procedures for characterising these sera by ELISA, and by radioimmunoprecipitation are described, as well as their use in studying the stability and intracellular location of the transforming proteins.

Protocol 7. Preparation of tumour-bearing hamster sera containing antibodies against Ad5-transforming proteins[a]

Equipment and reagents
- Four-day-old Syrian hamsters
- Ad5-transformed hamster (H14b) cells

Method

1. Ensure that permission has been obtained to conduct the procedures below on live animals, and that high standards of animal care are available.

2. Inoculate four-day-old Syrian hamsters by subcutaneous injection with 2×10^6 H14b cells.

3. After approximately four weeks, collect serum from tumour-bearing animals by cardiac puncture under terminal anaesthesia. The antibody response between animals may vary, and each animal's serum should therefore be analysed. This is most convincingly performed by ELISA (which gives an indication of total antibody levels), or immunoprecipitation (which shows the spectrum of antibodies recognising particular virus transforming proteins). These analyses could also be performed to test the antibody response of other animals to virus oncoprotein immunogens. The results of such analyses are shown in *Figure 1*.

[a] Described in ref. 24.

Protocol 8. Comparison of hamster H14b tumour sera by ELISA

Equipment and reagents
- Cytosine arabinoside (10 mg ml^{-1} in PBS)
- Human Hep-2 cells (available from the American Type Culture Collection, ATCC no CCL-23)
- Adenovirus 5 virus stock (ATCC no VR-5, 10^{11} plaque-forming units ml^{-1})[a]
- Horseradish peroxidase-conjugated rabbit or goat anti-hamster immunoglobulins (ICN)

8: Analysis of DNA virus proteins involved in neoplastic transformation

- 100 mm tissue-culture dishes
- PBS
- Methanol (at $-20\,°C$)
- Probe sonicator
- PBS plus 0.1% gelatin and 0.05% (v/v) Tween 20
- Forceps, and pieces of sterile rubber tubing
- 96-well ELISA microtitre plates
- Substrate solution: 10 mg o-phenylenediamine dissolved in 0.1 M citrate buffer, pH 4.5, 4 µl 30% (w/v) hydrogen peroxide added just before use
- 8 M sulfuric acid

Method

1. Grow Hep2 cells in 100 mm tissue culture dishes, as described above. When the cells are almost confluent, infect them with adenovirus 5 at a multiplicity of infection of 10 plaque-forming units cell^{-1} in the presence of AraC, which prevents synthesis of virus DNA, resulting in only the synthesis of virus transforming (and other early) proteins. In parallel, mock-infect Hep2 cells with an equivalent volume of DMEM. Incubate for 22 h in a CO_2 incubator.

2. At 22 h post-infection, remove the culture medium, and harvest infected and mock-infected cells by scraping off the monolayers with a small piece of rubber tubing.

3. Wash the cells three times in PBS at $4\,°C$, and resuspend at a concentration of 10^8 cells in 0.5 ml of PBS. Lyse the cells by sonicating three times for 10 s on ice, using a probe sonicator.

4. Dilute the lysates in PBS to give approximately 10^6 cell equivalents per ml.

5. Add 100 µl of this dilution to each well of a 96-well ELISA microtitre plate, and adsorb at $4\,°C$ for 18 h in a humidified box.

6. Remove the lysate, and add 200 µl of methanol chilled to $-20\,°C$ to each well. Place the plate at $-20\,°C$ for 10 min.

7. Remove the methanol, and wash the wells three times with PBS containing 0.1% gelatin and 0.05% Tween 20.

8. Serially dilute each tumour serum in PBS from 1:50 to 1:6400 along the plate. Incubate the plates for 1 h at $37\,°C$ in a humidified box.

9. Wash the wells three times in PBS/Tween/gelatin as before. To each well, add 100 µl of freshly prepared horseradish peroxidase-conjugated anti-hamster IgG (using the manufacturer's recommended dilution).

10. Incubate the plates at $37\,°C$ in a humidified box, and wash the wells six times with PBS/Tween/ gelatin. Add 100 µl of freshly prepared ELISA substrate to each well, and allow the colour to develop. Stop the reaction by the addition of 8 M sulfuric acid, and record the results by photography.

[a] Described in ref. 25.

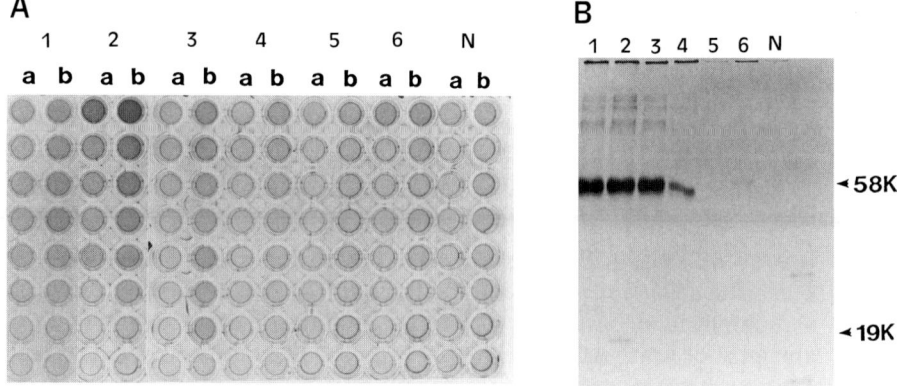

Figure 1. Analysis of sera from tumour-bearing hamsters by ELISA or radioimmunoprecipitation. Sera from six hamsters bearing tumours induced by adenovirus 5-transformed hamster cells (H14b cells) were analysed by ELISA (A) using extracts of uninfected (a) or Ad5-infected Hep2 cells (b); normal hamster serum (N) was used as a control. Sera were tested at 1:50 (top wells) to 1:6400 (bottom wells). In (B), the same six sera were tested by immunoprecipitation, using extracts of [^{35}S]methionine-labelled H14b cells. In general, there is good agreement between the results obtained by each technique. In particular, serum 2 has a relatively high ELISA titre, and also contains antibodies against the E1B-19 kDa as well as the E1B-58 kDa protein.

3.1 Analysis of virus oncoproteins by radiolabelling of transformed cells and immunoprecipitation

Immunoprecipitation, followed by SDS-gel electrophoresis, provides a simple and direct method of analysis of virus oncoproteins in infected or transformed cells when specific antibodies are available against the virus protein. The technique can also be used to study protein–protein interactions between virus and cellular proteins (by co-immunoprecipitation), the stability of the virus protein (by pulse–chase analysis) (*Figure 2*), and the intracellular location of the protein (*Figure 3*). A key factor in the results obtained by immunoprecipitation is the buffer used for solubilization of radiolabelled cells and for washing immunoprecipitates. In our laboratory, two buffers (RIPA and NETN, described below) are routinely used. RIPA often gives cleaner backgrounds in SDS-PAGE, but may not be so appropriate for preserving protein–protein interactions, for which NETN may be preferable. It is therefore worth experimenting with different solubilization and washing buffers, and noting the results obtained.

Immune complexes can be efficiently collected by adsorption to *Staphylococcus aureus* protein A. Most classes of animal immunoglobulins can bind to protein A, which is used either as the purified protein linked to Sepharose, or as *S. aureus* membrane fragments. (It is also possible to use Sepharose-linked protein G for immunoprecipitation; this protein has a wider

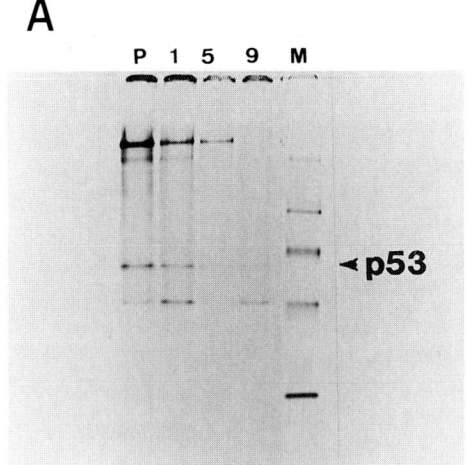
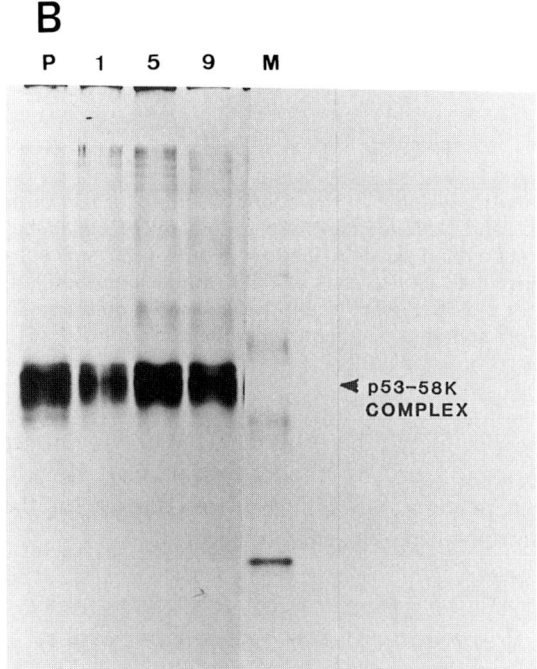

Figure 2. Pulse-chase analysis of the p53 tumour suppressor protein. In (A), primary baby rat kidney (BRK) cells were radiolabelled with [^{35}S]methionine for 1 h, washed, and incubated for the indicated number of hours (*Protocol 14*), then subjected to immuno-precipitation with the monoclonal anti-p53 antibody PAb421. In (B), Ad5-transformed rat cells were subjected to a similar procedure. Note that a complex of E1B-58 kDa and p53 is co-precipitated from Ad5-transformed cells that is stable after 9 h of chase. In contrast, p53 from BRK cells has a short half-life of less than 1 h.

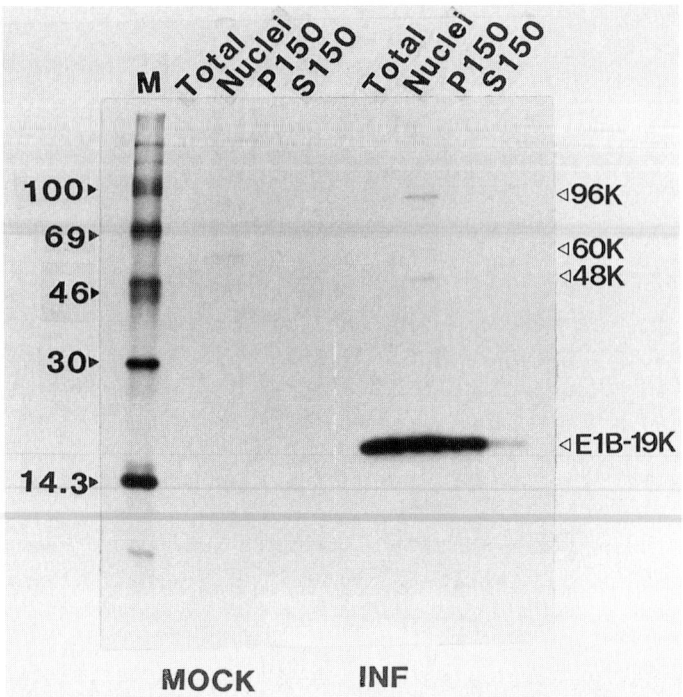

Figure 3. Sub-cellular localization of the adenovirus E1B-19 kDa protein in Ad5-infected Hep-2 cells. Mock- and Ad5-infected Hep-2 cells were radiolabelled for 2 h with [^{35}S]methionine, fractionated as described in *Protocol 13*, and subjected to immunoprecipitation, as described in *Protocol 11*, with monoclonal anti-E1B-19 kDa antibody. Note the enrichment of E1B-19 kDa in the nuclear and membrane (P150) fractions, and the evidence of co-precipitating polypeptides in the nuclear fraction.

antibody-binding specificity than protein A.) While membrane fragments are cheaper to buy (and can be readily prepared), they can give higher backgrounds than protein A-Sepharose. Procedures for the preparation of both types of immunosorbent are given below.

Protocol 9. Preparation of *Staphylococcus aureus* immunosorbent

Equipment and reagents

- *Staphylococcus aureus* (Cowan 1 strain), available from the American Type Culture Collection (ATCC no. 12598)
- Sterile Luria broth (LB): 1% bactotryptone, 0.5% yeast extract, 0.5% NaCl, 0.1% glucose
- PBS plus 0.05% sodium azide
- PBS plus 0.05% sodium azide and 1.5% (w/v) formaldehyde
- Water bath at 80°C
- 1.5 ml sterile microcentrifuge tubes

8: Analysis of DNA virus proteins involved in neoplastic transformation

Method

1. Prepare an inoculum of *Staphylococcus aureus* Cowan I strain (SAC) by adding a few drops of glycerol culture to 10 ml LB, and shaking at 37 °C overnight.
2. Add 8 ml of inoculum to 800 ml LB, and shake for 24 h at 37 °C.
3. Harvest cells by centrifugation at 7000 g for 10 min at 20 °C, and resuspend the cell pellet in 10 ml of 0.05% sodium azide in PBS.
4. Add 190 ml of 0.05% sodium azide in PBS, stir, and centrifuge. Repeat washing the cells.
5. Weigh the cells, and resuspend in 1.5% formaldehyde in 0.05% sodium azide in PBS to approximately 10% w/v.
6. Shake at 23 °C for 90 min, and collect cells by centrifugation.
7. Resuspend the cell pellet in PBS-azide to 10% (w/v), and heat the cell suspension, with occasional shaking, at 80 °C for 5 min.
8. Harvest cells by centrifugation, and wash in 0.05% sodium azide in PBS as before.
9. Prepare a 10% (w/v) suspension of cells in 0.05% sodium azide in PBS, and store in aliquots in sterile 1.5 ml microcentrifuge tubes at −70 °C.

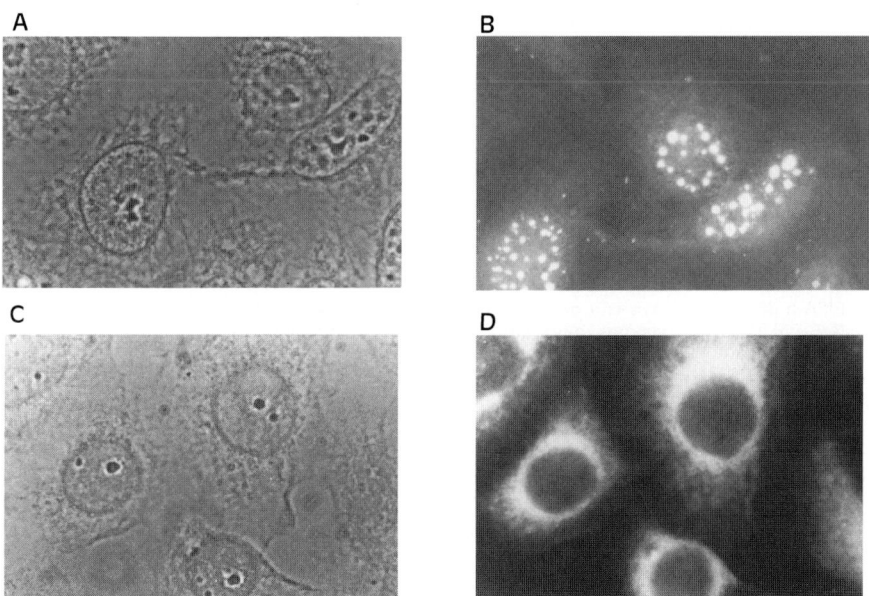

Figure 4. Intracellular localization of the adenovirus E1B-19 kDa and E1B-58 kDa proteins in Ad5-infected Hep-2 cells. Ad5-infected Hep-2 cells were fixed, and subjected to immunofluorescent antibody staining, using monoclonal anti-E1B-58 kDa (A and B) and anti-E1B-19 kDa (C and D) antibodies as described in *Protocol 14*. Slides were viewed under phase-contrast (A and C) and UV epifluorescence (B and D) optics. Note the intranuclear location of E1B-58 kDa, and the cytoplasmic and perinuclear location of E1B-19 kDa.

Protocol 10. Preparation of protein A-Sepharose for immunoprecipitation

Equipment and reagents

- Protein A immobilized on Sepharose beads (Sigma). This is produced as a 5 ml bed volume in a 6.25 ml suspension of 20% ethanol.
- NETN buffer: 50 mM Tris-HCl pH 7.5, 150 mM NaCl, 1mM EDTA, 0.5% NP-40, 1mM PMSF
- RIPA buffer: 20 mM Tris-HCl pH 7.5, 0.1 M NaCl, 0.5% (w/v) sodium deoxycholate, 1.0% (v/v) Triton X-100, 0.1% SDS, 1 mM PMSF

Method

1. Centrifuge 250 μl of the suspension in a microcentrifuge, remove the ethanol, and wash several times in RIPA or NETN buffer, depending on the buffer to be used for immunoprecipitation.

2. Resuspend the pellet in 1 ml (or up to 1 ml) of immunoprecipitation buffer, either RIPA or NETN depending on the buffer to be used for immunoprecipitation, to make a 20% suspension.

3. Leave for 30 min at 4°C to allow the protein A-Sepharose (PAS) to swell again in the buffer.

Protocol 11. Radiolabelling of transformed cells, and immunoprecipitation

Equipment and reagents

- RIPA buffer: 20 mM Tris-HCl pH 7.5, 0.1 M NaCl, 0.5% (w/v) sodium deoxycholate, 1.0% (v/v) Triton X-100, 0.1% SDS, 1 mM PMSF
- ^{35}S-Translabel (ICN)
- DMEM without cysteine and methionine (ICN)
- NETN buffer: 50 mM Tris-HCl pH 7.5, 150 mM NaCl, 1 mM EDTA, 0.5% NP-40, 1 mM PMSF
- Serum-free DMEM
- SDS-PAGE sample buffer
- Dimethylsulfoxide (DMSO, Analar, BDH)
- 20% diphenyloxazole (PPO, Analar, BDH) in DMSO
- Methanol–acetic acid fixation fluid (33% methanol, 13% glacial acetic acid)
- Bath-type sonicator
- Rotary blood mixer

Method

1. Grow transformed cells to 70% confluence in 100 mm tissue-culture dishes.

2. Label cells for 2 h with 50 μCi ml^{-1} of ^{35}S-Trans label (a mixture of [^{35}S] methionine and [^{35}S] cysteine) in 3 ml of methionine- and cysteine-free DMEM.

8: Analysis of DNA virus proteins involved in neoplastic transformation

3. Remove the labelling medium, and wash the cells twice in warmed serum-free DMEM.
4. Scrape the cells, wash by centrifugation in PBS, and resuspend in ice-cold immunoprecipitation buffer (either RIPA or NETN), approximately 10^7 cells in 0.5 ml buffer.
5. Sonicate cells in a bath-type sonicator, and leave on ice for 1 h.
6. (a) Centrifuge the cell lysates by centrifugation at 15 000 g for 5 min at 4°C in a refrigerated microcentrifuge.

    At this point, the supernatant can be pre-cleared of labelled proteins, which form non-specific background bands in SDS-PAGE. (If this is not required, proceed to Step 7).

    (b) Add 10 μl of normal serum (if a polyclonal serum is to be used in Step 7), or 100 μl DMEM plus 10% FCS (if monoclonal antibody-containing culture medium is to be used in Step 7) to 0.5 ml of labelled supernatant for 1 h at 4°C. Then add either 50 μl of 10% SAC or 20% PAS, and rotate for 2 h at 4°C on a rotary blood mixer. Centrifuge for 5 min at 15 000 g in a microcentrifuge.
7. Collect the supernatant. Incubate with culture medium containing monoclonal antibody (100 μl) or immune serum (10 μl) for 18 h at 4°C.
8. Collect immune complexes by addition of 50 μl of 10% SAC or 20% PAS for 1 h at 4°C on a rotary blood mixer. Centrifuge for 5 min at 15 000 g in a microcentrifuge.
9. Wash immunoprecipitates five times by repeated cycles of vortexing the pellet to a homogeneous suspension, and collecting the pellet by centrifugation.

    SAC pellets are more difficult to disperse than PAS pellets, and should therefore be sonicated in a bath sonicator to disperse clumps of immunosorbent.
10. Resuspend the final pellet in 20 μl of SDS-PAGE sample buffer, boil for 3 min, and cool quickly on ice.
11. Centrifuge at 15 000 g for 2 min in a microcentrifuge, and load the supernatant on to an SDS-polyacrylamide gel.
12. Perform electrophoresis.
13. Perform fluorography by fixing the gel in methanol–acetic acid fixation fluid for 30 min at room temperature. Carefully remove all the fixation fluid, and dehydrate the gel in DMSO (two changes, 30 min each), then treat with 20% PPO in DMSO for 1 h.
14. Wash the gel in running tap water for 1 h, dry under vacuum, and expose to X-ray film for 1–3 days at –70°C.

Protocol 11. *Continued*

15. Commercially available reagents can be used in place of Steps 13 and 14.

Protocol 12. Pulse-chase analysis: determination of the stability of virus oncoproteins and interacting cellular proteins

Equipment and reagents

- 100 mm tissue-culture dishes
- ^{35}S-Translabel (ICN)
- DMEM without cysteine and methionine (ICN)
- Serum-free DMEM
- Immunoprecipitation buffer (RIPA or NETN, see *Protocol 11*)

Method

1. Label sub-confluent 100 mm dishes of cells with ^{35}S-Trans label, as described in *Protocol 11*, Step 2.
2. Remove the labelling medium, and wash the monolayer twice in warmed serum-free DMEM.
3. Scrape the cells from one plate (the pulse sample $t = 0$), and add 10 ml of DMEM/10% FCS to each of the remaining plates.
4. Scrape cells from the plates at intervals, wash, and perform immunoprecipitation with appropriate antibodies, followed by SDS-PAGE as described in *Protocol 11*.

3.2 Sub-cellular distribution of virus oncoproteins

The intracellular location of virus oncoproteins can be studied both by biochemical, cell fractionation techniques (*Figure 3*), and by cell-biological, immunocytochemical techniques (*Figure 4*). Procedures for both techniques are given below. In the cell fractionation technique, cells are radioactively labelled, and the virus protein detected by immunoprecipitation; however, this could also be done by Western blotting (*Protocol 15* and *Figure 5*) of unlabelled cell fractions prepared in an identical fashion. It is a good strategy to use both biochemical and immunocytochemical techniques to determine the intracellular localization of a virus protein, to guard against possible limitations of each method (e.g. impurity of cell fractions, possible redistribution of proteins on cell fixation for immunocytochemistry).

8: Analysis of DNA virus proteins involved in neoplastic transformation

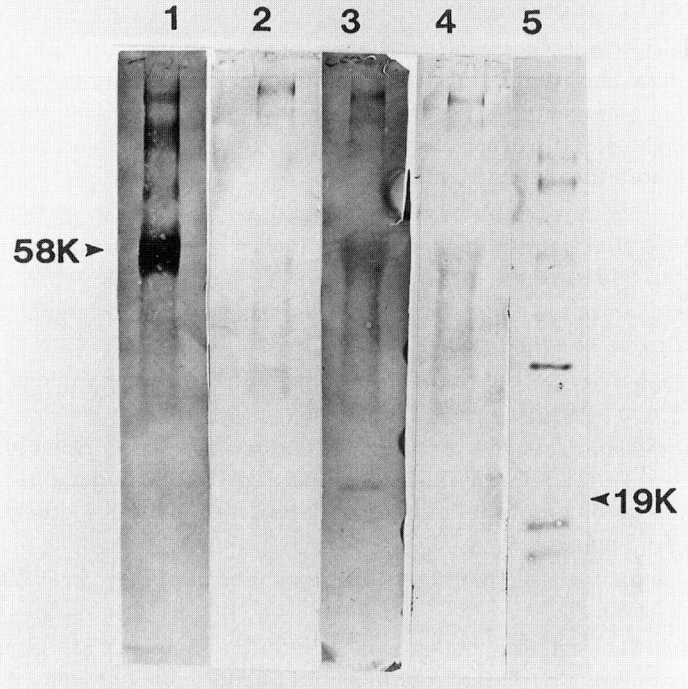

Figure 5. Western blot analysis of adenovirus E1B-58 kDa and E1B-19 kDa proteins in Ad5-transformed cells. Extracts of Ad5-transformed rat cells were subjected to Western blot analysis, as described in *Protocol 17*, using monoclonal anti- E1B-58 kDa (lane 1) and anti-E1B-19 kDa (lane 3) antibodies, as well as control antibody (lanes 2 and 4). Lane 5 contains pre-stained protein molecular weight markers.

Protocol 13. Cell fractionation

Equipment and reagents

- 100 mm tissue-culture dishes
- ^{35}S-Translabel (ICN)
- DMEM without cysteine and methionine (ICN)
- Immunoprecipitation buffer (RIPA or NETN, see *Protocol 11*)
- Lysis buffer: 10 mM Tris-HCl pH 7.5, 1.5 mM MgCl$_2$, 15 mM KCl, 6 mM 2-mercaptoethanol, 1 mM PMSF
- Serum-free DMEM
- 10 × M buffer: 200 mM Tris-HCl pH 7.5, 50 mM MgCl$_2$, 800 mM KCl, 60 mM 2-mercaptoethanol, 10 mM PMSF
- Nuclear wash buffer: 20 mM Tris-HCl pH 7.5, 0.2% (v/v) Triton X-100, 6 mM 2-mercaptoethanol
- 60% sucrose in 1 × M buffer

Method

1. Label six sub-confluent 100 mm plates of transformed or virus-infected cells with ^{35}S-Translabel, as described in *Protocol 11*.

Protocol 13. *Continued*

2. Collect cells by scraping, and wash twice with ice-cold PBS. Prior to the final centrifugation, divide the cell suspension into two portions, one containing 10% of the cells for immunoprecipitation analysis of the whole cell extract. Centrifuge this portion in a microcentrifuge, and store the pelleted cells at −20 °C.
3. Centrifuge the remainder of the cell suspension in a bench-top refrigerated centrifuge, resuspend the cell pellet in 4.5 ml of lysis buffer, and incubate for 10 min on ice.
4. Homogenize the cells in a sterile, all-glass Dounce homogenizer with 20 strokes of a tight pestle.
5. Add 0.5 ml of 10 × M buffer, and centrifuge the homogenate at 1000 g for 5 min at 4 °C.
6. Remove and save the supernatant, and resuspend the pelleted nuclei in 5 ml 1 × M buffer in the Dounce homogenizer, using a loose-fitting pestle, and centrifuge as before. Retain the final pellet (the nuclear fraction) on ice.
7. Combine the two supernatant fractions, and centrifuge at 30 000 g for 2 h at 4 °C.
8. Store the resulting pellet (the membrane fraction) at −20 °C, and dialyse the supernatant (cytoplasmic fraction) against water for 18 h at 4 °C, and freeze-dry.
9. Resuspend the nuclear pellet (from Step 6) in 1 ml of nuclear wash buffer, and keep on ice for 15 min with occasional mixing.
10. Layer the nuclear suspension onto a 4 ml cushion of 60% sucrose in 1 × M buffer, and centrifuge at 20 000 g for 1 h at 4 °C. Collect the pellet, and store at −20 °C.
11. Solubilize the whole cell pellet (from Step 2) and cell fractions in immunoprecipitation buffer (RIPA or NETN), and immunoprecipitate with appropriate antibodies, as described in *Protocol 11*.

Protocol 14. Immunofluorescent antibody staining of virus oncoproteins

Equipment and reagents
- Sterile 2 × 2 cm glass coverslips
- PBS
- 60 mm tissue-culture dishes
- Methanol (at −20 °C)
- Watchmaker's forceps
- Square bacterial Petri dishes, containing moist filter paper
- Primary antibody
- 37 °C incubator
- FITC-conjugated secondary antibody
- Glass microscope slides
- Clear nail varnish
- Vectashield™ (Vector Laboratories)
- Light microscope with epifluorescence optics and camera

8: Analysis of DNA virus proteins involved in neoplastic transformation

Method

1. Grow virus-transformed cells on glass coverslips in 60 mm tissue-culture dishes. Cells grown on coverslips can also be infected with virus. Ideally, when antibody reactions are performed, the cells should not be confluent, but should be growing in well separated colonies, so that cell morphology can be easily discerned.

2. Remove the medium by aspiration. Wash cell monolayers twice with PBS, aspirating the PBS from the dishes.

3. Add methanol, kept at −20°C, and place the dishes in the −20°C freezer for 10 min.

4. Wash monolayers three times with PBS. Holding the coverslips in watchmaker's forceps, carefully drain the excess PBS on to soft tissue.

5. Place the coverslips, cell side up, in the Petri dishes on moistened filter paper. Mark the position of each coverslip on the lid (or the paper, before moistening).

6. Pipette 25 µl of the primary antibody against the virus oncoprotein on to the coverslip, and gently spread it with the tip of the micropipette. Normally, a 1:100 dilution in PBS of a polyclonal antibody is used, or undiluted culture medium containing a monoclonal antibody. Incubate at 37°C for 1–2 h. Include suitable control antibody reactions, e.g. preimmune serum (in the case of polyclonal sera, an unrelated, isotype-matched antibody; in the case of monoclonal antibodies, the secondary antibody alone)

7. Wash the coverslips by holding them in forceps and dipping them up and down several times in each of three beakers of PBS. Drain off the traces of PBS on soft tissue. Place the coverslips back into the humidified dish.

8. Pipette 25 µl of the appropriate FITC-conjugated secondary antibody at the dilution recommended by the manufacturer (normally 1:50 in PBS) on to the coverslips, and incubate for 1–2 h at 37°C.

9. Wash the coverslips as before, drain the remaining PBS on to soft tissue, and mount each coverslip, cell side down, on a drop of Vectashield™ mounting medium on a glass slide. Seal the coverslip on to the slide with nail varnish.

10. When the nail varnish has set, gently clean the coverslip with moistened soft tissue, and view under phase-contrast and epifluorescence optics. Make photographs using a suitable photomicrography system.

Protocol 15. Western blotting to detect virus oncoproteins[a]

Equipment and reagents
- PBS
- RIPA buffer (see *Protocol 11*)
- SDS-PAGE sample buffer
- Dried milk powder
- Nitrocellulose membrane (0.45 μm)
- 10 × transfer buffer: for 1 litre of 10 × buffer, dissolve 30.3 g Tris base and 144 g glycine in 800 ml water, and bring to a final volume of 1000 ml)
- 0.2% Ponceau S in PBS
- Methanol
- Peroxidase enzyme substrate solution: 12 ml of a 3 mg ml^{-1} solution of α-chloronaphthol dissolved in methanol, 48 ml PBS, 20 μl 30% (w/v) H$_2$O$_2$
- Sterile pieces of rubber tubing
- Transformed cells grown to 80% confluence in 100 mm tissue-culture dishes

Method

1. Wash the cells in PBS, and remove them from the surface of the culture dish by gentle scraping with a sterile piece of rubber tubing held in forceps.

2. Remove the cell suspension, and transfer to a sterile Eppendorf tube. Pellet the cells by centrifugation at 15 000 *g* for 2 min in a refrigerated microcentrifuge.

3. Remove the supernatant, and resuspend the cell pellet in 0.5 ml RIPA buffer per dish. Leave the lysate on ice for 1 h.

4. Pellet the cellular debris by centrifugation in a microcentrifuge for 5 min at 4°C. Remove the supernatant, and determine the protein concentration in the supernatant using a standard protein assay system (e.g. from Bio-Rad).

5. For each sample, dilute 10 μg of protein in SDS-PAGE sample buffer, and separate the proteins by SDS-PAGE. At the end of electrophoresis, equilibrate the gel in transfer buffer.

6. Pre-treat a nitrocellulose membrane by soaking in methanol for 3 s, distilled water for 2 min, and transfer buffer for 5 min.

7. Transfer the proteins from the polyacrylamide gel to the pre-treated nitrocellulose membrane at 0.3 A for 3 h, or 0.11 A overnight.

8. When electroblotting is complete, remove the nitrocellulose membrane from the blotting apparatus. At this point, the membrane can be stained in 0.2% Ponceau S solution to show the presence of proteins. The membrane can be destained in water.

9. Incubate the membrane in PBS containing 0.3% dried milk powder at 37°C for at least 1 h. This incubation can be carried out overnight.

10. Wash the nitrocellulose membrane three times in PBS before incubating with the primary antibody against a virus oncoprotein

(diluted to a working concentration in PBS) at 37°C for 1 h, or at 4°C overnight. Typically, dilutions of 1:100 to 1:1000 should be tested for polyclonal antibodies; culture media containing monoclonal antibodies should be used undiluted. Suitable controls should be included, as described in the procedure for immunofluorescent antibody staining above.

11. After incubation with the primary antibody, wash the membrane four times in PBS before adding the secondary antibody diluted to a working concentration. Incubate at 37°C for 1 h.

12. After incubation with the secondary antibody, wash the membrane four times in PBS.

13. Pour the enzyme substrate solution onto the blot, and rock gently at room temperature or 37°C until coloured bands appear on the blot.

14. When coloured bands have appeared, wash the blot four times in PBS.

15. Dry the blot on Whatman 3MM paper, and wrap the blot in foil to preserve the bands, as they may be light-sensitive.

[a] An example of a Western blot is shown in *Figure 5*, where the adenovirus E1B-19 kDa and E1B-58 kDa proteins are detected using monoclonal antibodies.

4. Biological activity of virus oncoproteins

Virus oncoproteins are known to have a number of different biological activities, including the ability to interfere with the expression of cellular genes through interactions with cellular transcription factors. Oncoproteins such as SV40 large T, adenovirus E1A, and HPV E7 function as transactivators, and assays of activation of target promoters linked to reporter genes are useful in studying the biological activity of such virus oncoproteins. The most common reporter gene used is bacterial chloramphenicol acetyl transferase (CAT), which modifies chloramphenicol by mono- and diacetylation. CAT activity can be measured by the conversion of [^{14}C]-chloramphenicol to its acetylated derivatives. These derivatives are separated from unmodified [^{14}C]-chloramphenicol by thin-layer chromatography on silica gels (*Figure 6*). To normalize for DNA uptake into cells, another plasmid is included in the transfection, usually encoding β-galactosidase under the control of a constitutive promoter such as the SV40 early promoter. After transfection, each cell lysate is first assayed for β-galactosidase activity, then volumes of cell extract containing equivalent numbers of units of β-galactosidase are assayed for CAT activity.

Protocol 16. Transactivation assays: preparation of cell extracts

Equipment and reagents
- Reagents for transfection
- PBS
- 0.25 M Tris-HCl pH8.0
- Sonicating water bath

Method
1. Transfect cells with an expression plasmid encoding a virus transactivator, and a suitable plasmid containing a CAT reporter gene linked to a promoter that forms a target for the transactivator. Include also a β-galactosidase expression plasmid in the transfection to normalize for DNA uptake. The proportions of each plasmid in the transfection need to be determined experimentally, but a good guide is to use equal masses of virus transactivator and CAT plasmid, with around one-tenth of the combined plasmid mass as β-galactosidase expression plasmid.
2. Approximately 40 h after transfection, remove the medium from the cells, and collect cells by scraping them from the surface of the tissue-culture dish in cold PBS.
3. Pellet the cells by centrifugation in a refrigerated bench-top centrifuge at 1500 g for 5 min at 4°C.
4. Resuspend the cell pellet in 100 μl 0.25 M Tris-HCl pH 8.0.
5. Lyse the cells by sonication in a sonicating water bath, and then by three cycles of freeze-thawing.
6. Remove cell debris by centrifugation at 15000 g for 3 min at 4°C in a microcentrifuge.
7. Measure the protein concentration of each cell extract, using a commercial protein assay kit.

Protocol 17. Assay of β-galactosidase activity in transfected cell extracts

Equipment and reagents
- o-nitrophenylgalactoside: ONPG, 4 mg ml^{-1} in 0.1 M sodium phosphate buffer pH 7.5
- Z buffer: Na$_2$HPO$_4$.7H$_2$O (16.1 g l^{-1}), Na$_2$HPO$_4$.H$_2$O (5.5 g l^{-1}), KCl (0.75 g l^{-1}), MgSO$_4$.7H$_2$O (0.246 g l^{-1}).
- 1 M Na$_2$CO$_3$
- Standard β-galactosidase solution (Sigma)
- Visible light spectrophotometer

Method
1. Incubate the 50 μl of cell extract with 200 μl ONPG and 500 μl of Z buffer

8: Analysis of DNA virus proteins involved in neoplastic transformation

2. Incubate the reactions at 37°C until a detectable yellow colour is visible.
3. Stop the reaction by quenching with 500 μl of 1 M Na_2CO_3, prior to reading the absorbance at 420 nm.
4. A standard curve can be constructed using 0.002–0.014 units of a standard β-galactosidase solution.

Protocol 18. Assay of chloramphenicol acetyl transferase (CAT) in transfected cell extracts[a,b]

Equipment and reagents

- [^{14}C]-chloramphenicol (ICN)
- 4 mM acetyl-CoA
- 0.25 M Tris-HCl pH 7.8
- Centrifugal evaporator
- Silica gel thin-layer chromatography plates (Merck silica gel 60 plates, 20 × 20 cm)
- Thin-layer chromatography tank, equilibrated with chloroform:methanol (95:5)
- Scintillation counter
- Scintillation vials
- Scintillation fluid

Method

1. Heat-treat aliquots of cell lysate at 68°C for 10 min to inactivate endogenous deacetylase activity in the cell lysates
2. Based on the protein concentration and the number of units of β-galactosidase in each extract of transfected cells, incubate aliquots of cell extract (0.5–20 μl) in a final volume of 150 μl in 0.25 M Tris-HCl, pH 7.5, at 37°C for 1 h in the presence of 0.2 μCi [^{14}C]-chloramphenicol, and acetyl-CoA at a concentration of 0.53 mM.
3. Extract the acetylated products plus labelled chloramphenicol in 1 ml of ethyl acetate, dry in the centrifugal evaporator, and resuspend in 30 μl of ethyl acetate.
4. Spot the samples onto a silica gel thin-layer chromatography plate, and separate the products using ascending thin-layer chromatography in chloroform:methanol.
5. Air-dry the plate, and expose to X-ray film.
6. The percentage acetylation of chloramphenicol can be quantified by scintillation counting of appropriate regions of the thin-layer chromatography plate. To do this, use the autoradiograph as a guide, and rule out sectors with a pencil corresponding to the acetylated spots and the chloramphenicol substrate. Using a scalpel, carefully scrape off the silica from each sector in each assay on to a piece of aluminium foil, and transfer to a scintillation vial. Add scintillation fluid, and count in a scintillation counter. Take care during this

Protocol 18. *Continued*

> process: wear a face mask, and do not inhale radioactive silica particles. The quantification of CAT activity can be less hazardously estimated by using a phosphorimager or other radioactivity imaging device.
>
> [a] % CAT conversion = $\dfrac{\text{c.p.m. of acetylated products}}{\text{total c.p.m.}} \times 100$
>
> [b] Adapted from ref. 28.

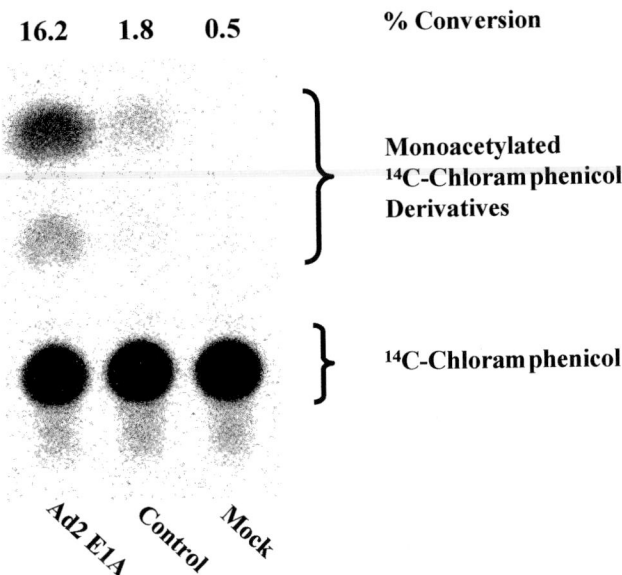

Figure 6. Transactivation of the adenovirus E2 promoter by E1A. Rat fibroblast 3Y1 cells were co-transfected (using the calcium phosphate method) with an Ad2 E1A expression plasmid, and a plasmid containing the Ad2 E2 promoter linked to CAT. CAT enzyme activity was assayed as described in *Protocols 16 and 18*. Note the 8–10-fold transactivation of the E2 promoter by E1A, compared with the control (without E1A). Percentage CAT conversion was determined by phosphorimaging, using a Fuji BAS1000 phosphorimager.

4.1 Studies of interactions between virus oncoproteins and cellular proteins

A number of methods have been used to investigate the interactions between virus oncoproteins and cellular proteins. Co-precipitation of a cellular protein with a virus protein has been frequently used to demonstrate such associations. An example of this is shown in *Figure 2*, where the cellular tumour suppressor protein p53 co-precipitates the adenovirus 5 E1B-58 kDa protein.

8: Analysis of DNA virus proteins involved in neoplastic transformation

Other approaches to the study of such interactions include 'pull-down' assays, where the cellular protein is bound *in vitro* to a tagged virus protein (commonly a bacterial fusion protein). Another powerful protein–protein interaction technique is the yeast two-hybrid assay (28–30).

Whether the 'pull-down' or the two-hybrid system is used, it should be stressed that proposed interactions between virus and cellular proteins should be confirmed in cultured cells (for example, by showing that a complex exists by co-immunoprecipitation).

4.1.1 Application of GST-fusions to study interactions between virus oncoproteins and cellular proteins

The interaction between a virus oncoprotein and a cellular protein can be studied by making a fusion of the virus oncoprotein with the glutathione-S-transferase (GST) protein. The fusion protein can then be bound to Sepharose beads on which glutathione has been immobilized. Cellular extracts are then incubated with the beads carrying the bound fusion protein, and complexes formed between the virus protein and cellular proteins can be purified by centrifuging to pellet the beads. An example of how such a 'GST-pull down' assay has been used is the use of GST-E1A fusion proteins in the study of the role of adenovirus E1A protein domains in complex formation (31).

Protocol 19. Preparation of GST fusion proteins[a]

Equipment and reagents

- 250 ml of LB medium (*Protocol 9*)
- Isopropylthiogalactoside (IPTG) 1M
- Glutathione-Sepharose beads (Sigma)
- Reduced glutathione: 5 mM in 50 mM Tris-HCl pH 8.0
- Probe sonicator

Method

1. Induce overnight cultures for 3 h with IPTG at a final concentration of 100 μM.
2. Centrifuge the cells at 3000 g for 30 min at 4°C.
3. Lyse the cell pellet by sonication in PBS.
4. Pellet the cell debris by centrifugation at 3000 g for 10 min at 4°C.
5. Incubate with glutathione-Sepharose 4B beads for 1 h at 4°C.
6. Wash the bound material with PBS, and elute with 5 mM free reduced glutathione in 50 mM Tris-HCl pH 8.0.
7. Dialyse overnight at 4°C against 50 mM Tris-HCl pH 8.0 containing 150 mM NaCl.

[a] Modified from ref. 33.

> **Protocol 20.** Affinity binding assays with GST-E1A
>
> *Equipment and reagents*
> - Radiolabelled cell extracts (*Protocol 11*)
> - Glutathione-Sepharose beads
> - PBS
> - Reagents for SDS-PAGE (*Protocol 11*)
>
> *Method*
> 1. Mix the GST fusion proteins with glutathione-Sepharose 4B beads. As a control, glutathione-Sepharose 4B beads should also be mixed with GST alone.
> 2. Incubate [^{35}S]methionine-labelled cell extracts with glutathione-Sepharose 4B beads and 27 µg of GST. This step acts to 'pre-clear' the cell extracts by removing non-specific binding activity.
> 3. Centrifuge in a microcentrifuge for 5 min. Remove the supernatant fraction.
> 4. Now that the labelled cell extracts have been 'pre-cleared', they can be incubated either with the Sepharose-bound fusion proteins, or with Sepharose-bound GST (as control) at room temperature for 30 min.
> 5. Centrifuge in a microcentrifuge for 5 min.
> 6. Add SDS-PAGE sample buffer, and analyse by electrophoresis and fluorography (*Protocol 11*).

4.2 Use of the yeast two-hybrid screen to identify interactions between virus oncoproteins and cellular proteins

The yeast two-hybrid assay is a method which allows the identification of proteins which form complexes in yeast cells. In this method, two proteins of interest are fused as hybrid proteins, one to a DNA-binding domain, and the other to a transcription-activating domain (27). The original method developed by Fields and Song (28) utilized the yeast GAL4 transcriptional activator protein to identify an interaction between the yeast SNF1 protein, a serine-threonine specific protein kinase, and the SNF4 protein which is required for maximal activity of SNF1. The native yeast GAL4 protein contains an N-terminal domain which binds DNA in a sequence-specific manner, but is unable to activate transcription, and a C-terminal region which contains transcriptional activation domains. Two hybrid proteins were created, by fusing one of two cDNAs encoding either SNF1 or SNF4 to the GAL4 DNA-binding domain or the GAL4 activating domain, respectively. These two plasmids were introduced into a yeast strain that is deleted for the *GAL4* and *GAL80* genes, and contained two reporter genes *HIS3* and *lacZ* that are

8: Analysis of DNA virus proteins involved in neoplastic transformation

controlled by GAL4-binding sites. The GAL4 DNA-binding domain and activating domain are unable to activate transcription of the *HIS3* and *lacZ* genes unless the two candidate proteins in each fusion are able to interact. If the two candidate proteins are able to interact, then the activating region of GAL4 is brought to its normal site of action, and the transcriptional activity of the GAL4 protein is reconstituted, allowing transcription of the reporter gene. If an interaction is observed, then deletions in each, and point mutations in each one, of the two candidate cDNAs can be introduced to define the minimum domain necessary for interaction. This method has since been modified (29) so that one of the candidate proteins is fused to the yeast LexA DNA-binding domain, and the other protein of interest is fused to the activating domain of the VP16 protein. The two plasmids are then introduced into a yeast strain which contains two reporter genes *HIS3* and *lacZ* that are under the control of LexA binding sites.

The yeast two-hybrid system has been used as a genetic method of studying the interactions between virus oncoprotein and cellular proteins. One particular example is the use of a yeast two-hybrid method to analyse the interactions of the adenovirus E1B 19K protein (31). The E1B 19K protein prevents apoptosis stimulated by the virus E1A protein. Using the yeast two-hybrid system, cDNAs encoding three different proteins designated Nip1, Nip2, and Nip3 were identified, and shown to interact with the E1B 19K protein.

For each protein tested, it is important to include the appropriate controls. An essential control is to test that each fusion cannot activate reporter gene expression autonomously without the presence of the other fusion construct. In order to carry out this control, each of the two fusion plasmids should be introduced into yeast with a control plasmid such as a DNA-binding domain–lamin fusion, or an activating domain–lamin fusion. Prior to starting the two-hybrid screen, it is also advisable to verify that a functional fusion protein is being made. One way to do this is to test for interaction with a protein that is known to interact with the candidate protein or, if no such protein is available, antibodies specific to the DBD or LBD can be used in a Western blot analysis. A final consideration is that the candidate proteins may contain signal sequences that direct them to subcellular compartments other than the nucleus. In this case, modification of these signal sequences to allow nuclear localization may be required.

The first step in the yeast two-hybrid assay system is the construction of the fusion constructs containing the two proteins of interest, and the DBD and AD of either GAL4 or LexA. To make these constructs, the candidate cDNAs are cloned into the polylinker of the plasmid carrying the GAL4 DBD (or LexA DBD) and the GAL4 AD (or the LexA AD). The inserts can be sequenced using either DBD or AD primers, to verify that they are in frame. The fusion vectors can now be introduced into the yeast tester strain L40.

Protocol 21. Preparation of media

Equipment and reagents

- For YPD medium: Bacto yeast extract (Difco), Bacto-peptone (Difco), 40% glucose (sterile)
- For SD medium: Bacto-yeast nitrogen base without amino acids (Difco)
- Drop-out mix contains the following ingredients, minus the appropriate supplement: adenine 0.5 g, alanine 2.0 g, arginine 2.0 g, asparagine 2.0 g, aspartic acid 2.0 g, cysteine 2.0 g, glutamine 2.0 g, glutamic acid 2.0 g, glycine 2.0 g, histidine 2.0 g, inositol 2.0 g, isoleucine 2.0 g, leucine 10.0 g, lysine 2.0 g, methionine 2.0 g, phenylalanine 2.0 g, proline 2.0 g, serine 2.0 g, threonine 2.0 g, tryptophan 2.0 g, tyrosine 2.0 g, uracil 2.0 g, valine 2.0 g
- In some cases, the addition of 3-amino-1,2,4-triazole (3-AT) (20–30 mM) to the SD medium is required.
- 0.1 M lithium acetate
- TE: 10 mM Tris-HCl, 1 mM EDTA, pH 7.5
- 40% PEG 3300
- 10 mg ml^{-1} carrier DNA: can be obtained by purchasing sheared and denatured herring testis DNA, or prepared by dissolving 1 g of salmon sperm DNA in 100 ml TE pH 8.0.
- Z buffer: *Protocol 17*
- X-gal stock solution: dissolve 20 mg of 5-bromo-4-chloro-3-indolyl-β-D-galactoside in 1 ml N,N-dimethylformamide (DMF)
- o-nitrophenyl β-D-galactopyranoside (ONPG): 4 mg ml^{-1} of ONPG in water
- 0.1% (w/v) SDS
- 1 M Na$_2$CO$_3$
- Yeast-lysis buffer: 2% Triton X-100, 1% SDS, 100 mM NaCl, 10 mM Tris-HCl pH 8.0, 1.0 mM EDTA
- Phenol:chloroform:isoamyl alcohol (25:24:1, v/v)
- Acid-washed glass beads
- 3 M sodium acetate

Method

1. YPD medium: dissolve 10 g of Bacto-yeast extract and 20 g of Bacto-peptone (if making plates, then add 20 g Bacto-agar) in 950 ml of water. Adjust the pH to 6.0, autoclave, and cool to 55°C. Add 50 ml of a sterile stock solution of 40% glucose.

2. SD medium: dissolve 6.7 g Bacto-yeast nitrogen base without amino acids, and 2 g drop-out mix in 950 ml water (include 20 g Bacto-agar if making plates). Adjust the pH to 6.0, autoclave, and cool to 55°C. Add 50 ml of a sterile stock solution of 40% glucose.

3. M9 medium: prepare 10 × M9 salts by dissolving the following salts in water to a final volume of 1 litre: 58 g Na$_2$HPO$_4$, 30 g KH$_2$PO$_4$, 5.0 g NaCl, and 10.0 g NH$_4$Cl. Dissolve 10 × M9 salts in 887 ml water, and add 15 g agar. Autoclave, then add 10 ml of 20% glucose, 1 ml of 100 mM CaCl$_2$, 1 ml of 1 M MgSO$_4$, and 1 ml 1 M thiamine-HCl.

4. LB medium (*Protocol 9*), and LB-Ampicillin (50 μg ml^{-1}).

Protocol 22. Preparation of competent yeast cells

Equipment and reagents
- See *Protocol 21*

Method

1. Inoculate a yeast colony into 10 ml of YPD, and grow overnight at 30°C with shaking.

8: *Analysis of DNA virus proteins involved in neoplastic transformation*

2. Transfer the overnight culture into 100 ml YPD, and incubate the culture at 30°C with shaking at 230 r.p.m. until an OD_{600} of 0.5–0.8 is reached.
3. Centrifuge the cells at 1500 *g* for 5 min at room temperature.
4. Discard the supernatants, and resuspend the pellet in 25–50 ml of 0.1 M lithium acetate in TE.
5. Centrifuge the cells at 1500 *g* for 5 min at room temperature, and resuspend the washed cells in 1 ml of 0.1 M lithium acetate in TE.
6. Incubate the cells for 1 h at 30°C with shaking at 230 r.p.m. The cells are now competent for transformation.

Protocol 23. Transformation of competent yeast cells

Equipment and reagents
- See *Protocol 21*

Method
1. Add 100 µl of competent yeast cells (*Protocol 22*) for each transformation into a 1.5 ml microcentrifuge tube.
2. Add the plasmid DNAs (approximately 0.5 to 2 µg of each plasmid) and 100 µg of sheared, denatured salmon-sperm DNA to 100 µl of competent yeast cells, and mix.
3. Add 600 µl of sterile PEG-lithium acetate solution to each tube, and mix well by inversion.
4. Incubate at 30°C for 30–60 min (shaking is not required).
5. Heat shock for 15–30 min in a 42°C waterbath.
6. Pellet the cells by centrifugation for 15 s at 15 000 *g* in a microcentrifuge.
7. Remove the supernatants, and resuspend the cells in 100 µl of sterile TE
8. Spread on 100 × 150 mm plates containing the appropriate SD medium (SD-Leu-Trp).
9. Incubate plates at 30°C until colonies appear (this usually takes 2–4 days).

Protocol 24. Measuring the transcriptional activity of the reporter gene

Equipment and reagents
- See *Protocol 21*

Protocol 24. *Continued*

Method

1. Replica plate the yeast colonies directly (or pick colonies with sterile toothpicks from transformation plates, and spread as small patches) on SD-Leu-Trp plates, then replica plate on SD-Leu-Trp-His plates, and finally on filter-paper circles (90 mm Whatman #50) placed on SD-Leu-Trp plates.
2. Incubate the plates at 30 °C for 1–2 days. Monitor growth on SD-Leu-Trp-His plates, and perform a β-galactosidase assay (described in Steps 4–6 below).
3. Allow the replica plated yeast colonies to grow overnight.
4. Remove the filter paper from the plate, and permeabilize the yeast cells by freezing the filter paper in liquid nitrogen.
5. Place the filter paper carrying the permeabilized yeast cells (yeast cells facing up) into a Petri dish containing a second filter paper circle (90 mm Whatman #3) that has been soaked in Z-buffer/X-Gal solution (2.5 ml Z buffer/X-gal solution per Petri dish).
6. Incubate at 30 °C, and check periodically for the appearance of blue colonies. The formation of blue colonies may take between 30 min and 10 h.

Acknowledgements

Work in the authors' laboratory is supported by Yorkshire Cancer Research. We acknowledge the late Sara Dixon and Nikolaos Georgopoulos for provision of figures.

References

1. Fanning, E. (ed.) (1994). *Seminars Virol.*, **5**, 1.
2. Boulanger, P. A., and Blair, G. E. (1991). *Biochem. J.*, **275**, 281.
3. Blair, G. E., and Hall, K. T. (1998). *Seminars Virol.*, **8**, 387.
4. Chinnadurai, G. (1998). *Seminars Virol.*, **8**, 399.
5. Flint, J., and Shenk, T. (1997). *Annu. Rev. Genet.*, **31**, 177.
6. Manfredi, J. J., and Prives, C. (1994). *Biochim. Biophys. Acta*, **1198**, 65.
7. Brizuela, L., Olcese, L. M., and Courtneidge, S. A. (1994). *Seminars Virol.*, **5**, 381.
8. Phillips, A. C., and Vousden, K. H. (1998). In *Viruses and human cancer* (ed. J. R. Arrand, and D. R. Harper). p. 39. BIOS Scientific Publishers, Oxford.
9. Tommasino, M., and Crawford, L. (1995). *BioEssays*, **17**, 509.
10. Hesketh, R. (1994). *The oncogene handbook.* Academic Press, San Diego, CA.
11. Spencer, T. E., Jenster, G., Burcin, M. M., Allis, C. D., Zhou, J., Mizzen, C. A., McKenna, N. J., Onate, S. A., Tsai, S. Y., Tsai, M-J., and O'Malley, B. W. (1997). *Nature*, **389**, 194.

12. Prives, C. (1998). *Cell,* **95,** 5.
13. Brodsky, J. L., and Pipas, J. M. (1998). *J. Virol.,***72,** 5329.
14. Conzen, S. D., Snay, C. A., and Cole, C. N. (1997). *J. Virol.,* **71,** 4536.
15. Lill, N. L., Tevethia, M. J., Eckner, R., Livingstone, D. M., and Modjtahedi, N. (1997). *J. Virol.,* **71,** 129.
16. Summers, S. A., Lipfert, L., and Birnbaum, M. (1998). *Biochem. Biophys. Res. Commun.,* **246,** 76.
17. Yi, X., Peterson, J., and Freund, R. (1997). *J. Virol.,* **71,** 6279.
18. Su, W., Liu, W., Schaffhausen, B. S., and Roberts, T. M. (1995). *J. Biol. Chem.,* **270,** 12331.
19. Cullere, X., Rose, P., Thathamangalam, U., Chatterjee, A., Mullane, K. P., Pallas, D. C., Benjamin, T. L., Roberts, T. M., and Schaffhausen, B. S. (1998). *J. Virol.,* **72,** 558.
20. Zehbe, I., Wilander, E., Delius, H., and Tommasino, M. (1998). *Cancer Res.,* **58,** 829.
21. Zhu, J., Rice, P. W., Abate, M., and Cole, C. (1992). *J. Virol.,* **66,** 2780.
22. Lin, J-Y., and Simmons, D. T. (1991). *J. Virol.,* **65,** 6447.
23. DeFeo-Jones, D., Vuocolo, G. A., Haskell, K. M., Hanobik, M. G., Kiefer, D. M., McAvoy, E. M., Ivey-Hoyle, M., Brandsma, J. L., Oliff, A., and Jones, R. (1993). *J. Virol.,* **67,** 716.
24. Williams, J. F. (1973). *Nature,* **243,** 162.
25. Precious, B., and Russell, W. C. (1985). In *Virology: a practical approach* (ed. B. W. J. Mahy), p. 193. IRL Press, Oxford.
26. Gorman, C. (1985). In *DNA cloning: a practical approach* (ed. D. M. Glover). Vol. 2, p. 143. IRL Press, Oxford.
27. Van Aelst, L. (1998). In *Transmembrane signaling protocols.* (ed. D. Bar-Sagi). p. 201. Humana Press, Totowa, NJ.
28. Fields, S., and Song, O. (1989). *Nature,* **340,** 245.
29. Hollenberg, S. M., Sternglanz, R., Cheng, P. F., and Weintraub, H. (1995). *Mol. Cell. Biol.,* **15,** 3813.
30. Barbeau, D., Charbonneau, R., Whalen, S. G., Bayley, S. T., and Branton, P. E. (1994). *Oncogene,* **9,** 373.
31. Boyd, J. M., Malstrom, S., Subramanian, T., Venkatesh, L. K., Schaeper, U., Elangovan, B., D'Sa-Eipper, C., and Chinnadurai, G. (1994). *Cell,* **79,** 341.

9

Chemotherapy of DNA virus infections

PATRICIA A. CANE and DEENAN PILLAY

1. Introduction

In recent years there have been considerable advances in the development of effective antiviral drugs. Much of this progress has been driven by the need to combat human immunodeficiency virus (HIV) infection, but there are currently nearly thirty antiviral agents in clinical development or licensed for use in humans against DNA viruses. These are listed, together with their principal activities, in *Table 1*. In theory, any of the stages of the replication cycle of a virus (attachment, penetration, uncoating, replication, maturation, and release) are potential targets for antiviral drugs, but in the case of DNA viruses most of the effective drugs act by inhibiting DNA replication. Such drugs are mainly either nucleoside analogues such as acyclovir, ganciclovir, and lamivudine, or phosphate analogues such as foscarnet.

2. Antivirals effective against herpesviruses

2.1 Aciclovir

Aciclovir is highly active against herpesviruses, and is used clinically for herpes simplex virus (HSV) and varicella-zoster virus (VZV). The target for aciclovir is the virus-encoded DNA polymerase. Both HSV and VZV encode a thymidine kinase (TK) capable of phosphorylating, in addition to thymidine, uridine, cytidine, thymidylate (dTMP), and a variety of nucleoside analogues, including compounds such as aciclovir which do not have a pyrimidine base. Thus, HSV- and VZV-infected cells generate aciclovir monophosphate far in excess of that produced by uninfected cells. Cellular enzymes mediate further phosphorylation. Inhibition of the viral DNA polymerase by this compound may occur as follows:

(a) aciclo-GTP competes with dGTP in a reversible manner
(b) aciclo-GTP acts as a substrate, and is incorporated into the growing DNA chain

Table 1. Antiviral agents active against DNA viruses

Generic name	Systematic name	Principal activities
aciclovir	9-(2-hydroxyethoxymethyl)guanine	HSV-1, HSV-2, VZV, EBV, CMV
adefovir dipivoxil	bis(pivaloyloxymethyl)-9-[2-(phosphonomethoxy)ethyl]adenine	HSV-1, HSV-2, HHV-6, CMV, EBV, HBV
benzimidavir	5,6-dichloro-2-(isopropylamino)-1-(β-L-ribofuranosyl)benzimidazole	CMV
brivudin	(E)-5-(2-bromovinyl)-2'-deoxyuridine	HSV-1, VZV, EBV
cidofovir	(S)-1-(3-hydroxy-2-phosphonyl-methoxypropyl)cytosine	CMV, HSV-1, HSV-2, VZV, EBV, HHV-6, HPV, ADV
cyclic HPMPC	1-[((5)-2-hydroxy-2-oxo-1,4,2-dioxaphosphorinan-5-yl)methyl] cytosine	CMV, HSV-1, HSV-2, VZV, EBV, HHV-6, HPV, ADV
n-docosanol	1-docosanol	HSV-1, HSV-2, CMV, VZV, HHV-6
fam/ciclovir	diacetyl 6-deoxy-9-(4-hydroxy-3-hydroxymethyl-but-l-yl)guanine	HSV-1, HSV-2, VZV, EBV, HBV
fiacitabine	1-(2'-deoxy-2'-fluoro-β-D-arabino-furanosyl)-5-iodocytosine	HSV-1, HSV-2, CMV, HBV
fialuridine	1-(2'-deoxy-2'-fluoro-β-D-arabino-furanosyl)-5-iodouracil	HSV-1, HSV-2, HBV
fomivirsen sodium	ISIS 2922 (antisense)	CMV
foscarnet	trisodium phosphonoformate hexahydrate	HSV-1, HSV-2, VZV, CMV, EBV, HHV-6, HBV
(-) - FTC	2',3'-dideoxy-5-fluoro-3'-thiacytidine	HBV
ganciclovir	9-(1,3-dihydroxy-2-propoxymethyl) guanine	HSV-1, HSV-2, VZV, CMV, EBV, HHV-6, HBV
GEM 132	antisense	CMV
idoxuridine	5-iodo-2'-deoxyuridine	HSV-1, HSV-2, VZV
lamivudine	(-)-β-L-2',3'-dideoxy-3'-thiacytidine	HBV
lobucavir	(R)-9-[2,3-bis(hydroxymethyl) cyclo-butyl] guanine	HSV-1, HSV-2, VZV, EBV, CMV, HBV
netivudine	1-(β-D-arabinofuranosyl)-5-(1-propynyl)uracil	HSV-1, VZV
penciclovir	9-(4-hydroxy-3-hydroxymethyl-but-1-yl) guanine	HSV-1, HSV-2, VZV, EBV, HBV
sorivudine	1-β-D-arabinofuranosyl-(E)-5-(2-bromovinyl)uracil	HSV-1, HBV, VZV
trifluridine	5-trifluoromethyl-2'-deoxyuridine	HSV-1, HSV-2
valaciclovir	L-valine, 2-[(2-amino-1,6-dihydro-6-oxo-9H- purin-9-yl)methoxy]-ethyl-ester	HSV-1, HSV-2, VZV, EBV, CMV
vidarabine	9-β-D-arabinofuranosyladenine monohydrate	HSV-1, HSV-2, VZV, CMV, HBV

(c) the polymerase cannot continue chain elongation, since acyclo-GTP does not contain a 3' hydroxyl group.

At this point, it is thought that the polymerase becomes inactivated in a complex with the subsequent deoxynucleoside triphosphate. The cellular DNA polymerase α is relatively insensitive to ACV-triphosphate, which further enhances the drug selectivity.

9: Chemotherapy of DNA virus infections

Resistance to aciclovir may occur by mutations either in the TK or the DNA polymerase genes (1). Three TK phenotypes are recognized:

- TK negative
- TK partial, in which the enzyme expressed has sufficiently low activity to represent aciclovir resistance
- TK altered, in which TK is capable of phosphorylating natural substrates but not aciclovir

A number of mutations within the TK gene have been identified. Of interest, although TK is non-essential for HSV replication in cell culture, there is evidence that TK activity is required *in vivo* for reactivation and pathogenicity. It is possible that so-called TK-negative drug-resistant viruses *in vivo* do represent mixtures of phenotypes, with a small component of TK-expressing virus.

2.2 Penciclovir

Penciclovir has a similar mode of action to that of aciclovir. It is preferentially monophosphorylated by HSV and VZV TKs, with minimal activation by host-cell polymerases. It appears that the intracellular half-life of penciclovir triphosphate is longer than that of aciclovir triphosphate. By contrast, the aciclovir species may have a higher affinity for the viral polymerase than penciclovir triphosphate. The clinical implications of these differences are unclear, since the clinical benefit of the two drugs appears similar. Valaciclovir and famciclovir represent the pro-drugs of aciclovir and penciclovir, respectively, providing a higher bioavailability when taken orally.

Not surprisingly, the majority of aciclovir-resistant clinical isolates of HSV and VZV are cross-resistant to penciclovir, since the mechanisms of action of the two drugs are very similar.

2.3 Foscarnet

Foscarnet is a pyrophosphate analogue with activity against HSV, VZV, and cytomegalovirus (CMV), as well as retroviruses and influenza. It reversibly and non-competitively inhibits the herpesvirus DNA polymerase by binding to the site normally occupied by pyrophosphate, which is a product of nucleic acid polymerization (2). Selectivity arises from the viral polymerase being more sensitive than cellular polymerases to this drug. Foscarnet does not therefore require prior phosphorylation for activity. As would be expected, foscarnet resistance maps to the viral polymerase gene and, in addition, virtually all HSV isolates resistant to foscarnet are cross-resistant to aciclovir. This implies that aciclovir triphosphate and foscarnet interact with polymerase at similar sites.

2.4 Cidofovir

Cidofovir is a cytidine nucleotide analogue with *in vitro* activity against a broad spectrum of herpesviruses, including CMV, HSV, VZV, EBV, HHV-6, and HHV-8, as well as adenovirus, human papillomavirus, polyomaviruses, and poxviruses. The phosphorylation of cidofovir is not dependent on viral enzymes. Following diphosphorylation by cellular enzymes, the compound acts both as a potent inhibitor of and as an alternative substrate for the viral polymerase, in competition with dCTP (3). Incorporation of cidofovir leads to reduced DNA synthesis, with cessation of synthesis following the incorporation of two molecules of CDVpp. The specificity of this drug for viral DNA synthesis results from a much higher affinity of CDVpp for viral DNA polymerase than for host-cell enzymes.

The prevalence of aciclovir-resistant HSV is estimated at 5–10% in immunocompromized patients, but is much rarer in those with normal immunity. Controlled trials have demonstrated efficacy of foscarnet and topical cidofovir for treatment of aciclovir-resistant mucocutaneous HSV infection. Little is known regarding cidofovir drug-resistant viruses.

2.5 Testing susceptibility of HSV to antiviral drugs

The traditional method for testing susceptibility of HSV to antiviral drugs is by a plaque reduction assay as outlined in *Protocols 1–3*. The principle of this test is the detection of inhibition of plaque formation at a range of concentrations of the test drug.

Protocol 1. Growing stocks of HSV

Equipment and reagents

- MEM medium supplemented with Earle's salts, L-glutamine, 25 mM Hepes, 2% FCS, 100 IU ml^{-1} penicillin, and 100 µg ml^{-1} streptomycin
- Vero or MRC5 cell monolayer freshly confluent in a 25 cm^2 or 75 cm^2 flask

Method

1. Remove the medium from the monolayer.
2. Inoculate the flask with 100 µl virus in 5 ml MEM for a 75 cm^2 flask, 20 µl virus in 2 ml MEM for a 25 cm^2 flask, and swirl the inoculum over the cell sheet.
3. Incubate at 35°C for 1 h
4. Add further medium, 15 ml for a 75 cm^2 flask, 4 ml for a 25 cm^2 flask.
5. Incubate at 35°C.
6. Examine daily; when the maximum CPE has developed, scrape off the cells into the medium, and repeatedly pipette to disperse.

9: Chemotherapy of DNA virus infections

7. Freeze–thaw rapidly to disrupt cells, or sonicate for 30–60 s.
8. Remove cell debris by centrifuging at 2000 g for 10 min.
9. Aliquot the supernatant in small volumes (0.1–0.2 ml), and freeze at −70 °C.

Protocol 2. Plaque assay for HSV

Equipment and reagents

- Dissecting microscope
- Vero cells
- MEM medium supplemented with Earle's salts, L-glutamine, 25 mM Hepes, 2% FCS, 100 IU ml^{-1} penicillin, and 100 μg ml^{-1} streptomycin
- 24-well tissue-culture plates
- 4% w/v carboxymethyl cellulose (CMC) in PBS; autoclave
- Overlay: 1 part 4% CMC and 3 parts medium
- 10% formalin in PBS
- Crystal violet (0.5% in 20% methanol)

Method

1. Seed 24-well tissue-culture plates with Vero cells (2×10^5 cells per well), and grow to confluence, normally overnight at 37 °C in 5% CO_2.
2. Make tenfold dilutions of virus in medium, up to 10^{-6}.
3. Remove medium from wells, and inoculate with 0.2 ml of virus dilution per well.
4. Incubate at 37 °C in 5% CO_2 for 1 h.
5. Remove the inoculum, and add 1 ml overlay to each well.
6. Incubate at 37 °C for 48 h.
7. Aspirate off the overlay gently, and flood the wells with 10% formalin in PBS. Leave at room temperature for 30 min.
8. Remove the formalin, and rinse the wells with water.
9. Flood the wells with Crystal violet, leave for 5–10 min, rinse with tap water, drain, and allow to air-dry.
10. Count the plaques, using a dissecting microscope.

Protocol 3. Plaque reduction assay for testing antivirals against HSV

Equipment and reagents

- Vero cells
- MEM medium supplemented with Earle's salts, L-glutamine, 25 mM Hepes, 2% FCS, 100 IU ml^{-1} penicillin, and 100 μg ml^{-1} streptomycin
- 24-well tissue-culture plates
- 4% w/v carboxymethyl cellulose (CMC) in PBS; autoclave
- Overlay: 1 part 4% CMC and 3 parts medium

Protocol 3. *Continued*

- Drugs made up to appropriate serial twofold dilutions in CMC overlay medium (the pure forms of the drugs need to be used, and are usually available direct from the manufacturer).
- Laboratory strain of HSV such as SC16 (4).
- 10% formalin in PBS
- Crystal violet: 0.5% w/v in 20% methanol

Method

1. Prepare 24-well plates as in *Protocol 2*.
2. Remove the medium from the wells, and inoculate the wells of half the plate with 0.2 ml of virus, diluted to give approximately 50–100 plaques per well.
3. Add 0.2 ml of medium to the remaining wells. These will provide controls for cellular toxicity of the drugs.
4. Incubate at 37 °C in 5% CO_2 for 1 h
5. Remove the inoculum, and add 1 ml overlay per well. For each drug concentration tested, use 2 virus-infected wells and 2 mock-infected wells. Include 'no drug' controls.
6. Incubate at 37 °C for 48 h.
7. Aspirate off the overlay gently, and flood the wells with 10% formalin in PBS. Leave at room temperature for 30 min.
8. Remove the formalin, and rinse the wells with water.
9. Flood the wells with Crystal violet, leave for 5–10 min, rinse, drain, and allow to air-dry.
10. Check the cell morphology in control wells to estimate if any drug toxicity is occurring. Count the plaques using a dissecting microscope.
11. Calculate the concentration of the drug that reduces the plaque number by 50% compared with the no drug control (IC_{50}). This can be done by the plotting number of plaques against drug concentration.

2.6 Resistance assays for therapeutic antivirals used for HSV

In practice, it is necessary to distinguish *in vitro* drug resistance from clinical resistance, in which a viral infection fails to respond to treatment, which may be due to the presence of drug-resistant virus, but could also be due to poor patient compliance or pharmacokinetic and metabolic effects. True antiviral resistance is shown where there is a decrease in susceptibility to an antiviral drug that can be shown by *in vitro* testing. As mentioned above, aciclovir (and penciclovir) resistance occurs as a result of mutations in either the viral TK or DNA polymerase genes. Resistance to foscarnet is due to mutation of the

9: Chemotherapy of DNA virus infections

viral DNA polymerase gene. Resistance of HSV to antivirals is only very rarely observed in isolates from immunocompetent individuals.

It is difficult to standardize methods of antiviral susceptibility testing, because many variables can influence the final result. These include the cell line, the viral inoculum titre, the assay method, the endpoint criteria, and the method of calculation of the endpoint. Antiviral susceptibility assays in use include plaque reduction, dye uptake, enzyme immunoassay, and DNA hybridization. The plaque reduction assay is the standard method for measuring antiviral susceptibility, and is considered the 'gold standard' against which other assays should be interpreted. This assay is described in detail in *Protocol 4*.

Interpretation of endpoints for HSV testing varies according to whether an isolate is HSV 1 or 2. Thus it is necessary first to type isolates. Susceptibility results are traditionally expressed as IC_{50} values, because of the greater mathematical precision of the 50% endpoint compared with a 90% or 99% endpoint. Many variables can effect plaque reduction assays, so there can be considerable day-to-day variation in absolute IC_{50} values. It is important to use first- or low-passage clinical isolates, as these are more likely to be representative of the original population infecting the patient. The interpretation of IC_{50} values currently used in our laboratory is shown in *Table 2*. Isolates giving IC_{50} values that lie in the intermediate range may either be mixtures of highly resistant and sensitive viruses, or true intermediate-resistant viruses.

Finally, the phosphorylation of nucleoside analogue drugs may differ between different cell lines. This may explain the higher IC_{50} for penciclovir than for aciclovir for inhibition of HSV in Vero cells. More similar results are observed in MRC-5 cells. Thus, absolute differences between drugs in IC_{50} values generated *in vitro* may not have a bearing on comparative efficacy *in vivo*.

Protocol 4. Assaying susceptibility of HSV isolates to therapeutic antivirals by plaque reduction assay

Equipment and reagents

- Vero cells
- HSV typing kit (e.g. PathoDx Herpes Typing kit, Diagnostic Products Corporation)
- Control standard sensitive and resistant HSV strains such as SC16 (4) and DM21 (5)
- MEM medium supplemented with Earle's salts, L-glutamine, 25 mM Hepes, 2% FCS, 100 IU ml^{-1} penicillin, and 100 µg ml^{-1} streptomycin
- 24-well tissue-culture plates
- 4% w/v carboxymethyl cellulose (CMC) in PBS; autoclave
- Overlay: 1 part 4% CMC and 3 parts medium
- 10% formalin in PBS
- Antiviral drugs made up to an appropriate concentration in CMC overlay medium (see below) (the pure forms of the drugs need to be used, and are usually available direct from the manufacturer).
- Crystal violet (0.5% in 20% methanol)
- Test drug: the drugs currently available for the treatment of HSV include aciclovir, penciclovir, foscarnet, and cidofovir. Stock solutions of 4 mM aciclovir, penciclovir, and cidofovir, and 20 mM foscarnet, should be made up in water. Drugs are tested at fourfold dilutions, starting with 40 µM for aciclovir, 160 µM for penciclovir, 400 µM for foscarnet, and 400 µM for cidofovir.

Protocol 4. *Continued*

Method

1. Prepare 24-well plates as in *Protocol 2* (1 plate per isolate).
2. Inoculate each well with 0.2 ml of virus, diluted to give approximately 50–100 plaques per well. Include plates for standard sensitive and resistant virus strains. The plate layout should be as shown in *Figure 1*. 'No drug' and cell controls should be included.
3. Follow Steps 4–11 of *Protocol 3*.
4. Interpretation: see *Table 2*.

3. Resistance to antivirals of human cytomegalovirus (HCMV)

HCMV is a major cause of morbidity and mortality in immunocompromised patients; for example, CMV retinitis is observed in AIDS patients, and pneumonitis in organ-transplant patients. Ganciclovir and foscarnet have been shown to be effective against HCMV in clinical trials.

Ganciclovir is an acyclic guanosine analogue, similar in structure to aciclovir. It has *in vitro* activity against a range of herpesviruses and, like aciclovir, is an

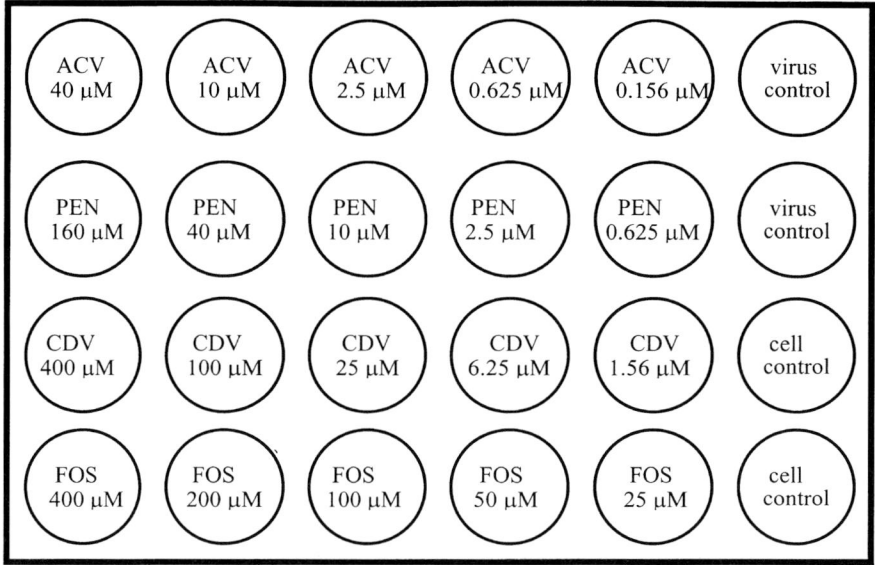

Figure 1. Diagram to show the layout of a 24-well plate for Vero cell HSV plaque reduction assay for antiviral drug susceptibility. ACV, aciclovir; PEN, penciclovir; CDV, cidofovir; FOS, foscarnet.

9: Chemotherapy of DNA virus infections

Table 2. Interpretation of HSV plaque reduction assay

Virus	Drug	IC_{50}	Interpretation
HSV 1	aciclovir	<3 μM	sensitive
		3–40 μM	intermediate
		>40 μM	resistant
	penciclovir	<7.5 μM	sensitive
		7.5–40 μM	intermediate
		>40 μM	resistant
	foscarnet	<250 μM	sensitive
		250–400 μM	intermediate
		>400 μM	resistant
HSV 2	aciclovir	<6.5 μM	sensitive
		6.5–40 μM	intermediate
		>40 μM	resistant
	penciclovir	<38 μM	sensitive
		38–40 μM	intermediate
		>40 μM	resistant
	foscarnet	<250 μM	sensitive
		250–400 μM	intermediate
		>400 μM	resistant

excellent substrate for HSV thymidine kinase. It is converted to the triphosphate, which inhibits the viral polymerase. Unlike aciclovir, however, it is also a good inhibitor of human cytomegalovirus. The protein kinase product of CMV gene UL 97 monophosphorylates ganciclovir. A cellular guanylate kinase adds a second phosphate, and other cellular kinases add the third. Both GCV triphosphate and ACV triphosphate competitively inhibit incorporation of dGTP into the viral DNA polymerase. The relative efficacy of GCV over ACV in CMV-infected cells, as well as *in vivo*, may be due to the shorter intracellular half life of aciclovir triphosphate. It is possible that GCV triphosphate is not as efficient a chain terminator as the ACV equivalent, since the 3' hydroxyl group remains for further elongation. This GCV monophosphate can be internally incorporated within a growing DNA chain, and partially explains the bone marrow toxicity of this drug. Ganciclovir is available for intravenous and oral use, and a pro-drug, val-ganciclovir, is undergoing clinical trials at present.

Ganciclovir strains of CMV usually have the phosphorylation-deficient phenotype resulting from mutations in UL97 which fall within specific domains of the enzyme (6). Mutations in the viral DNA polymerase can also lead to drug resistance (7). Drug resistance appears most common in patients receiving anti-CMV treatment and maintenance therapy for more than 3 months.

3.1 Phenotypic CMV drug susceptibility assay

As with HSV, the gold standard for determining susceptibility of HCMV to antivirals is the plaque reduction assay, and this is described in *Protocols 5 and 6*, which have been adapted from the DAIDS Virology Manual for HIV Laboratories (8).

Protocol 5. Preparation of cytomegalovirus stocks

Equipment and reagents
- MEM medium supplemented with Earle's salts, L-glutamine, 25 mM Hepes, 5% FCS, 100 IU ml^{-1} penicillin, and 100 μg ml^{-1} streptomycin
- MRC5 cells
- Cryoprotective medium containing 20% FCS and 10% DMSO

Method
1. Obtain a CMV isolate, usually a culture in a screw-capped tube.
2. Trypsinize the cells, and transfer the cell suspension to a 25 cm^2 flask of MRC5 cells. Incubate at 37 °C.
3. When CPE is observed, trypsinize the cells, and redistribute them to a fresh flask, adding fresh cells if necessary.
4. Observe the flask daily, and repeat the trypsinization and re-distribution every 3–4 days until the CPE reaches at least 50–75%.
5. When the CPE involves 50–75% of the monolayer, trypsinize the cells, and gently resuspend them in medium, pipetting up and down to obtain a single-cell suspension.
6. Count the cells in a haemocytometer, and estimate the proportion of viable infected cells to determine the number of plaque-forming cells (PFC) in the suspension.
7. Adjust the cell concentration to 400 PFC ml^{-1}, to provide an inoculum dose of 60–80 PFC per 0.2 ml.
8. Carry out the drug sensitivity assay the same day (see *Protocol 6*).
9. Freeze aliquots of the remaining virus stock by resuspending pelleted cells in cryoprotective medium, and freezing slowly at –80 °C.

Protocol 6. CMV drug susceptibility assay, plaque reduction method

Equipment and reagents
- Dissecting microscope
- MRC5 cells
- Control sensitive and resistant CMV strains
- 24-well tissue culture plates

9: Chemotherapy of DNA virus infections

- CMC overlay: 1 part CMC with 3 parts medium, containing an appropriate concentration of the drug. For ganciclovir, the following final drug concentrations should be used in the assay: 0, 1.5, 3, 6, 12, 24, 48, and 96 μM. For foscarnet, the final drug concentrations should be 0, 25, 50, 100, 200, 400, 800, and 1000 μM. The drug concentrates should be kept frozen, and the overlays made up fresh.
- MEM medium supplemented with Earle's salts, L-glutamine, 25 mM Hepes, 5% FCS, 100 IU ml^{-1} penicillin, and 100 μg ml^{-1} streptomycin
- 4% CMC made up in PBS

Method

1. Prepare 24-well plates of MRC5 cells. For each virus tested, use one 24-well plate. Drug-sensitive and drug-resistant controls must be included in each assay, so at least 3 plates (2 controls and 1 unknown) need to be prepared for each assay.
2. Inoculate each well with 0.2 ml of cell suspension containing 60–80 PFC of cell-associated virus. Incubate for 90 min in a 37°C 5% CO_2 incubator to allow adsorption.
3. Prepare drug concentrations in the overlay solutions.
4. Carefully remove the inoculum and medium from the wells. Add overlay containing an appropriate concentration of the drug (3 wells per drug concentration).
5. Incubate plates at 37°C in a 5% CO_2 incubator for 7–10 days, examining daily for plaque formation.
6. When plaque formation is well defined, aspirate the overlay, and fix the plates in 10% formalin in PBS for 30 min. Decant off the formalin, and rinse the plates in water. Stain the plates with Crystal violet for 5–10 min, rinse, and allow to dry.
7. Count plaques using a dissecting microscope.
8. Calculate IC_{50}, as for *Protocol 3*.
9. Interpretation:

Drug	IC_{50}	Interpretation
Ganciclovir	<6 μM	sensitive
	>12 μM	resistant
Foscarnet	<400 μM	sensitive

3.2 Genotypic assays for detection of ganciclovir resistance-associated mutations in CMV

It is often difficult and time consuming to grow clinical isolates of HCMV, so genotypic assays have been developed to detect the most commonly observed mutations in the UL97 gene. The assay outlined in *Protocol 7* is an adaptation of one originally described by Chou *et al.* (6), and can be carried out directly on a blood sample from a patient, thus eliminating the need for virus isolation. This test uses restriction fragment analysis (RFLP) of PCR products, derived by semi-nested PCR, to detect mutations at codons 460, 594, and 595. *Protocol 8* is a point mutation assay (PMA) provided by Dr Vince Emery, Royal Free Hospital, London, which detects mutations at codons 460, 520, and 594/595 of the UL97 gene.

Protocol 7. Assay for detection of ganciclovir resistance-associated mutations in the UL97 gene of HCMV using RFLP

Equipment and reagents

- Control strain of virus such as AD169 (ATCC #VR538)
- QIAamp Blood Kit (Qiagen 29104) or similar, for extraction of DNA from whole blood
- 10 × *Taq* buffer: 200 mM Tris-HCl pH 8.4, 500 mM KCl
- 25 mM MgCl$_2$
- DMSO
- Primers: use at 2 pmol ml^{-1} for the first round, and 5 pmol ml^{-1} for the second round.
  Primers CPT1088 and 1619 yield a 532 bp amplicon covering codon V460, in which mutants can be detected using digestion with *Nla*III.
  Primers CPT1713 and 1830M yield a 118 bp fragment in which mutations at V594, F595, and S595 can be detected by digestion with *Hha*I, *Mse*I, and *Taq*I respectively.
- CPT1088: (5′ ACGGTGCTCACGGTCTGGA-T 3′)
- CPT1619: (5′ AAACGCGCGTGCGGGTCGC-AGA 3′)
- CPT1713: (5′ CGGTCTGGACGAGGTGCGC-AT 3′)
- CPT1830M: (5′ AATGAGCAGACAGGCGT-CGAAGCAGTGCGTGAGCTT-GCCGTTCTT 3′)
- 5 mM dNTP mixture
- *Taq* DNA polymerase (5 units ml^{-1})
- Mineral oil
- QIAquick PCR Purification Kit (Qiagen)
- Restriction enzymes: *Nla*III, *Hha*I, *Taq*I, and *Mse*I
- Agarose suitable for resolution of small fragments, e.g. GibcoBRL agarose-1000
- DNA ladder suitable for sizing small fragments, e.g. GibcoBRL 25 bp ladder

Method

1. Extract DNA from blood or virus stock according to the kit instructions.
2. For each sample, mix the following for the first round of PCR. Make up a master mix of all reagents except the template, and dispense 96 μl to each tube: 10 × *Taq* buffer 10 μl; DMSO 10 μl; MgCl$_2$ 6 μl; dNTP mix 4 μl; primer CPT1088 4 μl; primer CPT1830M 4 μl; *Taq* polymerase 0.4 μl; template DNA 5 μl; distilled water 63.6 μl
3. Briefly mix by vortexing, and overlay with 40 μl of mineral oil.
4. Amplify by 30 cycles of 95°C for 45 s, 54°C for 45 s, and 72°C for 45 s.
5. Transfer 5 μl of the product of the first round to nested PCR, made up as above, but use primers CPT1088 and CPT1619 for analysis of codon 460, and CPT1713 and CPT1830M for analysis of codons 594 and 595.
6. Amplify by 30 cycles of 95°C for 45 s, 54°C for 45 s, and 72°C for 45 s.

9: Chemotherapy of DNA virus infections

7. Analyse 10 µl of sample by gel electrophoresis, and if amplification is successful, purify the PCR product using the QIAquick PCR purification kit, according to the manufacturer's instructions.
8. Digest the PCR products with restriction enzymes, and analyse fragments using a 3% agarose-1000 gel, with a 25 bp ladder as size marker.

Protocol 8. Point mutation assay for detection of resistance mutations in CMV UL97 gene

Equipment and reagents

- Wash buffer/TTA diluent (10 ×): 10 mM Tris-HCl with 0.5% Tween 20, and 1.0% sodium azide (store as 10 × concentrate at room temperature).
- PMA diluent: 40 mM Tris-HCl pH 7.8, with 20 mM $MgCl_2$, and 50 mM NaCl. Store at 4°C
- Annealing mix (for 96 wells): 2475 µl PMA diluent, 25 µl probe stock (make up immediately before use)
- Labelling mix (24 wells): 48 µl Klenow polymerase stock (48 µl per test), 168 µl 0.1 M DTT, 24 µl [^{35}S]dNTP dilution (make up immediately before use).
- Probe stock: 66 µg ml^{-1} (16.5 µl of 1µg ml^{-1} probe + 8.5 µl water)
- Klenow polymerase stock: 2 µl Klenow polymerase, 200 µl PMA diluent

- [^{35}S]dNTP dilution: 1 in 10 dilution in water of NEN Dupont 1000–1500 Ci $mmol^{-1}$ dATP, dCTP, dGTP, and dTTP. These labels should be purchased at the same time to have the same activity. Store the 1:10 dilution at −20°C
- PCR reagents, as in *Protocol 7*
- UL97 Primers:
  Outer **NUL1** 5' CAA CGT CAC GGT ACA TCG ACG TTT 3'
  SC2 5' GCC ATG CTC GCC CAG GAG ACA GG 3'
  Inner **NUL3** 5' CAT CGA CAG CTA CCG ACG TGC 3'
  NUL4 5' GTA GCT CAT TTG CGC CGC CAG 3'
- Probes as illustrated in *Figure 2*.

A. *PCR*

1. For ten reactions, mix the following for the first round of PCR. Make up a master mix of all reagents except the template, and dispense 95 µl to each tube: 10 × *Taq* buffer 100 µl, $MgCl_2$ 80 µl (2 mM final concentration), dNTP mix 30 µl (200 µM final concentration), outer primers 1 µl each (1 µg ml^{-1}), *Taq* polymerase 2 µl, template DNA 5 µl, distilled water 736 µl.
2. Briefly mix by vortexing, and overlay with 40 µl mineral oil.
3. Heat at 95°C for 12 min, then amplify by 39 cycles of 94°C for 1 min, 55°C for 1 min, and 72°C for 2 min, and one cycle of 94°C for 1 min, 55°C for 1 min, and 72°C for 10 min.
4. Transfer 2 µl of the product of the first round to nested PCR, made up as above, but use either sense primers (1 µl Nul3 (biotinylated) + 1 µl Nul3 + 2 µl Nul4), or antisense primers (1 µl Nul4 (biotinylated) + 1 µl Nul4 + 2 µl Nul3).
5. Heat at 95°C for 10 min, then amplify by 25 cycles of 94°C for 1 min, 55°C for 1 min, and 72°C for 1 min, followed by 72°C for 10 min.

> **Protocol 8.** *Continued*
>
> B. *Microtitre plate point mutation assay*
>
> The method below is for the assay of 22 samples and 2 controls using a complete 96-well plate.
>
> 1. Allow four wells for each sample or control to be assayed.
> 2. Add 15 µl per well of 1 × TTA buffer.
> 3. Add 10 µl of each sample to the four wells (labelled A, C, T, G), and mix with the 1 × TTA diluent, then incubate at 37°C for 60 min.
> 4. Wash the wells three times with 1 × TTA.
> 5. Add 40 µl per well of 0.15 M NaOH, and incubate at room temperature for 5 min.
> 6. Wash the wells four times with 1 × TTA.
> 7. Add 25 µl of annealing mix to each well, seal the wells, and incubate the microtitre plate for 3 min at 55°C. (Use a shallow water bath or floating plate.)
> 8. Allow the microtitre plate to cool slowly to room temperature over 30 min (use a metal heating block heated to 55°C, and allow it to cool).
> 9. Add 9.8 µl of the labelling mix to each well (i.e. labelling mix A, C, T, or G to one well for each sample), and incubate at room temperature for 3 min.
> 10. Wash the wells four times with 1 × TTA. Due to the short incubation time for Step 9, it is convenient to do Steps 9 and 10 sequentially for each labelling mix.
> 11. Add 40 µl of 0.15 M NaOH to each well.
> 12. Incubate at room temperature for a minimum of 5 min.
> 13. Transfer the 40 µl of NaOH solution to a 10 ml scintillation vial containing 5 ml of a scintillation cocktail compatible with an aqueous sample.
> 14. Count the vials for 1 min in a scintillation counter.

4. Antivirals active against hepatitis B virus

Treatment of disease caused by hepatitis B virus (HBV) has been revolutionized by the introduction of antiviral therapies. A number of nucleoside and nucleotide analogues are now known to inhibit HBV replication *in vitro* and *in vivo*, including lamivudine, penciclovir, lobucavir, and adefovir. No specific viral kinase activity has been identified, and therefore activation of these drugs is mediated by cellular kinases. Lamivudine triphosphate is thought to inhibit the elongation of minus strand DNA synthesis, whereas famciclovir

```
460
                              G   A
5'    B-CTG CAC TTT GAC ATT ACA CCC ATG AAC GTG CTC ATC GAC GTG AAC 3'
3'      GAC GTG AAA CTG TAA TGT GGG TAC TTG CAC GAG TAG CTG CTC TTG 5'
5'(6)   TGC CAC TTT GAR ATT ACA CCC A
                                    G
   (5)                            C TTG CAC GAG TAG CTG CAC TTG 5'
                                  A

520
                                G
5'      C CGT CTC CGC GAA TGT TAC CAC CCT GCT TTC CGA CCC ATG
        G GCA GAC GCG CTT ACA ATG GTG GGA CGA AAG GCT GGG TAG-B
5'(4)   C CGT CTG CGC GAA TGT TAC CAC
                                    G

594/595
                         T   CT
5'    B-GCG GCC TGC CCC CCC TTG GAG AAC GGT AAG CTC ACG CAC T

(3)                      GC AAC CTC TTG CCR TTC GAG TG 5'
                         A
(2)                         AC CTC TTG CCR TTC GAG TGC G 5'
                          G
(1)                       C CTC TTG CCR TTC GAG TGC GTG A 5'
                          A
```

Figure 2. Design of probes for HCMV point mutation assay. The probes are numbered 1–6. Probes 1–3 and 5 bind to the biotinylated sense strand of the PCR product, whilst probes 4 and 6 bind to the biotinylated anti-sense DNA strand. At the 3' end of each probe, the base indicates the wild type, and the base beneath indicates the mutant. R indicates a mixed base to take into account the silent mutation.

may act to inhibit the priming reaction. It is of note that lamivudine is also a potent inhibitor of HIV replication, another virus with reverse transcriptase activity.

4.1 Genotypic assay for resistance associated mutations in HBV using nucleotide sequencing.

Resistance to lamivudine and famciclovir in HBV has been documented *in vivo*, associated with specific mutations in the polymerase gene. In the case of lamivudine resistance, mutations leading to M550V or M550I within the viral polymerase are always observed, and are variably associated with other mutations (9). The situation with mutations conferring resistance to famciclovir is less clear. In the absence of a standardized *in vitro* drug susceptibility assay for HBV, genotypic assays remain the most convenient method of analysis for drug resistance.

Protocols 9 and 10 give methods for extracting HBV DNA from serum, PCR conditions for amplification of the appropriate region of the polymerase, and nucleotide sequencing. Some HBV positive sera have very high viral loads, so extreme care must be taken to avoid cross-contamination between samples. A negative control serum should be included in all tests, and handling of more than a few sera simultaneously should be avoided.

Protocol 9. Rapid extraction of HBV DNA from serum for PCR

Equipment and reagents
- 0.2 M NaOH
- 0.12 M Tris-HCl pH 7.5
- Mineral oil

Method

1. Add 10 µl 0.2 M NaOH to a 10 µl serum sample.
2. Add 2 drops mineral oil.
3. Heat at 40°C for 1 h.
4. Add 30 µl of 0.2 M Tris-HCl pH 7.5.
5. Heat at 95°C for 10 min.
6. Preferably use the same day, or store for a maximum of one week at 4°C.

Protocol 10. PCR and sequencing of HBV to detect lamivudine and famciclovir resistance-associated mutations

Equipment and reagents
- 10 × *Taq* buffer: 200 mM Tris-HCl pH 8.4, 500 mM KCl
- 25 mM $MgCl_2$
- Forward primer: GCCCGTTTGTCCTCTAAT at 5 pmol ml^{-1}
- Reverse primer: TAACCCCATCTTTTTGTTT-TG at 5 pmol ml^{-1}
- These primers amplify a region of the polymerase between nucleotides 468 and 865
- 5 mM dNTP mixture
- *Taq* DNA polymerase, 5 units ml^{-1}
- Mineral oil
- QIAquick PCR Purification Kit (Qiagen)

Method

1. For each sample, mix the following. Make up a master mix of all reagents except the template, and dispense 96 µl to each tube: 10 × *Taq* buffer 10 µl, $MgCl_2$ 8 µl, dNTP mix 4 µl, forward primer 4 µl, reverse primer 4 µl, *Taq* polymerase 0.2 µl, template DNA 4 µl, distilled water 66 µl.
2. Briefly mix by vortexing, and overlay with 40 µl mineral oil.
3. Amplify by 40 cycles of 95°C for 45 s, 54°C for 45 s, and 72°C for 45 s.
4. Analyse 10 µl of the sample by gel electrophoresis and, if amplification is successful, purify the PCR product using QIAquick PCR purification kit, according to the manufacturer's instructions.

9: Chemotherapy of DNA virus infections

5. The PCR product is now suitable for sequencing by standard automatic and manual methods, using the same primers as for PCR. When doing manual sequencing, we currently use Amersham Thermo Sequenase radiolabelled terminator cycle sequencing kit with ^{33}P-labelled ddNTPs.

5. Real time PCR and fluorimetry for detection of mutations

The detection of specific mutations is a key approach to the analysis of drug resistance for many viruses, and this also has applications in the more general study of molecular mechanisms of viral replication and pathogenesis. Traditionally, mutations have been detected using sequencing, point mutation assays, and hybridization assays. Recently, new technologies have been developed that allow extremely rapid PCRs (10–20 minutes for a 40-cycle reaction). These have been combined with generic detection of PCR product using fluorescent dyes such as ethidium bromide or Sybr Green 1, or specific oligonucleotide probes labelled with dyes such as fluorescein or Cy5. The detection of PCR product takes place concurrently with the thermal cycling. Equipment for this type of test includes the LightCycler manufactured by Idaho Technology (10–12).

At the time of writing, the hardware for these tests is expensive, but the reduction in time, and in labour and consumable costs, makes it seem likely that this technology will become more readily available in the near future. In *Protocol 11*, a method for detection of lamivudine resistance-associated mutations in HBV using a LightCycler is described as an example of the application of this technology (13).

Reaction conditions for PCRs in the LightCycler are very different from those used in conventional PCRs. This is because the reaction is done in a very small volume (1–10 μl) in glass capillaries, where the surface area of glass is high relative to the reaction volume. Concentrations of magnesium are usually higher than in conventional PCRs, and must be optimized for each reaction. Bovine serum albumin (BSA) must be included in the reactions, to prevent inactivation of the *Taq* polymerase on the surface of the glass.

In our experience, we have found that amplification using the LightCycler is most efficient for products less than 300 base pairs in length, so we try to aim for amplicons of about 100 base pairs in length. The PCR primers described in *Protocol 11* amplify a 103 bp fragment covering codon 550 of the HBV polymerase, which is invariably mutated in lamivudine-resistant virus. The mutations observed are either M550V (ATG→GTG) or M550I (ATG→ATT). The mutations are detected by hybridization of labelled probes to the PCR product, and then determining the temperature at which the probes melt from the product. Two labelled probes are used in the reactions: one is 3'-labelled with fluorescein, while the other is 5'-labelled with Cy5. The

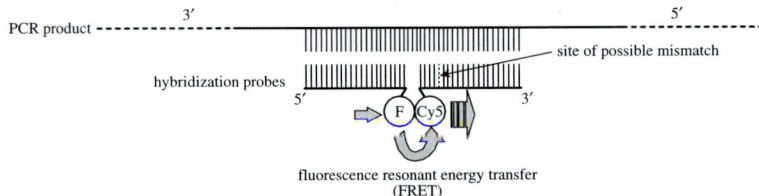

Figure 3. Design of fluorescent probes to detect mutations in codon 550 of HBV polymerase, using a LightCycler. PCRs each contain one Cy5-labelled probe and the Fluorescein-labelled probe. The Cy5 molecule fluoresces only when adjacent to the Fluorescein molecule (F). The probes are designed so that the Cy5 label will detach at a lower temperature than the Fluorescein-labelled probe.

probes are designed to anneal so that they are one base apart. When both probes are binding to the PCR product, there is transfer of fluorescence resonant energy (FRET) from the fluorescein molecule to the Cy5 molecule. The probes are designed such that the melting temperature of the Cy5-labelled probe, which covers the mutation area, is lower than that of the fluorescein-labelled probe. In addition, the Cy5-labelled probe needs to be blocked at its 3' end, either by biotinylation or phosphorylation, so that it does not interfere with the amplification. The principle of design of probes is illustrated in *Figure 3*.

Protocol 11. Use of the LightCycler to detect lamivudine resistance-associated mutations in HBV

Equipment and reagents

- LightCycler (Roche Molecular Biochemicals)
- Glass capillary cuvettes and holders.
- PCR mastermix (2 × concentrated to give final concentrations of 50 mM Tris pH 8.5, 3 mM MgCl$_2$, 200 μM of each dNTP, 250 μg ml^{-1} bovine serum albumin, and 0.4 units of *Taq* polymerase per 10 μl reaction mix). N.B. 'Usual' PCR mixes are not suitable for use in the LightCycler; higher Mg^{2+} concentrations are required, and BSA (non-acetylated) must be present to prevent adsorption and denaturation of polymerase on the surface of the glass capillary tubes.
- Forward primer: 5' TACTAGTGCCATTTGT-TCAGTGG 3' (5 μM)
- Reverse primer: 5' CACGATGCTGTACAGA-CTTGG 3' (5 μM)

These primers amplify the polymerase gene between nucleotides 681 and 783.
- Probes: 3'-fluorescein-labelled probe: CCC-CCACTGTTTGGCTTTCAG (2 μM)
- 5'-Cy5 3'-biotin (or phosphorylated)-labelled probes (2 μM):
  1. wild type: TATATGGATGATGTGGTATT
  2. Mutant 1: TATGTGGATGATGTGGTATT
  3. Mutant 2: TATATTGATGATGTGGTATT
- All probes must be reverse-phase cartridge-purified. N.B. Only one Cy5-labelled probe should be included in each reaction. The temperature at which the probe melts from the PCR product is diagnostic of whether there are mutations present in the target codon.

Method

1. For each 100 μl of PCR mix, add the following: 3 mM MgCl$_2$ master mix 50 μl, primers 10 μl, Fluorescein-labelled probe 10 μl, Cy5-labelled probe 10 μl, water 10 μl.

2. Dispense 9 µl to each small Eppendorf tube.
3. Add 1 µl template to each tube.
4. Dispense 2–5 µl into the plastic top of each glass capillary cuvette.
5. Lightly cap the cuvettes.
6. Spin cuvettes in holders at 1000 r.p.m. for 10–15 s.
7. Firmly cap the cuvettes.
8. Insert cuvettes in LightCycler.
9. Heat at 95°C for 45 s, then 40 cycles of 95°C for 1 s, 50°C for 2 s, and 72°C for 6 s, followed by a melt programme of 40–75°C at 0.2°C s^{-1} with continuous monitoring of fluorescence with acquisition set at 32. The ratio of Cy5 fluorescence (F2) to fluorescein fluorescence (F1) should be measured.

Acknowledgements

Many thanks to colleagues in the PHLS Antiviral Susceptibility Reference Unit, and to Dr V. Emery, Royal Free Hospital, London, for providing *Protocol 8*.

References

1. Gaudreau, A., Hill, E., Balfour, H. H., Erice, A., and Boivin, G. (1998). *J. Infect. Dis.*, **178**, 297.
2. Erikkson, B., Larsson, A., Helgstrand, E., Johansson, N-G., and Oberg, B. (1980). *Biochem. Biophys. Acta,* **607**, 53.
3. Safrin, S., Cherrington, J., and Jaffe, H. S. (1997). *Rev. Med. Virol.*, **7**, 145.
4. Hill, T. J., Field, H. J., and Blyth, W. A. (1975). *J. Gen. Virol.*, **28**, 341.
5. Efstathiou, S., Kemp, S., Darby, G., and Minson, A. C. (1989). *J. Gen. Virol.*, **70**, 869.
6. Chou, S., Erice, A., Jordan, M. C., Vercellotti, G. M., Michels, K. R., Talarico, C. L., Stanat, S. C., and Biron, K. K. (1995). *J. Infect. Dis.*, **171**, 576.
7. Smith, I. L., Cherrington, J. M., Jiles, R. E., Fuller, M. D., Freeman, W. R., and Spector, S. A. (1997). *J. Infect. Dis.*, **176**, 69.
8. Division of AIDS, National Institute of Allergy and Infectious Diseases. (1997). *DAIDS virology manual for HIV laboratories.* Publication NIH-97-3828. US Department of Health and Human Services, Washington, DC.
9. Pillay, D., Bartholomeusz, A., Cane, P., Mutimer, D., Schinazi, R. F., and Locarnini, S. (1998). *Int. Antiviral News*, **6**, 167.
10. Bernard, P. S., Lay, M. J., and Wittwer, C. T. (1998). *Anal. Biochem.*, **255**, 101.
11. Lay, M. J., and Wittwer, C. T. (1997). *Clin. Chem.*, **43**, 2262.

12. Wittwer, C. T., Herrmann, M. G., Moss, A. A., and Rasmussen, R. P. (1997). *BioTechniques,* **22**, 130.
13. Cane, P. A., Cook, P., Ratcliffe, D., Mutimer, D., and Pillay, D. (1999). *Antimicrob. Agents Chemother.* **43**, In press.

10

Herpes simplex virus and adenovirus vectors

CINZIA SCARPINI, JANE ARTHUR, STACEY EFSTATHIOU,
YVONNE MCGRATH and GAVIN WILKINSON

1. Introduction

Gene therapy has the potential to treat a wide variety of acquired and inherited disorders, including inborn errors of metabolism, infectious disease, and cancer. For each of these applications an appropriate vector system is required to deliver the therapeutic gene efficiently to the nucleus of the target cell, in order to mediate either transient or long-term expression of the required gene product. In this chapter we consider the biological properties and methods associated with the construction and propagation of two large DNA virus vector systems: herpes simplex virus and adenovirus.

2. Herpes simplex virus

2.1 Biological properties

Herpes simplex virus type 1 (HSV-1) is a large enveloped DNA virus capable of replicating in a wide variety of cell types in culture. Virions are composed of an icosadeltahedral capsid 100 nm in diameter, containing a GC-rich, double-stranded genome of about 152 kb, which is surrounded by a protein-rich region called the tegument, and a glycoprotein-containing lipid envelope. The genome has been fully sequenced (1), and consists of two covalently linked components, designated as L (long) and S (short). Each component consists of unique sequences flanked by inverted repeats (2). The genome codes for at least 74 genes, three of which are diploid. Of these genes, at least 38 are known to be non-essential for growth in tissue culture. The number of non-essential genes present, and therefore the amount of DNA that can be substituted without impairing the capacity of the virus to grow *in vitro*, makes herpes virus an ideal vector for the insertion of large DNA sequences, or for the co-expression of multiple genes. Following natural infection with HSV-1, the virus replicates within cells of the mucosal epithelium, gaining access to sensory nerve endings, through which the viruses can be retrogradely transported to sensory neurones. Within these neurones, HSV can either enter another productive replication cycle, presumably causing the death of the infected neurone, or establish a latent state of infection.

2.2 Gene expression during lytic infection

HSV-1 lytic gene expression has been well characterized in both epithelial and fibroblast-derived cells *in vitro* (3, 4). The virus enters the cell by fusion of its envelope with the cellular membrane, so that the capsid covered by the tegument is released into the cell. The capsid is then transported to the nuclear membrane, where the viral DNA and at least some of the tegument proteins migrate to the nucleus. Viral transcription occurs in three regulated temporal classes: immediate-early (IE), early (E) and late (L). IE gene products (ICP0, ICP4, ICP27, ICP22, and ICP47) are expressed in the absence of prior viral protein synthesis, and are necessary for the transactivation of early and late gene expression. A complex between the cellular factors Oct-1, ancillary cellular protein (HCF), and a viral tegument protein, VP16, forms and binds to octamer motif-TAATGARAT sequences, which are located upstream of all IE genes (5, 6). This complex greatly enhances the activation of IE transcription during the initiation of lytic infection. Once synthesized, IE gene products result in the expression of the E genes, seven of which are necessary and sufficient for viral DNA replication. This is then followed by the transcription of L genes, which being mainly structural proteins, will in turn allow assembly and release of mature virions.

2.3 The latent state

The capacity to establish latency is a characteristic feature of all herpesviruses, and involves three separable phases: establishment, maintenance, and reactivation. Following natural infection, establishment of HSV latency occurs within sensory neurones innervating the site of primary infection. A lack of permissivity of at least a proportion of sensory neurones results in failure of productive cycle gene expression, and failure of entry into the lytic cycle. The neurones in which herpes establishes latency reside primarily in the sensory ganglia, although there is evidence for the presence of latent virus also in the central nervous system (CNS) (7–9). Transcription during HSV latency occurs from a very restricted portion of the viral genome, which maps to the repeats flanking the unique long region of the viral genome, and is driven by a single viral promoter. The activity of this promoter leads to the generation of a number of nuclear RNAs, which have been designated latency-associated transcripts (LATs). Two of those, termed major LATs, of 2.0 and 1.5 kb respectively, are highly abundant non-polyadenylated transcripts, which map in antisense direction to part of one of the IE genes, ICP0 (10). These two transcripts appear to be introns derived from splicing of a less abundant 8.5 kb polyadenylated precursor RNA, termed minor LAT (11, 12). The function of LATs is still not fully understood. Some LAT deletion mutants display a slow reactivation phenotype (13), or appear to establish latency with reduced efficiency (14).

Whatever function LATs may perform, the capacity of the LAT promoter to function during latency, which can last a lifetime, indicates that this promoter has potential for a lifelong expression of therapeutic genes. While transcription from homologous and heterologous promoters is partially or totally switched off during latency (15), insertion of the reporter gene LacZ linked to the encephalomyocarditis virus internal ribosomal entry site (IRES) 1.5 kb downstream of the LAT transcriptional start site has been shown to result in the expression of the reporter gene in sensory neurones for up to 5 months (16). Although further studies are necessary to establish the minimal elements necessary for long term transcription, these findings indicate that LAT-driven expression of therapeutic genes in HSV could indeed provide us with an effective gene therapy vector.

2.4 Basic techniques of virus handling

In our laboratory, wild-type stocks of virus are grown on baby hamster kidney (BHK) cells, which can be easily maintained as adherent monolayers. Protocols for the passaging, storage, and recovery of BHK cells have been described in Chapter 1, and in a previous volume of this series (17).

2.4.1 Growth, purification, and titration of virus

(i) Wild type viruses

Wild type HSV is a hazard group 2 pathogen, and as such is handled under category 2 conditions (Advisory Committee on Dangerous Pathogens, 1995. *Categorisation of pathogens according to hazard and categories of containment.* 4th edn. London, HMSO). Our methods for growth, purification, and titration of HSV are described in *Protocols 1–3*. Most of the vectors which have been developed in our laboratory are derived from the HSV-1 strain 16 (SC16) (18). Although the sequence of SC16 has not been determined, manipulation and analysis of recombinant viral genome is based on restriction fragment sizes of the sequenced HSV-1 strain 17.

When preparing virus stocks, there are a few points to bear in mind. First, the multiplicity at which the infections are done is important; stocks should always be prepared by infecting at low multiplicity (<0.01 PFU per cell), as infection at high multiplicity results in the production of populations of defective particles containing incomplete genomes. Moreover, multiple passages of a virus stock results in the introduction of heterogeneity of the viral population, due to the random accumulation of mutations. Thus, for each new virus recombinant, master, submaster, and working stocks are maintained. The master stock is the original virus stock, which has been characterized thoroughly both *in vitro,* and by restriction pattern and hybridization analysis. This is generally a small stock, as it is usually derived from a propagated plaque pick, and consists of about 1 ml of virus suspension at a titre of 10^8–10^9 PFU ml^{-1}. To increase the quantity of virus available, a much larger submaster stock is propagated from the master. The submaster is then used to prepare all the working stocks used to perform experimental work, so that there are never more than two passages between the original stock and the virus used for a particular experimental procedure.

(ii) Replication-defective viruses
Other issues arise when the virus has been manipulated to become replication-defective by the introduction of mutations or the deletion of essential genes. Some manipulations do not require the preparation of special reagents to produce *in vitro* high-titre stocks. For instance, viruses temperature-sensitive (ts) for the essential immediate-early ICP4 gene can be grown *in vitro* at the permissive temperature of 31–32 °C, rather than the standard 37 °C. Deletion mutants can only be grown *in vitro* by engineering cell lines to express the essential gene in *trans*. Such complementing cell lines have been constructed for a number of essential genes, ranging from glycoproteins such as gH to IE genes such as ICP4, and ICP27 (19–22). When producing such complementing cell lines, care must be taken that no overlapping between the viral genome and the plasmid used to express the gene in *trans* occurs, as recombination may take place if homology exists, resulting in the production of revertant and therefore replication-competent virus.

Protocol 1. Growth of virus

Equipment and reagents

- Glasgow modification of Dulbecco's Minimum Essential Medium (GMEM, Life Technologies), supplemented with 2% or 10% (v/v) newborn calf serum (NCS), 5% (v/v) tryptose phosphate broth (TPB), 100 units ml^{-1} penicillin, and 100 mg ml^{-1} streptomycin (Life Technologies). Media supplemented with 2% or 10% are referred to as GMEM-2% and GMEM-10%, respectively.
- Roller bottles containing about 80% confluent monolayers of BHK, as assessed by microscopical analysis
- 200 ml centrifuge bottles (Falcon or Corning)
- Refrigerated benchtop centrifuge

Method

1. Remove the medium from the roller bottles, and add the virus inoculum at a multiplicity of infection (m.o.i.) of 0.001 plaque-forming units (PFU) per cell in 20 ml of GMEM-2%.

2. Incubate at 37 °C for 1 h (to allow virus absorption). Remove the inoculum, then add 100 ml of GMEM-10%. Leave to incubate for 2–3 days at 37 °C, by which time the cells will have rounded (100%

10: Herpes simplex virus and adenovirus vectors

cytopathic effect or CPE) and can be detached by gentle shaking of the bottle.
3. After shaking off adherent cells, transfer the cell suspension to a sterile 200 ml centrifuge bottle. Separate cells from the medium by centrifugation at 2000 r.p.m. (1400 g) for 10 min at 4°C.
4. Decant the medium into sterile 200 ml containers (e.g. Beckman rotor 19 buckets). The medium contains virus which has been released from the cells, and can be further processed to obtain a stock of pure virions. Further handling of this material to obtain a cell-free stock is described in *Protocol 2*.
5. Resuspend the cell pellet in 4 ml of GMEM-10% per roller bottle, transfer to a sterile 20 ml tube, and sonicate to disrupt the cells. Keep the sample in a water–ice mix while sonicating, to reduce inactivation of the virus by overheating. Store the sample on ice between sonications.
6. Dispense the cell-associated virus into small aliquots (220–250 µl) in cryotubes, and store at –70°C.
7. Leave at –70°C for at least 4 h before titration.

Although cell-associated virus is generally used for *in vitro* and *in vivo* studies, there are instances when it may be necessary to use purified virus. In particular, studies on cell viability of neuronal cultures after infection with replication defective viruses have shown that high cell toxicity can be observed after exposure of the cultures to extracts from uninfected cells (23). We have also observed increased neuronal cell death when non-purified virus is used to infect cultures of sensory ganglia neurones, and therefore always use purified stocks for our work on neurones. Purified stocks are also preferred when the virus is introduced by stereotaxic injection into the brain of rats or mice, in order to reduce both the toxicity and the intensity of the immune response.

Protocol 2. Virus purification

Equipment and reagents
- Beckman or equivalent ultracentrifuge and rotors (SW28 and 19 for Beckman)
- Beckman or equivalent ultraclear tubes for centrifugation
- Phosphate-buffered saline (PBS), endotoxin-free
- Ficoll 400 at 5, 7.5, 10, 12.5, and 15% (w/v) in PBS

Method
1. Pellet the supernatant obtained at Step 4 of *Protocol 1* for 2 h in a Beckman rotor 19 or equivalent at 18 000 r.p.m., 4°C.
2. Resuspend the pellet in 2–3 ml of medium (resuspension is made easier by leaving the suspension at 4°C overnight, and then resuspending). Sonicate for 20 s.

Protocol 2. *Continued*

3. Prepare a cold linear 5–15% Ficoll 400 gradient in PBS in Beckman ultraclear tubes (the gradient can be prepared the day before, and left overnight at 4°C). Layer the virus on top, and centrifuge in a Beckman SW28 rotor for 2 h at 12 000 r.p.m. (19 000 g), 4°C.

4. Using overhead illumination, two bands are clearly visible in the centrifuged tube: a clearly defined intense band containing the virions about halfway down the gradient, and a diffuse band above this.

5. Collect the virion band, and transfer it to a fresh ultracentrifuge tube; dilute to fill the tube with PBS.

6. Pellet the virions by centrifuging at 20 000 r.p.m. (65 000 g) in a Beckman SW28 rotor for 2 h, 4°C.

7. Remove the supernatant carefully, and resuspend the pellet in a small volume of sterile PBS. Aliquot the virus into 40 µl aliquots in cryotubes, and store at −70°C.

8. Leave at −70°C for at least 4 h before titration.

Titration of HSV can be done either on cell monolayers, or with cells in suspension. A protocol for virus titration on cell monolayers has been described in Chapter 1, and in a previous volume in this series (17). However, we find that the infection of cells in suspension is a more sensitive assay, and this method is described in *Protocol 3*. When accurate titres are critical, it is important that virus is re-titrated at the time of the experiment, as long-term storage can sometimes result in partial loss of infectivity of the viral stock.

Protocol 3. Virus titration in suspension

Equipment and reagents

- BHK cells freshly trypsinized, and resuspended in GMEM-10%
- GMEM-10% alone, or containing 1.5% (w/v) carboxymethyl cellulose (CMC) (GMEM-10%-CMC)
- 50 mm tissue-culture dishes (Falcon or Corning)
- 10% formalin solution in PBS
- 0.1% Toluidine blue solution

Method

1. Serially dilute virus stock by consecutive tenfold dilutions in GMEM-10%. To make the 10^{-1} dilution we normally use 200 µl from a 220–250 µl aliquot of cell-associated virus stock, or 30 µl from a 40 µl aliquot of purified virus. Testing of dilutions in the range of 10^{-5}–10^{-10} is appropriate for most virus stocks. Use a fresh pipette or tip for each dilution to prevent carry-over of virus, which will introduce error in the determination of the virus titre.

2. Add 2×10^6 cells in 200 μl of medium to each dilution of the virus to be tested. Incubate with shaking at 37 °C for 1 h.
3. Plate the cells in 50 mm tissue-culture dishes in 5 ml of GMEM-10%-CMC. Incubate the cells at 37 °C for 2–3 days.
4. Fix the cell sheet by removing the CMC-containing medium, and covering with 10% formalin solution for at least 15 min. The monolayer can then be stained by substituting the formalin solution with a Toluidine blue solution for 15 min.
5. Wash off the Toluidine blue solution, drain the plates, allow them to air-dry, then count the plaques using an inverted microscope.

Note: If a ts mutant is being titrated, incubate duplicate pairs of plates at the permissive temperature of 32 °C, and at the non-permissive temperature of 39 °C, to check for the presence of revertants.

2.5 Construction of recombinant virus genomes

Insertion of DNA fragments into the HSV genome is most commonly achieved by homologous recombination of shared sequences, or by direct ligation of the foreign DNA into the viral genome. Homologous recombination is the standard approach, and can be accomplished by cotransfection of viral DNA with a suitably constructed plasmid. This method can be used to insert DNA at any desired location in the genome, and is a relatively simple procedure; however recombinants are generated at a low frequency (generally <1%), and the isolation of pure recombinant progeny can be laborious. Use of a selection systems, such as the insertion of a reporter gene such as LacZ at the locus of choice to obtain a blue virus, and subsequent selection of white progeny once the gene of interest has been inserted at the same locus, greatly facilitates the process. Alternatively, insertion of the foreign gene at a locus such as the thymidine kinase (TK) gene, which is non-essential for virus replication *in vitro*, but is essential for the activity of the antiviral nucleotide analogue acyclovir, allows selection of recombinants by growing transfection progeny in the presence of the antiviral agent. This latter approach is not desirable if the virus obtained is to be used in pathogenesis studies, as it has been shown that TK-negative viruses have an impaired capacity to establish latency and to reactivate *in vivo* (24). When no other selection system can be used, the presence of the desired gene can be established by dot-blotting progeny virus through a manifold, and hybridising with a probe specific to the gene of interest. In some instances, where the required gene is already present in a second virus, it can be transferred to the recipient virus via homologous recombination, following co-infection at high m.o.i. (i.e. 5 PFU cell^{-1}).

A more recent approach for constructing recombinant viruses is the direct ligation of DNA fragments in unique restriction sites, either already present in the viral genome, or novel restriction sites introduced by site-directed mutagenesis. One example of the many variations of such an approach consists of a two-step method (25). In the first step, a reporter gene cassette flanked by *Pac*I restriction sites is inserted in the desired location of the viral genome by homologous recombination. As there are no other *Pac*I sites in viral genome, the reporter gene can then be removed by *Pac*I digestion, and the viral DNA repaired by homologous recombination with a transgene-containing plasmid. Since the pressure to repair the viral DNA is high, recombination occurs at a high frequency, resulting in the replacement of the lacZ cassette with the transgene in 20–65% of the progeny. This method can be used repeatedly for the sequential deletion of multiple HSV genes, allowing the insertion of multiple transgenes. Other methods in which the DNA is religated *in vitro* maintaining the *Pac*I site result in recombination frequencies of 100% (26), although using this approach only one foreign gene can be inserted.

Since homologous recombination is in many cases an appropriate strategy, we report here our methods to construct recombinant viruses using this technique (*Protocols 4–7*).

2.5.1 Preparation of virus DNA

Virus DNA can be prepared by various methods, such as caesium chloride density-gradient centrifugation of DNA isolated directly from cell-released virus, or by extraction of gradient-purified virions without further purification. We find that extraction of DNA from cell-associated virus preparations (*Protocol 4*) gives DNA of the quality required for efficient transfection and recombination protocols.

Protocol 4. Preparation of virus-infected cell DNA

Equipment and reagents
- BHK monolayers at 80–90% confluence, usually in a 175 cm^2 flask
- GMEM-2% and GMEM-10%
- TE: 10 mM Tris-HCl pH 8, 1 mM EDTA
- 10% (w/v) sodium dodecyl sulfate (SDS)
- 20 mg ml^{-1} proteinase K
- Phenol equilibrated with TE, phenol:chloroform 1:1, chloroform
- Ethanol
- 3 M sodium acetate pH 5.5
- Glass loop

Method

1. Infect cells as for the production of virus stock, but at an m.o.i. of 0.1 PFU cell^{-1} (*Protocol 1*).

2. After 24 h or when 100% CPE is evident, detach the cells from the plastic by gentle shaking. Transfer the cells to 200 ml centrifuge tubes. Spin to separate the cells from the supernatant at 2000 r.p.m. (1400 *g*) for 10 min at 4°C.

3. Remove the supernatant, and resuspend the cell pellet in 5 ml of TE.

10: Herpes simplex virus and adenovirus vectors

> Add 10% SDS and proteinase K at final concentrations of 0.5% and 50 µg ml^{-1} respectively.
>
> 3. Incubate overnight at 37 °C. Add 5 ml of TE, then add an equal volume of phenol (i.e. 10 ml), and invert the sample gently a few times. Care must be taken not to shake the sample, as the DNA can be easily sheared.
>
> 4. Centrifuge at 2000 r.p.m. (1400 g) for 5 min, and transfer the upper aqueous phase to a fresh tube. Repeat the phenol extraction once more, then extract the aqueous phase twice with phenol:chloroform, and finally twice again with chloroform.
>
> 5. Precipitate the DNA by adding to the last-extracted aqueous phase 0.1 volume of 3 M sodium acetate, and 2 volumes of ethanol. If the DNA-containing tube is gently inverted a few times, the DNA will form a thread-like white filament. This can be easily spooled using a glass loop.
>
> 6. Leave the DNA to air-dry on the glass loop, then transfer it to an Eppendorf tube containing about 500 µl of TE. The DNA will quickly rehydrate and fall off the loop.
>
> 7. Leave the DNA overnight at 4 °C to dissolve, then determine the DNA concentration using a spectrophotometer. The yield of DNA should be about 8–12 µg for each 175 cm^2 flask infected.
>
> 8. If the DNA is to be used for transfection, store at 4 °C, and use within a month.
>
> Note: If the DNA is going to be used for analysis of the genomic structure of the virus by restriction enzyme mapping, aliquotting in 20–50 µl portions and storing at –20 °C is recommended.

2.5.2 Insertion of a reporter gene by recombination

We normally insert a reporter gene to facilitate the screening of progeny virus. As already mentioned, this approach consists of inserting a reporter gene cassette in a locus of interest, such as the TK or LAT region. This plasmid may be constructed so that the reporter cassette is inserted at a specific point within the target gene, thus resulting in gene disruption, or the incorporation of the reporter cassette which results in the deletion of one or more target genes. There are different techniques available for transfection of DNA into cells. Transfection using lipidic mixtures (lipofection) has become widely used in recent years. Many companies are providing improved lipofection reagents, and optimized protocols. We find that standard calcium phosphate precipitation followed by a DMSO boost is an adequate technique to use for transfection, and describe this approach in *Protocol 5*.

Protocol 5. Co-transfection of viral and plasmid DNA with calcium phosphate

Equipment and reagents
- BHK cells plated at 2×10^6 in a 5 cm dish
- Hepes buffer: 21 mM Hepes, 0.7 mM Na_2HPO_4, 137 mM NaCl, 5 mM KCl, 5.6 mM glucose, pH 7.05
- Calcium chloride, 2 M in distilled water (autoclaved, then filtered)
- GMEM-2%, GMEM-10%,-and GMEM-2% containing 20% DMSO

Method

1. Gently mix 0.5 ml Hepes buffer with 10 µg of HSV DNA and 3 µg of linearized plasmid DNA. The volume of solutions of DNA added to Hepes buffer should not exceed 20 µl. Add 30 µl of 2 M $CaCl_2$. Mix by inversion, and stand for 12 min.

2. Wash cells 3 times with Hepes buffer, 3–5 ml per dish. Use a vacuum pump to remove the last few drops of Hepes.

3. Drip gently 0.5 ml of precipitate onto the cells. Leave for 10 min.

4. Carefully overlay with 4 ml of warm GMEM-2%, and incubate at 37 °C for 4–6 h.

5. Remove the medium and replace with 2 ml of pre-warmed 20% DMSO in GMEM-2%. Leave for 2 min.

6. Remove all DMSO medium quickly and carefully, and wash the cells with 4 ml of warm GMEM-2% 4 times, 3–5 ml for each wash, to remove any residual trace of DMSO. This is particularly important as DMSO is toxic to cells.

7. Overlay with 4 ml of GMEM-10%, and leave at 37 °C for 2–3 days for plaques to form.

Once the transfection progeny are obtained, selection to identify recombinant viruses is necessary. Isolation of single-virus derived progeny is achieved by either picking single plaques as described in *Protocol 6*, or by limiting dilution as described in *Protocol 7*. When a reporter gene such as LacZ driven by the CMV IE promoter is used to facilitate the selection of the recombinant progeny, inclusion of X-Gal in an agarose overlay is often used to select for the plaques expressing β-galactosidase. Confirmation of the purity of recombinant progeny can be then achieved by sensitive blue staining, as described in *Protocol 8*.

10: Herpes simplex virus and adenovirus vectors

Protocol 6. Selection and growth of individual virus clones by plaque picking

Equipment and reagents

- Agarose medium: 1% (w/v) Seakem LGT agarose in GMEM-10%. This medium is made by melting a sterile 2% stock of agarose in water at high temperature. When this reaches a temperature of 50°C, GMEM-20% is added. When required, 150 µg ml^{-1} X-Gal is added to agarose-containing medium at 45°C.
- Plugged Gilson P1000 pipette tips
- Cell scrapers

Method

1. Infect and overlay cells with GMEM-10%-CMC, as described in *Protocol 3*. When plaques have grown to a visible size, select plates containing a small number of well separated plaques by viewing under a microscope. Remove the CMC-containing medium carefully, and overlay cells with the agarose medium held at 45°C. If white/blue selection is required, add medium containing 150 µg ml^{-1} X-Gal.

2. When the agar has solidified, put plates at 37°C. Blue plaques should become visible within 4–24 h.

3. Pick blue/white plaques using a sterile plastic tip with a Gilson P1000 pipette for each plaque. Withdraw cells from the area of the plaque, and transfer them to a tube containing 2×10^6 BHK cells in suspension in 2 ml of GMEM-10%. Incubate with shaking at 37°C for 1 h.

4. Plate out the cell suspension in 50 mm plates in 5 ml of GMEM-10%. Incubate at 37°C for 2–3 days until CPE is complete.

5. Detach cells from the plate with a cell scraper, then transfer the medium containing the cells to a centrifuge tube. Separate cells from medium by spinning at 2000 r.p.m. (1400 *g*) for 10 min at 4°C.

6. Decant the medium, resuspend the pellet in 1–2 ml of GMEM-10%, and sonicate to disrupt the cells. Keep the sample in a water–ice mix while sonicating.

7. Dispense the virus into small aliquots (200–250 µl) in cryotubes, and store at −70°C.

8. Leave at −70°C for at least 4 h before titrating.

Note: Cells with agar overlays can be fixed with formalin for 1 h without removing the agar. After fixation the overlay can easily be removed by washing under a tap, and the cells can be stained as in *Protocol 3*.

Protocol 7. Selection and growth of virus clones identified by dot-blotting and hybridization

Equipment and reagents
- Reagent reservoir
- Multichannel pipette and tips
- Vacuum blotting manifold
- 10^7 BHK cells per 96-well plate
- GMEM-10% NCS
- 2 × SSC: 300 mM NaCl, 30 mM tri-sodium citrate
- Positively charged nylon membrane (Boehringer Mannheim, cat. no. 1417240)
- Whatman 3MM paper
- 10% w/v SDS
- Denaturing solution: 0.5 M NaOH, 1.5 M NaCl
- Neutralization solution: 0.5 M Tris-HCl pH 7.4, 1.5 M NaCl

Method

1. To a suspension of BHK cells in GMEM-10%, add a virus inoculum sufficient to infect the whole plate with a determined number of PFU per well, e.g. to infect each well with 30 PFU, add 3000 PFU. Make the volume up to 2 ml, and shake gently for 1 h to allow virus adsorption.

2. Add a further 8 ml of GMEM-10%, then dispense 0.1 ml per well using a reagent reservoir and a multichannel pipette.

3. After plaques have formed, freeze the plate at −70 °C and thaw. Scrape the cells into the medium of each well using tips on a multichannel pipette.

4. Set up the vacuum manifold with a nylon membrane. This is done by assembling the manifold after pre-wetting the membrane with 2 × SSC, then exposing it to vacuum so that the wells do not contain excess solution when the cell suspension is added. One corner well is usually left for a positive control.

5. Add 50 μl of the cell suspension to each well of the manifold. Freeze the remaining contents of the wells at −70 °C.

6. Once the suspension is sucked through onto the filter, remove the filter from the manifold and place on two sheets of 3MM paper soaked with 10% SDS for 2 min.

7. Transfer to 3MM paper soaked in denaturation solution, and leave for 2 min.

8. Transfer to 3MM paper soaked in neutralization solution, and leave for 2 min.

9. Finally transfer to 3 MM paper soaked in 2 × SSC, and leave for 2 min.

10. Air-dry, UV cross-link, and hybridize following a standard Southern blot protocol.

10: Herpes simplex virus and adenovirus vectors

> 11. The positive wells identified are used to make virus stocks, which are then used to infect 96-well plates with a lower number of PFU per well, and the procedure is repeated. Eventually plates are infected with a total of 20–30 PFU for 96 wells, to allow clonal purification of recombinant virus by limiting dilution assay.

2.5.3 Analysis and confirmation of the viral structure

Once a single-plaque-derived virus is obtained, confirmation of virus genome structure is achieved by digestion of the viral DNA and hybridization. In general, two or more digestion patterns should be analysed by hybridization with two or more relevant probes. The techniques used are standard in molecular biology, and are therefore not described here (see Chapter 2). Accurate prediction of the sizes of the digested DNA is possible for HSV strain 17, and partially for the strain SC16, whereas for other viruses, assessment of the pattern of digestion of the parental genome is necessary. However, inclusion of the parental virus DNA as a control in the analysis of a recombinant virus structure is always suggested.

Further confirmation of the structure of a recombinant virus can be obtained in some cases by biological assays. In particular, growth of a virus lacking an essential gene in a non-complementing cell line should not give rise to any viral progeny. Alternatively, a virus containing a reporter gene can be checked by looking at the expression of the reporter gene itself; for instance, sensitive X-Gal staining can be used for viruses expressing LacZ, or visualization of fluorescent plaques with a UV microscope can be used for viruses tagged with GFP. Here we include the method for the sensitive blue stain of LacZ-expressing viruses (*Protocol 8*). This method can also be used at any stage of the purification of a blue virus on a white background (or vice versa), to establish the ratio between wild type and recombinant virus.

Protocol 8. Sensitive white/blue staining of infected monolayers

Equipment and reagents

- 0.1 M Tris-HCl buffer pH 7.0
- Glutaraldehyde fixative: 0.5% v/v glutaraldehyde in PBS
- PBS-MgCl$_2$: 2 mM MgCl$_2$ in PBS
- Detergent solution: 2 mM MgCl$_2$, 0.02% NP-40, 0.01% deoxycholate in PBS
- Fe/Fe solution: 50 mM potassium ferricyanide, 50 mM potassium ferrocyanide
- X-Gal solution: 20 ml of detergent solution, 2 ml of Fe/Fe solution, 0.5 ml of X-Gal (40 mg ml^{-1} in dimethyl formamide)
- 0.1% w/v Neutral red

Method

1. Remove the medium from the cells, and wash with 0.1 M Tris buffer.
2. Cover with 0.5% glutaraldehyde. Leave at room temperature for 15 min.

Protocol 8. *Continued*

3. Wash 3 times with ice-cold PBS-MgCl$_2$.
4. Cover the cell sheet with PBS-MgCl$_2$, and leave on ice for 5 min.
5. Replace with detergent solution, and leave on ice for 5 min.
6. Cover with X-Gal solution, and incubate at 37°C for 4–24 h or overnight.
7. Remove X-Gal, and rinse monolayers with PBS, and then with water.
8. Counterstain with 0.1% Neutral red, rinse with water, and air-dry.

2.6 The use of wild-type and replication-defective viruses as vectors to deliver genes to the peripheral and central nervous system

The use of animals for experimental procedures is strictly regulated in the UK, and therefore the legally required Project and Personal Licences must be obtained from the Home Office before any experimental work is carried out. Moreover, training and supervision should be provided for each investigator working with animals. In other countries, appropriate legislation must be adhered to. Most animal models used for the study of HSV latency and gene delivery employ the mouse, although rabbit and guinea pigs are also used (27, 28).

An important issue is how much virus should be used in each model. This will depend mainly on the route of infection, the strain of virus, and the susceptibility of the animal itself. Replication-defective viruses can be used at higher dosages than wild-type virus, allowing the study of models not normally used for wild-type virus such as direct injection of the brain, since as little as 10 PFU of wild-type virus can be lethal in mice by this route.

2.6.1 Peripheral injection into ear/flank or footpad with wild-type viruses

The mouse ear model has been extensively characterized, both in the acute and in the latent stage of infection (18). Wild-type virus inoculated subcutaneously or by scarification into the ear pinna results in an acute infection, with a subsequent virus spread to the corresponding II, III, and IV cervical ganglia, spinal cord, and brain stem. Balb/c mice 5–6 weeks old are inoculated with up to 2×10^6 PFU of wild-type HSV-1 in the left ear pinna. One month after infection, following resolution of acute infection, mice are considered latently infected. At this time point or later, pooled or individual II, III, and IV cervical ganglia from a group of at least 5 mice can be tested for the presence of latent virus by *in situ* hybridization with a probe recognising the LAT transcripts, or for the expression of reporter genes.

Another widely used model uses the inoculation of wild-type virus by

10: Herpes simplex virus and adenovirus vectors

scarification of the flank skin. The thoracic region of the PNS is divided into 13 (T1 to T13) structural segments, each innervating corresponding segments of skin (dermatomes). Adjoining dermatomes of the flank overlap by 50%. In this model, infection with HSV results in zosteriform lesions at the site of inoculum during the acute phase of infection. These lesions resolve within 10–15 days, after which the virus is considered latent. Productive infection is restricted to the ganglia innervating the site of inoculation, usually T9, whereas latent infection is more widespread. Pooled or individual T8, T9, and T10 ganglia are usually chosen for the analysis of the presence of latent virus.

Inoculation of the footpad in mice is another well characterized model for HSV. Infection with wild-type virus in the rear footpad causes clinical signs during the acute phase, such as ipsilateral foot drop, and ipsilateral or bilateral paralysis. The viral infection results in the establishment of latent infection in the dorsal root ganglia (DRG) of lumbar segment 5 (L5), with some contribution to L4 and L6 (29). This model is particularly relevant to those who use replication-defective vectors, as it has been shown that injection in the footpad of viruses disabled for their capacity to replicate results in the establishment of latency at levels comparable to wild-type virus (30).

2.6.2 Removal of ganglia and staining for β-galactosidase

Dissection of animals usually requires the use of a dissecting microscope. As DRGs are located adjacent to the spinal column, removal of ganglia will involve opening the spine, removal of the spinal cord, and either pulling out the ganglia before fixation, or removal of a whole section of the spine followed by fixation and removal of the ganglia at later time. Here we describe a method for the staining of whole ganglia for β-galactosidase activity (*Protocol 9*).

Protocol 9. β-galactosidase staining of whole ganglia

Equipment and reagents

- Fixative solution: 2% w/v paraformaldehyde, 0.2% v/v glutaraldehyde in PBS.
- PBS-MgCl$_2$: 2 mM MgCl$_2$ in PBS
- Detergent solution: 2 mM MgCl$_2$, 0.02% NP-40, 0.01% deoxycholate in PBS
- 20% v/v and 50% v/v glycerol in PBS
- Fe/Fe solution: 50 mM potassium ferricyanide, 50 mM potassium ferrocyanide
- X-Gal solution: 20 ml of detergent solution, 2 ml of Fe/Fe solution, 0.5 ml of X-Gal (40 mg ml^{-1} in dimethylformamide)

Method

1. Remove the ganglia into fixative solution. Fix the ganglia for 1 h on ice (start timing when the last ganglion is added to the fixative).
2. Remove the ganglia from the fixative solution, and rinse twice in PBS-MgCl$_2$.

Protocol 9. *Continued*

3. Replace with detergent solution, and incubate the ganglia on ice for 30 min.
4. Replace with X-Gal solution, and incubate at 37°C for 3–8 h.
5. Remove the staining solution. Rinse in PBS, and clarify in 20% glycerol in PBS for 2 h at 4°C, and then in 50% glycerol overnight.
6. Transfer ganglia in 50 % glycerol to a glass slide. Cover with a coverslip, and count the stained blue neurones under the microscope by focusing throughout the whole thickness of the ganglia.
7. Optional: ganglia can then be dehydrated and paraffin-embedded. Sections can be cut, and the number of blue neuronal profiles per section can then be determined.

2.6.3 Analysis of CNS tissues

Defective viruses can also be studied by direct injection into the brain. The functional and structural anatomy of the brains of rat and mouse have been well characterized, and it is possible to identify any particular region within the brain by three-dimensional coordinates, called stereotaxic coordinates. The use of a ring allows the operator to inject in a precise area of the brain of an anaesthetized animal by setting the three coordinates. The choice of the area to infect will depend on the specific needs of the operator, such as the interest for a particular neurological disease. This model can be used to study the efficiency of the delivery of genes by a replication-defective virus, as well as its toxicity or the effects due to the delivery of a particular therapeutic gene. Two different methods to evaluate the expression of the reporter gene LacZ are reported in *Protocols 10 and 11*.

Protocol 10. β-galactosidase staining of CNS tissues

Equipment and reagents
- PBS
- 4% w/v paraformaldehyde in PBS
- 25% sucrose in PBS
- 4% w/v Seakem LGT agarose in distilled water
- Detergent solution: 2 mM $MgCl_2$, 0.02% NP-40, 0.01% deoxycholate in PBS
- 0.1% w/v Neutral red
- Fe/Fe solution: 50 mM potassium ferricyanide, 50 mM potassium ferrocyanide
- X-Gal solution: 20 ml of detergent solution, 2 ml of 50 mM Fe/Fe solution, 0.5 ml X-Gal (40 mg ml^{-1} in dimethylformamide)
- Citifluor aqueous mounting medium (Citifluor Ltd.)

Method

1. Terminally anaesthetize mice by intraperitoneal injection with 0.1 ml of sodium pentobarbitone per mouse.

10: Herpes simplex virus and adenovirus vectors

2. Perfuse with 20 ml of 4% paraformaldehyde in PBS via the left ventricle.
3. Dissect the brain/spinal cord.
4. Post-fix for 1–2 h at 4°C with 4% paraformaldehyde.
5. Equilibrate in 25% sucrose overnight on ice.
6. Make 4% w/v Seakem LGT agarose in distilled water by heating in a microwave oven. Allow to cool to 37°C. Place the CNS tissue in a well of an ice cube-making tray, and place the tray on ice. Slowly add LGT agarose to the CNS tissue, and allow to set.
7. Cut the embedded tissue on a vibratome. Sections can be of 150–300 μm. Collect sections into ice-cold detergent solution. Incubate at 4°C for 1–2 h.
8. Replace the detergent solution with X-Gal solution. Incubate at 37°C for 2–16 h with gentle shaking. The length of this incubation depends upon the strength of the promoter driving LacZ expression, as different promoters will drive accumulation of variable amounts of protein. For a strong promoter such as the CMV IE, a 2 h incubation is sufficient, as the amount of LacZ present is so high that the enzymatic conversion of X-Gal into a clearly distinguishable blue product will occur quickly. For weaker promoters such as LAP, an overnight incubation is recommended.
9. Wash twice with PBS, mount, and allow to dry for at least 24 h.
10. Rehydrate through descending grades of ethanols to the aqueous phase, and then counterstain with Neutral red.
11. Dehydrate through ascending graded ethanols, clear in xylene, and mount with DPX mountant.

Protocol 11. Immunohistochemical detection of LacZ

Equipment and reagents

- Paraffin-embedded brain or spinal cord tissue, 30–60 μm sections on slides
- PAP pen (DAKO)
- Quench solution: 10% v/v methanol, 3% v/v H_2O_2
- TBS (Tris-buffered saline): 50 mM Tris-HCl pH 7.4
- TXTBS: 0.2% v/v Triton X-100 in TBS pH 7.4
- Rabbit anti-β-galactosidase (Sigma)
- Biotinylated mouse anti-rabbit (DAKO)
- Blocking solution: 3% normal goat serum in TXTBS
- TNS: 50 mM Tris-HCl pH 7.4
- Streptavidin ABC kit (DAKO)
- Diaminobenzidine (DAB) solution: 20 mg DAB in 40 ml of TNS, to which 12 ml of 30% H_2O_2 is added just before use
- Haematoxylin 10% w/v
- DPX mounting solution (BDH, UK)
- Xylene

> **Protocol 11.** *Continued*
>
> *Method*
>
> 1. Dewax paraffin-embedded sections for 2 × 15 min in xylene.
> 2. Rehydrate through 100%, 90%, 70%, and 50% ethanol, leaving for 5 min in each ethanol solution, then transfer to PBS and leave for a further 5 min.
> 3. Incubate at room temperature for 15 min in quench solution to quench the endogenous peroxidase activity.
> 4. Remove the H_2O_2, and rinse in TBS 3 times, 5 min each.
> 5. Circle the section on the slide with a PAP pen. Block non-specific staining by incubating with blocking solution. Allow 50–100 µl per section. Incubate for 1 h at room temperature in a humidified chamber.
> 6. Remove the fluid and blot off the remaining excess, then add 100 µl per section of a 1:1000 dilution of antibody in TXTBS supplemented with 1% normal goat serum. Incubate overnight at room temperature in a humidified chamber.
> 7. Wash the slides 3 times with TBS, 5 min each.
> 8. Incubate with 100 µl per section of biotinylated secondary antibody, diluted 1:200 in TBS supplemented with 1% normal goat serum, for 3 h at room temperature in a humidified chamber.
> 9. Wash slides 3 times in TBS, 5 min each.
> 10. Incubate sections with a 1:200 dilution of the solutions A and B of the streptavidin ABC kit, in TBS supplemented with 1% normal goat serum, for 2 h at room temperature in a humidified chamber.
> 11. Wash slides 3 times in TBS, then twice in TNS, 5 min each.
> 12. Develop by incubating with the DAB solution for a few minutes in the dark, until the sections turn a light brown.
> 13. Rinse the slides 3 times in TNS, 5 min each.
> 14. Optional: counterstain with Haematoxylin, then rinse excess stain out in water.
> 15. Dehydrate in an ascending series of ethanol for 2 min each, clear in xylene for 5 min, and mount in DPX mounting solution.

2.7 *In vitro* culture of neurones to study the biology of HSV

Examining the basic characteristics of virus infectivity, vector delivery, and expression of gene constructs is most readily done *in vitro*. It has been shown previously that latent HSV-1 infections may be established, maintained, and

10: Herpes simplex virus and adenovirus vectors

reactivated in primary cultures of dorsal root and trigeminal neurones, and that these infections display many of the characteristics of latent infections in animal models and natural infections (31–35). Hence we use primary dorsal root neurone cultures to model delivery, expression, and stability of our vector gene constructs in neurones.

In our primary dorsal root neurone culture model included here (*Protocol 12*), we are able to establish latent infection, and analyse expression of reporter gene constructs throughout the course of infection. Generally we use the antibody staining method included (*Protocol 13*) for the dual detection of the neuronal markers (β-tubulin III) and the reporter gene β-galactosidase.

Protocol 12. Production of primary dorsal root neurone cultures

Equipment and reagents

- Dissection microscope with cold light source
- 5 cm glass Petri dish
- Fine forceps, strong surgical scissors, and fine scissors
- 21-gauge needles
- F-14$^+$: Ham's nutrient mixture F-14 medium (modified) (Imperial Laboratories, UK), supplemented with 100 units ml^{-1} penicillin, 100 μg ml^{-1} streptomycin, 0.5 μg ml^{-1} Fungizone, and 1% UltroserG (USG, Gibco BRL)
- Complete medium: F-14$^+$ supplemented with 20 μg ml^{-1} nerve growth factor 2.5S (Sigma), 0.5 × B27 supplement (Gibco BRL), and 50 μM fluorodeoxyuridine (FdU, Sigma)
- Neonatal rat pups (up to 3 days old)
- Plating medium: as for the complete medium, but containing 5% USG
- Collagen pre-plates: six-well tissue-culture dishes (Falcon) coated with a thin layer of 0.5 mg ml^{-1} rat-tail collagen type VII (Sigma), air-dried, and pre-adsorbed with 2 ml of F-14$^+$ for 15 min at 37 °C
- Matrigel-coated coverslips: sterile 13 mm diameter coverslips in 24-well tissue-culture dishes (Falcon), coated with a thin layer of 0.5 × Matrigel (Becton-Dickinson)
- Trypsin-EDTA (Gibco BRL)
- Reduced-bore siliconized glass pipettes (sterile)

Method

1. Kill rat pups by decapitation, using a pair of strong surgical scissors.
2. Dissect out the spine with hindquarters still attached, and pin out, dorsal side up.
3. Cut away the dorsal surface of the spine, and remove the spinal cord to expose the ganglia.
4. With the aid of a dissecting microscope, remove the ganglia using fine forceps, and place immediately in complete medium on ice.
5. In a glass Petri dish, trim the roots off the ganglia, and transfer to an Eppendorf tube containing F-14$^+$ on ice.
6. Collect the ganglia by centrifugation at 200 *g* for 5 min.
7. Resuspend the ganglia in trypsin-EDTA (1 ml per pup), and incubate at 37 °C for 20 min.
8. Mechanically dissociate the ganglia by gentle pipetting, using a siliconized reduced-bore pipette for 3 min. Do not attempt to disperse all the ganglia at this stage.

Protocol 12. *Continued*

9. Allow the fragmented ganglia to settle, and remove the supernatant to a 50 ml Falcon tube containing 3 volumes of F-14$^+$, and retrypsinize the ganglionic debris.
10. Repeat the dissociation. At this point a few lumps may remain, which may either be further trypsinized, or removed from the preparation.
11. Dilute the trypsin suspension in 3 volumes of F-14$^+$, and remove any clumps or threads of debris.
12. Spin the suspension at 200 g for 10 min to pellet the cells.
13. Resuspend the cells in 2 ml of complete medium per pup, and pre-plate on collagen-coated six-well dishes (1 ml per well) for 3 h at 37°C with 5% CO_2.
14. Remove the medium from the plates, and spin at 200 g for 10 min.
15. Apply 100 µl of complete medium to each Matrigel-coated coverslip, and incubate at 37°C in the presence of 5% CO_2 until required.
16. Resuspend the cells in 1 ml of plating medium, and respin at 200 g for 10 min.
17. Resuspend the cells at 2.5 × 10^3 neurones ml^{-1}, and distribute 10 µl (approximately 250 neurones) onto the Matrigel-coated coverslips.

Detection of the exogenous gene product, in this case β-galactosidase, may be done using immunofluorescence, with simultaneous detection of a neuronal markers such as β-tubulin III.

Protocol 13. Antibody staining of neuronal cultures

Materials
- PBS
- 4% w/v paraformaldehyde in PBS
- 0.2% v/v Triton X-100 in PBS
- Normal horse or goat serum (NS)
- Primary antibodies diluted 1:1000 in 2% NS in PBS: rabbit anti-β-galactosidase, and a 1:1000 dilution of mouse anti-β-tubulin III (Sigma)
- Secondary antibodies diluted 1:1000 in 2% NS in PBS: 100 ng ml^{-1} diamino-propidium iodide (DAPI, Sigma), goat anti-rabbit FITC conjugate (Sigma), and donkey anti-mouse Cy3 conjugate (Amersham)
- Citifluor (Citifluor Ltd.)

Method
1. Remove the medium from the culture, and rinse briefly in PBS.
2. Fix in 4% paraformaldehyde solution for 20 min at room temperature.
3. Rinse in PBS, 3 × 5 min.
4. Add the Triton X-100 solution, and permeabilize for 15 min.
5. Remove the Triton X-100 solution, replace with 10% NS in PBS, and incubate at 37°C for 1 h.

10: Herpes simplex virus and adenovirus vectors

6. Remove and replace with diluted primary antibodies, and incubate at 37°C for 1 h.
7. Rinse 3 × 5 min in 1% NS in PBS.
8. Add diluted secondary antibodies, and incubate at 37°C for 45 min.
9. Rinse 3 × 5 min in PBS.
10. Rinse in distilled water briefly, drain, and dry the back of the coverslip.
11. Mount the coverslip in Citifluor for fluorescence microscopy.
12. Bound antibodies can be visualized by fluorescence microscopy using filter cubes containing excitation (ex) and barrier filters (bf) specific for the chromophore used. For the FITC and GFP range, use ex 455–495 nm and bf 510–555 nm; for the Cy3 range, ex 550–570 nm and bf 582–650 nm; and for the DAPI range, ex 390–402 nm and bf 425–465 nm.

3. Adenovirus

3.1 Biological properties

Adenoviruses have been isolated from a wide range of mammalian and avian species, with some 49 distinct human adenovirus (Ad) serotypes having been identified and assigned to six different sub-groups (A–F). While adenovirus infection is commonly associated with the respiratory tract, keratoconjunctivitis, and gastroenteritis, specific serotypes tend to be more frequently associated with certain sites of infection. Occasionally, systemic adenovirus infections are reported, particularly in immunosuppressed individuals (36). The majority of adenovirus vectors are based on serotypes 2 (Ad2) and 5 (Ad5), which are encountered by most individuals as mild respiratory tract infections in early childhood.

The adenovirus particle (70–100 nm diameter) does not have an envelope, but is a large 20-sided icosahedron consisting of 240 hexon capsomeres, with additional penton capsomeres located at each of the 12 vertices. A trimeric fibre protein extends from each of the 12 vertices (attached to the penton base proteins), and is responsible for recognition and binding to the cellular receptor. A detailed three-dimensional structural model has been constructed for the virus particle, based on a combination of cryoelectron microscopy and X-ray crystallography (37–39). A globular domain at the end of the adenovirus fibre is responsible for recognition of the cellular receptor. The virus infects cells by receptor-mediated endocytosis (40) and, following rapid release from endosomal vesicles into the cytosol, the particle is transported to the nuclear envelope. Adenovirus transcription and DNA replication take place in the nucleus.

3.2 Pattern of gene expression during lytic adenovirus infection

Adenovirus has a linear double-stranded DNA genome of approximately 35 kb, flanked by short inverted terminal repeats which each contain an origin of DNA replication (*Figure 1*). Like herpesviruses, adenovirus operates a cascade system of gene regulation, in which early genes are expressed prior to replication of the virus genome, and late genes afterwards (41). There are four regions of early phase transcription (E1–E4), which each encode multiple gene products by means of alternate RNA splicing. The E1A gene is expressed first, and acts as a powerful transactivator to stimulate transcription from the other early promoters. Two additional genes encoding the IVa2 and IX proteins are expressed with delayed early/late kinetics. Late-phase transcription is driven primarily by the major late promoter. Although transcription from this promoter is complex, involving multiple polyadenylation signals and an elaborate usage of RNA splicing, five gene clusters can been defined (L1–L5). Late-phase gene expression is primarily concerned with the synthesis of virion proteins. Virus particles assemble and accumulate in the nucleus of the infected cell. Unlike HSV, there is no specific pathway for virus release, although prolonged infection will eventually lead to cell lysis, which may be facilitated by the adenovirus death protein (E3-11.6K) (42).

3.3 Adenovirus vectors

The adenovirus genome is transcribed in both directions, and alternative splicing occurs allowing multiple transcripts to be generated from the same

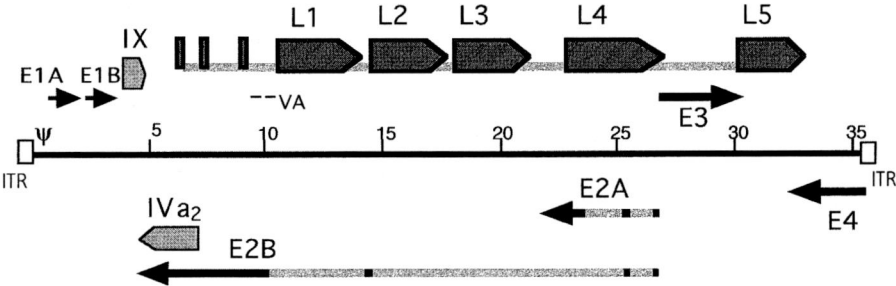

Figure 1. A simplified transcriptional map of the Ad5 genome. Transcriptional units active during the early phase of infection (E1–E4) each encode multiple gene products, primarily as a consequence of differential RNA splicing. The small polymerase III virus-associated (VA) RNAs modulate translation. The major late promoter drives the transcription of L1–L5. IX and IVa$_2$ are transcribed from independent promoters. Faint lines are used to represent longer intron sequences. The inverted terminal repeats (ITR) and *cis*-acting sequence required for DNA packaging (ψ) are indicated. The genome size is indicated in kilobase pairs.

10: Herpes simplex virus and adenovirus vectors

sequence. Any manipulation of the virus genome therefore requires the coding capacity of both DNA strands to be considered. Furthermore, adenovirus particles limit the amount of DNA that can be packaged to approximately 105% of the genome, only 1.8 kb more that the wild-type virus genome (43). The E3 gene region is dispensable for adenovirus replication *in vitro,* and can be deleted from both replication-competent and replication-deficient vectors to permit larger inserts. Adenovirus vectors may be either replication-competent or replication-deficient. Replication-competent adenovirus vectors have been advocated as potential immunization agents. In replication-competent adenovirus recombinants, transgenes are usually inserted in the E3 region, and very high levels of expression can be achieved with transcription driven by the adenovirus late promoter.

First generation replication-deficient adenovirus vectors are usually based on adenovirus $E1^-$, $E3^-$ deletion mutants. Deletion of the E1 gene region (both E1A and E1B) permits vectors to accommodate significantly larger inserts, removes a region of the genome associated with cellular transformation, and generates a vector that is not only replication-deficient but also cannot effectively activate early-phase gene expression. This defect in early-phase gene expression is extremely effective, generating a vector which can promote efficient transgene delivery and expression in the absence of significant vector gene expression. For many purposes, adenovirus $E1^-$, $E3^-$ vectors are perfectly adequate. However, breakthrough to early- and late-phase gene expression can occur, particularly following infection at high m.o.i., or if the target cell expresses an endogenous 'E1A-like' function.

Breakthrough expression from the vector genome has been associated with the induction of an immune response following *in vivo* gene delivery. Replication-deficient adenovirus vectors incorporating additional mutations or deletions have been generated to address this problem. The deletion of additional sequences from the vectors also permits the insertion of larger transgenes. The 2.8 kb E4 transcriptional unit encodes at least six proteins, and is required for efficient replication of virus DNA and processing of late-phase adenovirus mRNA (44). The E4 deletion reduces breakthrough vector gene expression, and thus also the immune response to the adenovirus vector *in vivo*. However, the inclusion of E4 in vectors has unexpectedly been found to help sustain active transgene expression for much longer periods (45, 46), so that for some applications deletion of E4 may be detrimental.

E2A encodes a ssDNA-binding protein which is required for adenovirus DNA replication, is expressed at high abundance, and is highly antigenic. Adenovirus $E1^-$ vectors have been constructed which contain a temperature-sensitive mutations in (47) or deletion of the E2A gene (48, 49). Defects in the E2A expression are also associated with reduced breakthrough to early-late gene expression. Impaired E2A expression was associated with a reduced cellular immune response to the adenovirus vector and prolonged transgene expression *in vivo* (50–52). However, other studies have been unable to

a)

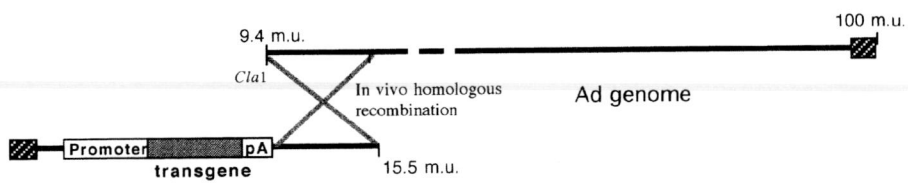

b)

Figure 2. Some methods for generating replication-deficient adenovirus recombinants. (a) Direct DNA ligation to adenovirus genomic DNA asymmetrically cleaved using restriction endonucleases. (b) Homologous recombination between adenovirus genomic DNA cleaved by a restriction endonuclease and a plasmid transfer vector. (c) Homologous recombination between a large plasmid (pJM17) containing adenovirus genomic and a plasmid transfer vector.

demonstrate enhanced persistence of the transgene following crippling of E2A function (49, 53). The results of such studies will depend on how the experiment is constructed. In situations where the host sees the expressed transgene as 'foreign', the predominant immune response tends to be directed against the expressed transgene (53–55).

The gene encoding pIX may also be deleted. Its function is to assist in the final stages of the assembly of the virus particle. Krougliak and co-workers calculate that an adenovirus E1, E3⁻, E4⁻, IX⁻ deletion mutant has the potential to accommodate an 11 kb insert (56). The generation of vectors with more extensive deletions is problematical, both because of the complexity of the virus genome and because of difficulties in providing complementing functions effectively in helper cell lines.

From the viewpoint of safety, size of transgene insertion, and absence of vector gene expression, the so-called 'gutless' adenovirus vectors are

10: Herpes simplex virus and adenovirus vectors

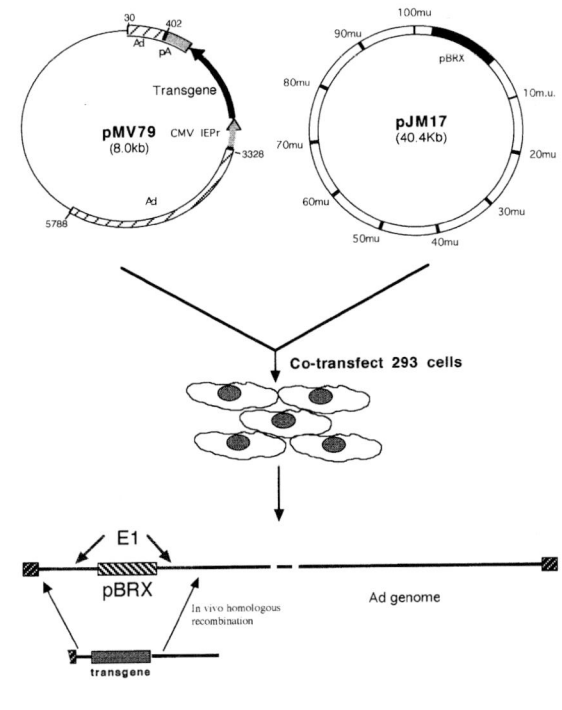

c)

extremely attractive. A number of groups have developed vector systems in which most or all adenovirus protein-coding sequences have been removed (57–64). The vectors contain the inverted terminal repeats (ITRs) and adjacent sequences responsible and essential for replicating the adenovirus genome, including the *cis*-acting signal sequence (ψ) which directs packaging of adenovirus DNA into virus particles (*Figure 2*). The gutless vector requires virtually all adenovirus gene functions to be provided for vector propagation and, since this cannot yet be achieved using a packaging cell line, they must be provided using a helper cell. An elegant system has been devised to minimize contamination of virus stocks with the helper virus. Helper viruses have been engineered so that the bacteriophage Cre-*lox*P recombinase system can be used to delete the *cis*-packaging signal ψ from the adenovirus helper virus during propagation of the gutless vector, to prevent helper virus encapsidation (60, 62). Although their potential is enormous, current gutless vectors are complex and technically difficult to use.

3.4 The helper cell line

Recombinant viruses based on adenovirus E1⁻ vectors are grown on a cell line which expresses E1 helper functions. 293 cells are a well established cell line widely used by adenovirologists and others. These cells were generated by transformation of human embryonic kidney cells with fragmented Ad5 genomic DNA, resulting in chromosomal insertion of bases 1 to 4344 (65, 66). 293 NS cells were subsequently selected for growth in suspension (67). Although the low adherence properties of 293 cells makes growth on roller bottles problematic, efficient scaling up of virus production on microcarriers can be achieved (68). 293 cells are currently available from Microbix Biosystems Inc. (Cat. No. PD-02-01), the European Centre for Animal Cell Cultures (ECACC Cat. No. 92052131), or the American Type Culture Collection (ATCC Cat. No. CRL 2573). 293 cells expressing not only Ad5 E1 but also additional adenovirus helper functions have been described (56).

Alternative helper cell lines for the growth of adenovirus E1⁻ vectors have been produced, of which perhaps the most commonly used are 911 cells (69, 70). 911 cells were derived by transformation of human embryonic retinoblasts with a plasmid containing the E1 gene region (bases 79–5789). Like 293 cells, 911 cells can be transfected with high efficiency, and can be used for the generation of adenovirus recombinants by homologous recombination. 911 cells grow more slowly than 293 cells, but provide slightly higher yields of recombinant virus. In our hands, 911 cells are particularly useful in performing plaque assays; the slower growth and contact inhibition exhibited by 911 cells is more amenable to the technique. 911 cells also tolerate agarose overlays well, and plaque formation is rapid (3–5 days). 911 cells are currently distributed by IntroGene, P.O. Box 2048, 2301 CA Leiden, The Netherlands.

3.5 Basic adenovirus handling techniques

Recombinant viruses should be plaque-purified three times when first generated to guarantee that the virus is derived from a single entity, before being expanded to produce a virus stock. Near-confluent 293 or 911 cell monolayers can be infected at either a low or high m.o.i. A low multiplicity of infection provides a selection for intact virus genomes capable of supporting productive infection, and thus helps maintain the integrity of the virus. Always keep separately an aliquot of the first plaque-purified virus stock produced as a master stock. Additional aliquots of this stock are designated as submasters, to be used for the generation of future virus stocks. Adenovirus recombinants are generally stable, but it is good practice to keep as close to the original, characterized virus stock as possible. The standard methods used for growing and purifying virus are given in *Protocols 14 and 15*.

10: Herpes simplex virus and adenovirus vectors

Protocol 14. Growth of virus

Equipment and reagents
- Dulbecco's Minimal Modified Essential Medium (MEM) (Sigma) supplemented with 10% (v/v) fetal calf serum, 100 units ml^{-1} penicillin, 100 μg ml^{-1} streptomycin, and 1 mM L-glutamine
- Tissue-culture flasks (175 cm^2)
- Arklone P (1,1,2-trichloro-1,2,2-trifluoro-ethane) (e.g. Basic Chemical Company)

Methods

1. Grow either 293 or 911 cells to 80% confluence in 175 cm^2 flasks before the addition of virus (low m.o.i., 0.001 PFU per cell; high m.o.i., 5 PFU per cell) in 15 ml of medium.

2. Cells are incubated with gentle rocking for 90 min before the inoculum is removed and fresh medium added. The infection will spread rapidly through the monolayer, and causes cells to become refractile, round up, and eventually detach from the monolayer. As soon as all cells exhibit a gross CPE (at high m.o.i. 40–48 h post-infection), harvest the monolayer together with detached cells. Infected 293 cells will readily detach if the flask is given a gentle tap, whereas it may be necessary to use a cell scraper to harvest infected 911 cells.

3. Cells are harvested in medium, transferred to 50 ml Falcon centrifuge tubes, and recovered by centrifugation at 2000 r.p.m. (1400 *g*) for 10 min at room temperature.

4. Decant the supernatant to waste. The medium contains significant amounts of virus released by cell lysis, but this is not normally recovered.

5. Cells are resuspended in 15 ml of PBS in a 50 ml Falcon tube, and extracted with an equal volume of Arklone P with vigorous mixing but not vortexing. The aqueous (top) phase is clarified by centrifugation at 2000 r.p.m. (1400 *g*) for 10 min at room temperature and harvested, carefully avoiding the interface.

6. The Arklone P phase can be re-extracted with a further 10 ml of PBS for efficient virus recovery. Following mixing and centrifugation as in Step 5, the supernatant is pooled with the first virus harvest. The virus can be aliquotted and stored at this stage. It is, however, still a fairly crude preparation, heavily contaminated with cellular protein. In general, avoid putting the virus through freeze–thaw cycles as much as possible. As is the case for HSV-1, further purification of the virus is often required. Adenoviruses can be further purified by CsCl gradient centrifugation.

Protocol 15. Virus purification

Equipment and reagents

- Beckman or equivalent ultracentrifuge and rotor (Beckman SW41)
- 14 × 89 mm Ultra-Clear centrifuge tubes (Beckman)
- CsCl Solution I: 60.3 g CsCl per 100 ml, in 5 mM Tris-HCl pH 7.8 (density = 1.45 g ml^{-1})
- CsCl Solution II: 44.2 g CsCl per 100 ml in 1 mM EDTA, 5 mM Tris-HCl pH 7.8 (density = 1.33 g ml^{-1})
- Dialysis tubing
- Buffer D (1 mM $MgCl_2$, 135 mM NaCl, 10 mM Tris pH 7.8, 10% glycerol)

Care should be taken to use sterile (autoclaved) solutions to ensure the virus stock is protected from microbial contamination or proteases. Centrifuge tubes may be disinfected by washing with ethanol, and the dialysis membrane should be boiled three times for 5 min in TE buffer immediately before use both to clean and disinfect it.

Method

1. Virus purified from infected cells by Arklone P extraction (*Protocol 14*) is used as the starting material. If the virus has been stored frozen, then it should be centrifuged at 2000 r.p.m. (1400 *g*) for 10 min (room temperature) on thawing to remove any precipitated material.

2. Prepare a step gradient by first pipetting 1.6 ml of CsCl Solution I into the centrifuge tube, and then 3 ml of CsCl Solution II gently layered on top. The tube is then filled to 2.5 mm from the top with the virus preparation. The number of tubes will depend on the volume of virus used. Centrifuge at 90 000 *g* for 2 h at room temperature.

3. Two bands should be clearly visible. An opalescent band between the high and low density CsCl solutions contains the virus, while a second band is present higher up the tube. Collect the virus carefully, using a 19-gauge needle and gently pulling the lower band into a 2 ml syringe. Great care must be taken to avoid a needle-stick injury.

4. The collected virus is pooled, and then diluted with an equal volume of 10 mM Tris-HCl pH 7.8.

5. Prepare a second CsCl gradient by first pipetting 2.0 ml of CsCl Solution I into the centrifuge tube, and then gently layering 3.0 ml of CsCl Solution II on top. Tubes are then filled to within 2.5 mm of the top with the diluted sample from Step 4. Centrifuge for 16 h at 100 000 *g* at room temperature.

6. Using overhead illumination, a single opalescent band containing the virus should be visible (CsCl density of 1.34–1.35 g ml^{-1}).

7. Collect the virus as before.

8. Dialyse the purified virus against 1 litre of buffer D for at least 4 h at

> 4°C. To ensure efficient removal of CsCl, the buffer is changed, and dialysis repeated twice more.
>
> 9. Aliquot the virus in cryotubes, and store at −70°C. Avoid multiple freeze–thaw cycles. Recovery from CsCl gradients is usually efficient, and results in a significant concentration of the virus. While extremely high titre virus may be useful for some procedures, it may also be appropriate to dilute the virus slowly with buffer D at this stage, to ensure its titre is not too high for routine use.

3.6 Quantification of adenovirus stocks

A number of methods have been devised for the quantification of virus stocks, including titration of the virus infectivity, quantitative assays of transgene expression, quantitative nucleic acid hybridization (71), particle count by electron microscopy, and the optical absorbance of purified virus using a spectrophotometer (72). A classical virus titration provides a measurement of the number of 'plaque forming units' (PFU) or of 'tissue-culture infectious doses' (TCID) in a given volume of a virus stock when added to a permissive cell. Clearly, replication-deficient adenovirus recombinants can only readily be titrated on a helper cell line, but this remains the most common method for quantifying recombinant virus. However, most target cell types appear to be less susceptible to adenovirus infection than the helper cell lines. To assess the relative 'infectability' of the target cell, an adenovirus recombinant encoding a reporter gene (usually LacZ or GFP) is commonly used. Titration of the replication-deficient adenovirus in the helper cell provides a benchmark for their use in non-permissive cell lines *in vitro* and *in vivo*.

A number of protocols for the titration of adenovirus recombinants both by plaque assay and $TCID_{50}$ determination have been published (73–75). In our hands, this method for performing plaque assays using 911 cells (adapted from (69)) has proved to be both rapid and reliable and is described in *Protocol 16*.

> **Protocol 16.** Virus titration
>
> *Material and reagents*
> - Six-well plates
> - 2% agarose gel (Gibco BRL)
> - 1% w/v Crystal violet in PBS (BDH)
> - Formaldehyde
> - 2 × medium: BME (2 ×) with Earle's salt without Phenol red (Gibco BRL), 4% fetal calf serum, 40 mM Hepes pH 7.4
>
> *Method*
> 1. Trypsinize 911 cells, and seed six-well dishes so that overnight growth will result in approximately 40% confluence. There should be

Protocol 16. *Continued*

good cell-to-cell contact throughout the monolayer. This will provide a sufficient density of cells to permit cell growth during the assay.

2. Perform tenfold serial dilutions of virus in growth medium. Typically 1 ml of the 10^{-8}, 10^{-9}, 10^{-10}, 10^{-11}, and 10^{-12} dilution is added to each of the wells, with a control well receiving medium alone. The inoculum is rocked gently over the cells at 37 °C for 90 min. If a rocking table is not available, the virus inoculum should be added for 4 h with regular periodic manual rocking.

3. Melt 2% agarose by immersion in a boiling water bath, and then cool to 65°C. Warm 2 × medium to 37 °C.

4. Remove the virus inoculum from the dishes, and wash the monolayer once with PBS, being careful to drain off all the buffer. Mix 12 ml of 2% agarose with 12 ml of 2 × medium quickly but gently, to avoid the generation of bubbles. The solution cools quickly with pipetting, and must be added almost immediately to cells before the agarose starts to set. Add 4 ml of agarose overlay to each well.

5. Allow to set for 5 min at room temperature, and then incubate at 37°C in a CO_2 incubator with a moist environment.

6. Plaques can be observed under the microscope within a few days, and are counted on day 5. Plaques can be counted directly or, alternatively, following staining as indicated below. The concentration of virus in the original stock is determined as the number of plaques multiplied by the reciprocal of the dilution. For example, twelve plaques at the 10^{-9} dilution is equivalent to a titre of 1.2×10^{10} PFU ml^{-1}.

7. Staining of cells can make the plaques more visible and easier to count. Fix the cells by adding 1 ml of formaldehyde to each well for 1 h.

8. The formaldehyde is poured off, and the agarose overlay is removed carefully with the aid of a spatula. Care must be taken not to disturb the monolayer, to avoid damaging plaques or creating holes in the monolayer that resemble plaques.

9. Add 1ml of 1% Crystal violet-PBS to each well for 30 s at room temperature.

10. Remove the stain, and rinse twice with PBS. Plaques should be visible either using a microscope or macroscopically.

3.7 Construction of replication-deficient adenovirus recombinants

Adenovirus vector technology is progressing rapidly, with vectors increasingly being modified or tailored to specific applications. Nevertheless, a number of elegant systems have been devised to make the insertion of a transgene into an adenovirus vector a simple and straightforward procedure. Some of the more common and efficient methods are described below.

3.7.1 Direct *in vitro* ligation with adenovirus genomic DNA

Adenovirus genomic DNA can readily be isolated from CsCl gradient-purified virus, and is infectious when transfected into cells. In some vectors, adenovirus genomic DNA has been modified to contain unique restriction endonuclease sites to facilitate the insertion of transgenes. Genomic adenovirus DNA can then be digested with a specific restriction endonuclease, and the transgene ligated directly to viral DNA. Recombinant virus is generated following transfection of permissive cells following DNA ligation.

Although a well established method (76), direct DNA ligation is now rarely used because of its low efficiency. However, the method has been revived and modified recently by Okada and co-workers (77) to provide an efficient system to generate adenovirus recombinants (*Figure 2a*). An adenovirus E1⁻ vector (AVC2null) was constructed with the strong constitutive HCMV IE promoter inserted in place of the E1 gene. Purified AVC2null DNA is then cleaved at unique cloning sites (*Xba*I and *Nsp*V) downstream of the HCMV major IE promoter. Cleavage with two non-compatible restriction enzymes prevents vector religation and permits directional insertion of the transgene. The transgene can be engineered by PCR to be flanked by the appropriate restriction endonuclease cleavage sites and ligated directly with adenovirus genomic DNA. A major advantage of this system is speed: there is no need to subclone the transgene into a transfer vector. The recombinant virus is generated following transfection of the helper cell line.

In addition to directional cloning, an important feature of the Okada procedure was the use of adenovirus genomic DNA which has the adenovirus terminal protein still covalently attached to both ITRs. The terminal protein has a role both in attachment of the virus genome to the nuclear matrix and in initiating DNA replication. Attachment of the terminal protein to the viral genome increased the efficiency of recombinant virus generation a hundredfold (77).

3.7.2 Homologous recombination with virus genomic DNA

This commonly used method depends on homologous recombination between purified adenovirus genomic DNA and a transfer vector to generate recombinant virus. Once more restriction endonuclease digestion is used to cleave the adenovirus, usually by digestion at the E1 locus, and the longer right arm DNA fragment is then purified following agarose gel electrophoresis. In a bacterial cloning experiment, the transgene is inserted into a transfer vector which contains adenovirus sequences necessary to repair the deleted left-hand end of the virus (*Figure 2b*). The transfer vector/transgene construct and the right-hand arm of the adenovirus genome are then co-transfected into the helper cell line and, following homologous recombination, the adenovirus recombinant virus is generated. This approach has been adapted for a number of adenovirus serotypes. This system is currently available commercially from Quantum Biotechnologies Inc., Quebec, Canada (also marketed by NBL Gene Sciences, Northumberland, UK).

3.7.3 Homologous recombination between two transfected plasmids

A bacterial plasmid containing the adenovirus genome is able to generate virus plaques when transfected into a permissive cell. Rather than purifying large quantities of adenovirus DNA, it is therefore possible simply to grow it up as plasmid in *E. coli*. Graham and co-workers developed an elegant method for the generation of adenovirus recombinant based on the co-transfection of two plasmids (78), and this method is now very widely used.

The size of insert that can be accommodated by an adenovirus vector is limited by the packaging limits of the virus particle. The plasmid pJM17 contains the complete Ad5 *dl*309 genome with a prokaryotic vector (pBRX) inserted in the E1 gene region. The size of pJM17 is larger than the adenovirus packaging limit, and hence transfection with pJM17 seldom generates virus plaques. However, co-transfection with pJM17 and a transfer vector containing the transgene frequently results in the generation of recombinant virus. Homologous recombination between the flanking adenovirus sequences in the transfer vector and pJM17 results in the E1 gene region and prokaryotic vector sequences being deleted and the transgene inserted (*Figure 2c*). The reduced size of the construct permits genome packaging and virus plaque formation in 293 or 911 helper cell lines.

More detailed protocols on using this vector system are published elsewhere (74, 75, 78), and vectors are currently available commercially from Microbix Biosystems Inc., 341 Bering Avenue Toronto, Ontario, Canada.

3.7.4 *E. coli* plasmid, cosmid, and YAC systems

Vector systems that facilitate the manipulation of large DNA virus genomes in *E. coli* have been developed, and recently they have been adapted for production of adenovirus recombinants (79, 80). Vogelstein and co-workers (80) produced vectors for adenovirus which enable all the DNA manipulation steps to be carried out in *E. coli*. The transgene is first sub-cloned into a transfer vector to provide flanking adenovirus sequence from either side of the E1 gene region (to be precise: Ad5 sequence 1–480 – *transgene* – Ad5 sequences 3534–5740). This plasmid is linearized and co-transfected into an appropriate *E. coli* strain (BJ5183) with a plasmid (pAdEasy-1) containing the 'right' arm of the Ad5 genome (from base 3534 onwards). Homologous recombination between the transfer vector and pADEasy-1 in *E. coli* generates a bacterial plasmid containing the adenovirus recombinant genome with the transgene inserted at the E1 locus. Antibiotics are used to select for the appropriate construct and plasmid DNA prepared. The adenovirus

10: Herpes simplex virus and adenovirus vectors

genome is then excised from the plasmid by digestion with the appropriate restriction endonuclease, and the adenovirus recombinant generated by transfection of 293 or 911 cells.

Cosmids are also based on a standard *E. coli* plasmid vectors, but contain a signal designed to package their DNA into bacteriophage lambda particles *in vitro*. This packaging step facilitates and selects for the cloning of very large DNA fragments (38–52 kb). Cosmid vectors containing the adenovirus genome have been constructed that permit the insertion of a transgene directly into the cosmid vector (81, 82). Similarly, recombinant adenovirus vector systems based on yeast artificial chromosomes have also been described (83).

3.8 Characterization of virus

Vector systems are designed so that the vast majority of the virus generated contain the appropriate insert. However, occasionally inappropriate recombination events take place which generate an aberrant product. The integrity of the adenovirus genome and the cloned insert should be confirmed by restriction endonuclease digestion of the recombinant virus genome. The production of sufficient adenovirus genomic DNA to visualize on an ethidium bromide-stained DNA gel is straightforward. The method described in *Protocol 17*, which is based on a commercial column system, is simple, rapid, and produces a good quality DNA preparation. An alternative method based on a simple phenol:chloroform extraction will also provide satisfactory results (84).

Protocol 17. Preparation of virus DNA

Equipment and reagents

- QIAamp Blood Kit (Qiagen Ltd)
- Analytical grade ethanol
- RNase A (Sigma): make up at a concentration of 20 mg ml^{-1} and incubate for 2 min in a boiling water bath to inactivate any contaminating DNase.
- A microfuge with a contained rotor. Alternatively a benchtop centrifuge with adapters for small tubes and sealed buckets may be used.
- 75 cm^2 tissue culture flasks

Method

1. Infect a 75 cm^2 flask of near-confluent 293 or 911 cells with the test virus. Infection is allowed to proceed until all cells in the flask exhibit a gross CPE. Cells are detached in tissue-culture supernatant, using a cell scraper if necessary, and recovered by centrifugation. The cell pellet is washed once with PBS, and all buffer carefully removed.

2. Cells are resuspended in 600 μl TE buffer, and transferred to a 1.8/2.0 ml screw-top centrifuge tube. An equal volume of Arklone P is added to the cell suspension, and mixed vigorously. The aqueous phase is cleared by centrifugation at maximum speed in a microfuge (2 min). The aqueous phase (approximately 500 μl) is collected.

Protocol 17. Continued

3. The sample is incubated with 1 mg of RNaseA for 30 min at 37°C.
4. First add 62.5 μl of protease solution, followed by 500 μl of buffer AL. Mix and incubate at 70°C for 10 min.
5. Add 525 μl of ethanol, and vortex.
6. Pass the sample through the Qiagen column by centrifugation at 8000 r.p.m. in a microfuge.
7. The sample bound in the column is washed by passing 500 μl of buffer AW through the column by centrifugation. Repeat this step.
8. Spin the column dry for 3 min at 13 000 r.p.m..
9. Add to the column 100 μl of buffer AE preheated to 70°C, and incubate for 5 min at 70°C.
10. Centrifuge for 1 min at 13 000 r.p.m. to recover the adenovirus DNA. Use 5 μl for a DNA digest.

Figure 3 illustrates a restriction endonuclease digest performed to test for the presence of an insert in an Ad5 E1⁻ recombinant produced according to the method of Graham and co-workers (78). The experiment compares DNA purified from a control virus with no insert (RAd60) and the test recombinant (RAd235). The RAd60 genome has an *Xba*I site at the position of the E1 gene deletion; this also corresponds to the normal site of transgene insertion.

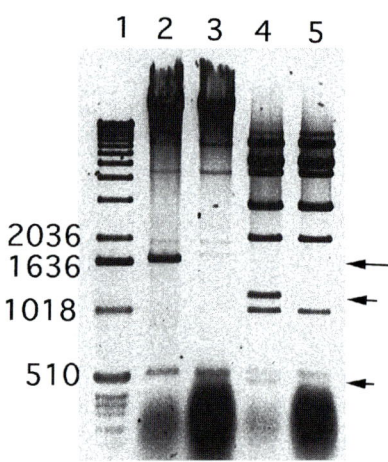

Figure 3. A restriction endonuclease digest of DNA purified from a recombinant virus. Lane 1. Molecular weight standard. Lane 2. RAd235 (1.6 kbp insert) DNA digested with *Xba*I. The excised insert is indicated by a long arrow. Lane 3. RAd60 (no insert) DNA digested with *Xba*I. Lane 4. RAd235 digested with *Hin*dIII. Excised inserts are indicated by short arrows. Lane 5. RAd60 DNA digested with *Hin*dIII.

*Xba*I cleavage of RAd60 DNA generates a short 500 bp fragment from the left end of the genome and a large fragment corresponding to the right arm of the virus genome (lane 2). *Xba*I cleavage also releases the 1.61 kbp expression cassette inserted in Rad235 (lane 3).

The adenovirus vector contains a number of *Hin*dIII sites, so the restriction endonuclease digestion pattern is complex. In this construct, *Hin*dIII also excises the transgene, but since there are also *Hin*dIII site within the transgene two additional bands of 1181 and 435 bp are detected (lane 4). The identity of the insert can be analysed by further restriction endonuclease digests, hybridization probing of Southern blots, PCR, or by direct DNA sequencing of the inserted fragment.

3.9 Infection of cells with adenovirus vectors

Adenoviruses infect cells by receptor-mediated endocytosis (40). Ad5 and Coxsackie B virus both bind to the same primary receptor on the cell surface (Coxsackie adenovirus receptor protein, CAR). Following the initial binding event, an RGD motif in the Ad5 penton base protein interacts with the α-integrin subunit to promote endocytosis (85). While efficient gene delivery to the target cell by an adenovirus vector is dependent on binding to the primary receptor, the presence of an appropriate integrin receptor appears to be an absolute requirement (86, 87).

Many protocols require modification of adenovirus vectors in order to enhance gene delivery to cells inefficiently infected by the virus, or to restrict virus infection to a specific cell type. Considerable research effort is going into modifying the tropism of adenovirus vectors. It has proved possible to genetically modify the fibre gene in the adenovirus genome (88–90), and adapter molecules which can bind both the virus and specific cell-surface components have also been exploited to target and enhance infection of specific cells (91–93).

There are, however, a number of simpler approaches to enhancing infection using conventional vectors. While adenovirus can infect a wide range of cell types, infection can be inefficient. Most target cells are less susceptible to infection with adenovirus vectors than the helper cell lines in which the virus was propagated and its titre determined. Increasing the amount of virus used can often overcome this barrier; m.o.i.s of 100–1000 PFU cell^{-1} can sometimes be needed. It should be noted that efficient infection is more dependent on the absolute concentration of virus surrounding the cell than the total virus-to-cell ratio (94). Infection can thus be substantially more efficient if performed in a small volume.

3.10 Enhanced infection

During a systematic analysis of factors controlling the efficiency of adenovirus infection, Trapnell and co-workers attempted to improve the efficiency of

Table 1. Reagents for enhancing the efficiency of adenovirus infection

	Source in inoculum	Final concentration
DEAE-dextran	Sigma	2 µg ml^{-1}
Polybrene	Sigma	4 µg ml^{-1}
Protamine	Sigma	5 µg ml^{-1}

adenovirus infection by increasing the effective concentration of virus in medium surrounding the target cell by the addition of a biocompatible polymer with the virus inoculum (94). Infection could be enhanced as much as tenfold by the addition of Poloxamer 407 (BASF, Parippany, NJ) to a final concentration of 15% (w/v) in the inoculating virus.

The efficiency of virus infection is also affected by the presence of charged molecules on the surface of the target cell, in particular negatively charged sialic acid residues. It has been recognized for some time that cations can enhance infection of a number of viruses, in particular retroviruses (95). In our hands, adenovirus-mediated gene delivery in the murine L929 cell line was dependent on the addition of DEAE-dextran to the virus inoculum (55). Although useful, DEAE-dextran is toxic to certain cells. In a systematic study, Arcasoy and co-workers demonstrated that a range of polycations, but not heparin (a polyanion), could facilitate adenovirus-mediated gene delivery to epithelial and endothelial cells (96). As an alternative, cationic liposomes have also been demonstrated to enhance adenovirus infection in some cell types, independently of binding to either the fibre receptor or the α-integrin on the target cell (97). The benefit of using polycations depends on the target cell type, and needs to be tested empirically. Based on the study by Arcasoy *et al.*(96), the reagents and concentrations given in *Table 1* are suggested for a preliminary screen for efficacy and toxicity.

References

1. McGeoch, D. J., Dalrymple, M. A., Davison, A. J., Dolan, A., Frame, M. C., McNab, D., Perry, L. J., Scott, J. E., and Taylor, P. (1988). *J. Gen. Virol.*, **69**, 1531.
2. Wadsworth, S., Jacob, R. J., and Roizman, B. (1975). *J. Virol.*, **15**, 1487.
3. Honess, R. W., and Roizman, B. (1974). *J. Virol.*, **14**, 8.
4. Wagner, E. K., Guzowski, J. F., and Singh, J. (1995). *Prog. Nucl. Acid Res. Mol. Biol.*, **51**, 123.
5. Stern, S., Tanaka, M., and Herr, W. (1989). *Nature*, **341**, 624.
6. Katan, M., Haigh, A., Verrijzer, C. P., and van der Vliet, P. C. (1990). *Nucl. Acids Res.*, **18**, 6871.
7. Kristensson, K. (1970). *Acta Neuropathol. Berlin*, **16**, 54.
8. Efstathiou, S., Minson, A. C., Field, H. J., Anderson, J. R., and Wildy, P. (1986). *J. Virol.*, **57**, 446.

10: Herpes simplex virus and adenovirus vectors

9. Cabrera, C. V., Wohlenberg, C., Openshaw, H., Rey-Mendez, M., Puga, A., and Notkins, A. L. (1980). *Nature,* **288**, 288.
10. Stevens, J. G., Wagner, E. K., Devi-Rao, G. B., Cook, M. L., and Feldman, L. T. (1987). *Science,* **235**, 1056.
11. Farrell, M. J., Dobson, A. T., and Feldman, L. T. (1991). *Proc. Natl. Acad. Sci. USA,* **88**, 790.
12. Zabolotny, J. M., Krummenacher, C., and Fraser, N. W. (1997). *J. Virol.,* **71**, 4199.
13. Steiner, I., Spivack, J. G., Linette, R. P., Brown, M. S., MacLean, A. R., Subak-Sharpe, J. H., and Fraser, N. W. (1989). *EMBO J.,* **8**, 505.
14. Sawtell, N. M., and Thompson, R. L. (1992). *J. Virol.,* **66**, 2157.
15. Bloom, D. C., Maidment, N. T., Tan, A., Dissette, V. B., Feldman, L. T., and Stevens, J. G. (1995). *Mol. Brain Res.,* **31**, 48.
16. Lachmann, R. H., and Efstathiou, S. (1997). *J. Virol.,* **71**, 3197.
17. Rixon, F. J., and McLauchlan, J. (1993). In *Molecular virology: a practical approach.* (ed. A. J. Davison, and R. M. Elliott), p. 285. IRL Press, Oxford.
18. Hill, T. J., Field, H. J., and Blyth, W. A. (1975). *J. Gen. Virol.,* **28**, 341.
19. Speck, P. G., Efstathiou, S., and Minson, A. C. (1996). *J. Gen. Virol.,* **77**, 2563.
20. Geller, A. I., Keyomarsi, K., Bryan, J., and Pardee, A. B. (1990). *Proc. Natl. Acad. Sci. USA,* **87**, 8950.
21. Samaniego, L. A., Webb, A. L., and DeLuca, N. A. (1995). *J. Virol.,* **69**, 5705.
22. Samaniego, L. A., Wu, N., and DeLuca, N. A. (1997). *J. Virol.,* **71**, 4614.
23. Ho, D. Y., Fink, S. L., Lawrence, M. S., Meier, T. J., Saydam, T. C., Dash, R., and Sapolsky, R. M. (1995). *J. Neurosci. Meth.,* **57**, 205.
24. Field, H. J., and Wildy, P. (1978). *J. Hygiene,* **81**, 267.
25. Krisky, D. M., Marconi, P. C., Oligino, T., Rouse, R. J., Fink, D. J., and Glorioso, J. C. (1997). *Gene Therapy* **4**, 1120.
26. Huang, Q. S., Deshmane, S. L., and Fraser, N. W. (1994). *Gene Therapy,* **1**, 300.
27. Stanberry, L. R., Kern, E. R., Richards, J. T., Abbott, T. M., and Overall, J. C. J. (1982). *J. Infect. Diseases,* **146**, 397.
28. Openshaw, H., McNeill, J. I., Lin, X. H., Niland, J., and Cantin, E. M. (1995). *J. Med. Virol.,* **46**, 75.
29. Stevens, J. G., and Cook, M. L. (1971). *Science,* **173**, 843.
30. Ecob-Prince, M. S., Rixon, F. J., Preston, C. M., Hassan, K., and Kennedy, P. G. (1993). *J. Gen. Virol.,* **74**, 995.
31. Doerig, C., Pizer, L. I., and Wilcox, C. L. (1991). *Virology,* **183**, 423.
32. Wilcox, C. L., and Johnson, E. M. (1987). *J. Virol.,* **61**, 2311.
33. Wilcox, C. L., and Johnson Jr., E. M. (1988). *J. Virol.,* **62**, 393.
34. Smith, I. L., Hardwicke, M. A., and Sandri-Goldin, R. M. (1992). *Virology,* **186**, 74.
35. Halford, W. P., Gebhardt, B. M., and Carr, D. J. J. (1996). *J. Virol.,* **70**, 5051.
36. Horwitz, M. S. (1996). In *Field's virology.* (ed. B. N. Fields, D. M. Knipe, and P. M. Howley), p. 2149. Lippincott-Raven, Philadelphia.
37. Xia, D., Henry, L. J., Gerard, R. D., and Deisenhofer. (1994). *Structure,* **2**, 1259.
38. Stewart, P. L., and Burnett, R. M. (1995). *Curr. Topics Microbiol. Immunol.,* **199**, 25.
39. Stewart, P. L., Fuller, S. D., and Burnett, R. M. (1993). *EMBO J,* **12**, 2589.
40. Varga, M. J., Weibull, C., and Everitt, E. (1991). *J. Virol.,* **65**, 6061.
41. Shenk, T. (1996). In *Field's virology.* (ed. B. N. Fields, D. M. Knipe, and P. M. Howley), p. 2111. Lippincott-Raven, Philadelphia.

42. Tollefson, A. E., Scaria, A., Hermiston, T. W., Ryerse, J. S., Wold, L. J., and Wold, W. S. (1996). *J. Virol.,* **70**, 2296.
43. Bett, A. J., Prevec, L., and Graham, F. L. (1993). *J. Virol.,* **67**, 5911.
44. Leppard, K. (1997). *J. Gen. Virol.,* **78**, 2131.
45. Armentano, D., Zabner, J., Sacks, C., Sookdeo, C. C., Smith, M. P., St George, J. A., Wadsworth, S. C., Smith, A. E., and Gregory, R. J. (1997). *J. Virol.,* **71**, 2408.
46. Brough, D. E., Hsu, C., Kulesa, V. A., Lee, G. M., Cantolupo, L. J., Lizonova, A., and Kovesdi, I. (1997). *J. Virol.,* **71**, 9206.
47. Yang, Y., Ertl, H. C. J., and Wilson, J. M. (1994). *Immunity,* **1**, 433.
48. Zhou, H., O'Neal, W., Morral, N., and Beudet, A. L. (1996). *J. Virol.,* **70**, 7030.
49. Lusky, M., Christ, M., Rittner, K., Dieterle, A., Dreyer, D., Mourot, B., Schultz, H., Stoeckel, F., Pavirani, A., and Mehtali, M. (1998). *J. Virol.,* **72**, 2022.
50. Englehardt, J. F., Litzky, L., and Wilson, J. M. (1994). *Hum. Gene Ther.,* **5**, 1217.
51. Englehardt, J. F., Ye, X., Doranz, B., and Wilson, J. M. (1994). *Proc. Natl. Acad. Sci. USA,* **91**, 6196.
52. Yang, Y., Nunes, F. A., Berencsi, K., Ganczol, E., Englehardt, J. F., and Wilson, J. M. (1994). *Nature Genet.,* **7**, 362.
53. Morral, N., O'Neal, W., Zhou, H., Langston, C., and Beaudet, A. (1997). *Hum. Gene Ther.,* **8**, 1275.
54. Jacobs, S. C., Stephenson, J. R., and Wilkinson., G. W. G. (1992). *J. Virol.,* **66**, 2086.
55. Fooks, A. R., Schadeck, E., Liebert, U. G., Dowsett, A. B., Rima, B. K., Steward, M., Stephenson, J. R. and Wilkinson, G. W. G. (1995). *Virology,* **210**, 456.
56. Krougliak, V., and Graham, F. L. (1995). *Hum. Gene Ther.,* **6**, 1575.
57. Fisher, K. J., Choi, H., Burda, J., Chen, S-J., and Wilson, J. M. (1996). *Virology,* **217**, 11.
58. Kochanek, S., Clemens, P. R., Mitani, K., Chen, H. H., Chan, S., and Caskey, C. T. (1996). *Proc. Natl. Acad. Sci. USA,* **93**, 5731.
59. Kumar-Singh, R., and Chamberlain, J. S. (1996). *Hum. Mol. Genet.,* **5**, 913.
60. Hardy, S., Kitamura, M., Harris-Sansil, T., Dai, Y., and Phipps, M. L. (1997). *J. Virol.,* **71**, 1842.
61. Mitani, K., Graham, F. L., Caskey, C. T., and Kochanek, S. (1995). *Proc. Natl. Acad. Sci. USA,* **92**, 3854.
62. Parks, R. J., Chen, L., Anton, M., Sankar, U., Rudnicki, M. A., and Graham, F. L. (1996). *Proc. Natl. Acad. Sci. USA,* **93**, 13565.
63. Morsy, M. A., Gu, M. C., Zhao, J. Z., Holder, D. J., Rogers, I. T., Pouch, W. J., Motzel, S. L., Klein, H. J., Gupta, S. K., Liang, X., Tota, M. R., Rosenblum, C. I., and Caskey, C. T. (1998). *Gene Ther.,* **5**, 8.
64. Lieber, A., He, C-Y., Kirollova, I., and Kay, M. A. (1996). *J. Virol.,* **70**, 8944.
65. Graham, F. L., Smiley, J., Russell, W. C., and Nairn, R. (1977). *J. Gen. Virol.,* **36**, 59.
66. Louis, N., Evelegh, C., and Graham, F. L. (1997). *Virology,* **233**, 423.
67. Graham, F. L. (1987). *J. Gen. Virol.,* **68**, 937.
68. Fooks, A. R., Warnes, A., Racher, A. J., Stephenson, J. R., Dowsett, A. B., and Wilkinson, G. W. G. (1995) In *Proceeding of 13th ESACT Meeting.* (ed. R.E. Spier, J. B., Griffiths and W. Berthold), p. 21. Butterworth-Heinemann.
69. Fallaux, F. J., Kranenburg, O., Cramer, S. J., Houweling, A., Van Ormondt, H., Hoeben, R. C., and Van der Eb, A. J. (1996). *Hum. Gene Ther.,* **7**, 215.

70. Fallaux, F. J., Bout, A., van der Velde, I., van den Wollenberg, D. J., Hehir, K. M., Keegan, J., Auger, C., Cramer, S. J., van Ormondt, H., van der Eb, A. J., Valerio, D., and Hoeben, R. C. (1998). *Hum. Gene Ther.,* **9**, 1909.
71. Atkinson, E. M., Debelak, D. J., Hart, L. A., and Reynolds, T. C. (1998). *Nucl. Acids Res.,* **26**, 2821.
72. Mittereder, N., March, K. L., and Trapnell, B. C. (1996). *J. Virol.,* **70**, 7498.
73. Precious, B., and Russell, W.C. (1985). In *Virology: a practical approach* (ed. B. W. J. Mahy), p. 193. IRL Press., Oxford.
74. Graham, F. L., and Prevec, L. (1991). In *Methods in molecular biology* (ed. E. J. Murray), p. 109. Humana Press, Clifton, NJ.
75. Lowenstein, P. R., Shering, A. F., Bain, D., Castro, M. G., and Wilkinson, G. W. G. (1996). In *Gene transfer into neurones: towards gene therapy of neurological disorders.* (ed. P. R. Lowenstein, and L. W. Enquist), p. 93. J. Wiley, Chichester.
76. Berkner, K. L., and Sharp, P. A. (1983). *Nucl. Acids Res.,* **11**, 6003.
77. Okada, T., Ramsey, W. J., Munir, J., Wildner, O., and Blaese, R. M. (1998). *Nucl. Acids Res.,* **26**, 1947.
78. McGrory, W. J., Bautista, D. S., and Graham, F. L. (1988). *Virology,* **163**, 614.
79. Chartier, C., Degryse, E., Gantzer, M., Dieterle, A., Pavirani, A., and Mehtali, M. (1996). *J. Virol.,* **70**, 4805.
80. He, T. C., Zhou, S., da Costa, L. T., Yu, J., Kinzler, K. W., and Vogelstein, B. (1998). *Proc. Natl. Acad. Sci. USA,* **95**, 2509.
81. Fu, S. (1997). *Hum. Gene Ther.,* **8**, 1321.
82. Kojima, H., Ohishi, N., and Yagi, K. (1998). *Biochem. Biophys. Res. Commun.,* **246**, 868.
83. Ketner, G., Spencer, F., Tugendreich, S., Connelly, C., and Hieter, P. (1994). *Proc. Natl. Acad. Sci. USA,* **91**, 6186.
84. Curtis, S., Wilkinson, G.W., and Westmoreland, D. (1998). *J Med. Microbiol.,* **47**, 91.
85. Wickham, T. J., Mathias, P., Cheresh, D. A., and Nemerow, G. R. (1993). *Cell,* **73**, 309.
86. Croyle, M. A., Walter, E., Janich, S., Roessler, B. J., and Amidon, G. L. (1998). *Hum. Gene Ther.,* **9**, 561.
87. Takayama, K., Ueno, H., Pei, X. H., Nakanishi, Y., Yatsunami, J., and Hara, N. (1998). *Gene Ther.,* **5**, 361.
88. Von Seggern, D. J., Kehler, J., Endo, R. I., and Nemerow, G. R. (1998). *J. Gen. Virol.,* **79**, 1461.
89. Krasnykh, V. N., Mikheeva, G. V., Douglas, J. T., and Curiel, D. T. (1996). *J. Virol.,* **70**, 6839.
90. Krasnykh, V., Dmitriev, I., Mikheeva, G., Miller, C. R., Belousova, N., and Curiel, D. T. (1998). *J. Virol.,* **72**, 1844.
91. Watkins, S. J., Mesyanzhinov, V. V., Kurochkina, L. P., and Hawkins, R. E. (1997). *Gene Ther.,* **4**, 1004.
92. Rogers, B. E., Douglas, J. T., Ahlem, C., Buchsbaum, D. J., Frincke, J., and Curiel, D. T. (1997). *Gene Ther.,* **4**, 1387.
93. Douglas, J. T., Rogers, B. E., Rosenfeld, M. E., Michael, S. I., Feng, M., and Curiel, D. T. (1996). *Nature Biotechnol.,* **14**, 1574.
94. March, K. L., Madison, J. E., and Trapnell, B. C. (1995). *Hum. Gene Ther.,* **6**, 41.

95. Duc-Nguyen, H. (1968). *J. Virol.,* **2**, 643.
96. Arcasoy, S., Latoche, J., Gondor, M., Pitt, B., and Pilewski, J. (1997). *Gene Ther.,* **4**, 32.
97. Qiu, C., De Young, M. B., Finn, A., and Dichek, D. A. (1998). *Hum. Gene Ther.,* **9**, 507.

A1

List of suppliers

Amersham Pharmacia Biotech:
http://www.apbiotech.com/
UK:
Amersham Pharmacia Biotech Limited, Amersham Place, Little Chalfont, Bucks HP7 9NA, UK.
USA:
Amersham Pharmacia Biotech Inc, 800 Centennial Avenue, P.O. Box 1327, Piscataway, NJ 08855-1327, USA.
AMICON
Millipore (UK) Ltd, The Boulevard, Blackmoor Lane, Watford, Hertfordshire WD1 8YW, UK.
Millipore Corporation, 80 Ashby Road, Bedford, Massachusetts, 01730-2271, USA.
American Type Culture Collection, PO Box 1549, Manassas, Virginia 20108, USA. (http://www.atcc.org/)
Anderman and Co. Ltd., 145 London Road, Kingston-upon-Thames, Surrey KT17 7NH, UK.
BDH Laboratory Supplies, Poole, Dorset BH15 1TD, UK. (http://www.bdh.com/)
Beckman Instruments
Beckman Instruments UK Ltd., Progress Road, Sands Industrial Estate, High Wycombe, Buckinghamshire HP12 4JL, UK.
Beckman Instruments Inc., PO Box 3100, 2500 Harbor Boulevard, Fullerton, CA 92634, USA. (http://www.beckman.com/)
Becton Dickinson
Becton Dickinson and Co., Between Towns Road, Cowley, Oxford OX4 3LY, UK.
Becton Dickinson and Co., 2 Bridgewater Lane, Lincoln Park, NJ 07035, USA. (http://www.bd.com/)
Bio
Bio 101 Inc., c/o Statech Scientific Ltd, 61–63 Dudley Street, Luton, Bedfordshire LU2 0HP, UK.
Bio 101 Inc., PO Box 2284, La Jolla, CA 92038–2284, USA.

List of suppliers

Bio-Rad Laboratories
Bio-Rad Laboratories Ltd., Bio-Rad House, Maylands Avenue, Hemel Hempstead HP2 7TD, UK.
Bio-Rad Laboratories, Division Headquarters, 3300 Regatta Boulevard, Richmond, CA 94804, USA. (http://www.biorad.com/)
Boehringer Mannheim
Boehringer Mannheim UK (Diagnostics and Biochemicals) Ltd., Bell Lane, Lewes, East Sussex BN17 1LG, UK.
Boehringer Mannheim Corporation, Biochemical Products, 9115 Hague Road, PO Box 504 Indianopolis, IN 46250–0414, USA.
Boehringer Mannheim Biochemica, GmbH, Sandhofer Str. 116, Postfach 310120 D-6800 Ma 31, Germany.
Citifluor Ltd, 18 Enfield Cloisters, Fanshaw St, London N1 6LD, UK.
CLONTECH Laboratories Inc., 1020 East Meadow Circle, Palo Alto, CA 94303, USA. (http://www.clontech.com/)
DAKO
DAKO Ltd, Denmark House, Angel Drove, Ely, Cambridgeshire CB7 4ET, UK.
DAKO Corporation, 6392 Via Real, Carpinteria, CA 93013 USA. (http://www.dako.com/)
Diagnostic Products Corporation
Diagnostic Products Corporation, 5700 West 96th Street, Los Angeles, CA 90045-5597, USA.
EURO/DPC Limited, Glyn Rhonwy, Llanberis, Caernarfon, Gwynedd LL55 4EL, UK. (http://www.dpcweb.com)
Difco Laboratories
Difco Laboratories Ltd., PO Box 14B, Central Avenue, West Molesey, Surrey KT8 2SE, UK.
Difco Laboratories, PO Box 331058, Detroit, MI 48232–7058, USA.
Du Pont
Du Pont (UK) Ltd., Industrial Products Division, Wedgwood Way, Stevenage, Herts, SG1 4Q, UK.
Du Pont Co. (Biotechnology Systems Division), PO Box 80024, Wilmington, DE 19880–002, USA.
Dynal UK, 10 Thursby Road, Croft Buisness Park, Bromborough, Wirral, Merseyside L62 3PW, UK. (http://www.dynal.no/)
EG&G Berthold, 100 Midland Road, Oak Ridge, TN 37830, USA. (http://www.berthold-us.com/)
European Collection of Animal Cell Culture, Division of Biologics, PHLS Centre for Applied Microbiology and Research, Porton Down, Salisbury, Wiltshire SP4 0JG, UK.
Falcon (Falcon is a registered trademark of Becton Dickinson and Co.)
Fisher Scientific Co., 711 Forbest Avenue, Pittsburgh, PA 15219–4785, USA.

List of suppliers

Flow Laboratories, Woodcock Hill, Harefield Road, Rickmansworth, Hertfordshire WD3 1PQ, UK.
Fluka
Fluka-Chemie AG, CH-9470, Buchs, Switzerland.
Fluka Chemicals Ltd., The Old Brickyard, New Road, Gillingham, Dorset SP8 4JL, UK.
Gibco BRL
Gibco BRL (Life Technologies Ltd.), Trident House, Renfrew Road, Paisley PA3 4EF, UK.
Gibco BRL (Life Technologies Inc.), 3175 Staler Road, Grand Island, NY 14072–0068, USA.
Arnold R. Horwell, 73 Maygrove Road, West Hampstead, London NW6 2BP, UK.
Hybaid
Hybaid Ltd., 111–113 Waldegrave Road, Teddington, Middlesex TW11 8LL, UK.
Hybaid, National Labnet Corporation, PO Box 841, Woodbridge, NJ 07095, USA.
HyClone Laboratories, 1725 South HyClone Road, Logan, UT 84321, USA.
Heat Systems Ultrasonics, Inc, 1938 New Farmingdale, NY 11735, USA.
ICN
ICN Biomedicals, 1 Elmwood, Chineham Business Park, Crockford Lane, Basingstoke, Hampshire RG24 8WG, UK.
ICN Pharmaceuticals Inc, 3300 Hyland Avenue, Costa Mesa, CA 92626, USA. (http://www.icnbiomed.com/)
Imperial Laboratories (Europe) Ltd, West Portway, Andover, Hampshire SP10 3LF, UK.
International Biotechnologies Inc., 25 Science Park, New Haven, Connecticut 06535, USA.
Invitrogen Corporation
Invitrogen Corporation, 3985 B Sorrenton Valley Building, San Diego, CA 92121, USA. (http://www.invitrogen.com/)
Invitrogen Corporation, c/o British Biotechnology Products Ltd., 4–10 The Quadrant, Barton Lane, Abingdon, OX14 3YS, UK.
Kodak
Eastman Fine Chemicals, 343 State Street, Rochester, NY, USA.
Life Technologies Inc., 8451 Helgerman Court, Gaithersburg, MN 20877, USA. (http://www.lifetech.com/)
Merck
Merck Industries Inc., 5 Skyline Drive, Nawthorne, NY 10532, USA. (http://www.merck.com/)
Millipore
Millipore (UK) Ltd., The Boulevard, Blackmoor Lane, Watford, Herts WD1 8YW, UK.
Millipore Corp./Biosearch, PO Box 255, 80 Ashby Road, Bedford, MA 01730, USA.

List of suppliers

MSD Sharp & Dohme GmbH:
Lindenplatz 1, D-85540 Haar, Germany.
MSE, see Sanyo Gallenkamp
Nalge Nunc
Nalge Nunc (Europe) Limited, Foxwood Court, Rotherwas Industrial Estate, Hereford HR2 6JQ, UK.
Nalge Nunc International, 75 Panorama Creek Drive, Rochester, NY 14625, USA. (http://nunc.nalgenunc.com/)
New England Biolabs (NBL)
New England Biolabs (NBL), 32 Tozer Road, Beverley, MA 01915–5510, USA. (http://www.neb.com/)
New England Biolabs (NBL), c/o CP Labs Ltd., PO Box 22, Bishops Stortford, Herts CM23 3DH, UK.
Nikon Corporation, Fuji Building, 2–3 Marunouchi 3-chome, Chiyoda-ku, Tokyo, Japan.
Perkin-Elmer
Perkin-Elmer Ltd., Maxwell Road, Beaconsfield, Bucks. HP9 1QA, UK.
Perkin-Elmer Ltd., Post Office Lane, Beaconsfield, Bucks. HP9 1QA, UK.
Perkin-Elmer-Cetus (The Perkin-Elmer Corporation), 761 Main Avenue, Norwalk, CT 0689, USA.
Pharmacia Biosystems
Pharmacia Biosystems Ltd., (Biotechnology Division), Davy Avenue, Knowlhill, Milton Keynes MK5 8PH, UK.
Pharmacia Biotech Europe, Procordia EuroCentre, Rue de la Fuse-e 62, B-1130 Brussels, Belgium.
Pharmacia LKB Biotechnology AB, Björngatan 30, S-75182 Uppsala, Sweden.
Photon Technology International, Inc, 1 Deerpark Drive, Suite F, Monmouth Junction, NJ 08852, USA. (http://www.pti-nj.com/)
Pierce Chemical Co, 3747 N. Meridian Rd, PO Box 117, Rockford, IL 61105, USA.
Promega
Promega Ltd., Delta House, Enterprise Road, Chilworth Research Centre, Southampton, UK.
Promega Corporation, 2800 Woods Hollow Road, Madison, WI 53711–5399, USA.
Qiagen
Qiagen Inc., c/o Hybaid, 111–113 Waldegrave Road, Teddington, Middlesex, TW11 8LL, UK.
Qiagen Inc., 9259 Eton Avenue, Chatsworth, CA 91311, USA.
R&D Systems
R&D Systems Europe Ltd, 4–10 The Quadrant Barton Lane Abingdon, Oxon OX14 3YS, UK.

List of suppliers

R&D Systems Inc., 614 McKinley Place N.E., Minneapolis, MN 55413, USA. (http://cytokine.rndsystems.com/)

Roche

F. Hoffmann-La Roche Ltd, CH-4070 Basel, Switzerland. (http://www.roche.com/)

Sanyo Gallenkamp plc, 1 Monarch Way, Belton Park Industrial Estate, Loughborough, Leicestershire LE11 5XG, UK.

Sarstedt

Sarstedt Ltd, 68 Boston Road, Beaumont Leys, Leicester LE4 1AW, UK.

Sarstedt Inc, PO Box 468, Newton, NC 28658-0468, USA. (http://www.sarstedt.com/)

Schleicher and Schuell

Schleicher and Schuell Inc., 10 Optical Ave, PO Box 2012, Keene, NH 03431, USA.

Schleicher and Schuell Inc., D-3354 Dassel, Germany.

Schleicher and Schuell Inc., c/o Anderman and Co. Ltd.

Scotlab Bioscience Ltd, (http://users.colloquium.co.uk/~SCOTLAB/)

Shandon Scientific Ltd., Chadwick Road, Astmoor, Runcorn, Cheshire WA7 1PR, UK.

Sigma Chemical Company

Sigma Chemical Company (UK), Fancy Road, Poole, Dorset BH17 7NH, UK.

Sigma Chemical Company, 3050 Spruce Street, PO Box 14508, St. Louis, MO 63178–9916. (http://www.sigma-aldrich.com/)

Sorvall

Kendro Laboratory Products Limited, International Centre, Boulton Road, Stevenage, Hertfordshire SG1 4QX, UK.

Kendro Laboratory Products, 31 Pecks Lane, Newtown, CT 06470-2337, USA. (http://www.sorvall.com/)

Stratagene

Stratagene Ltd., Unit 140, Cambridge Innovation Centre, Milton Road, Cambridge CB4 4FG, UK.

Strategene Inc., 11011 North Torrey Pines Road, La Jolla, CA 92037, USA.

Tropix

Applied Biosystems, Kelvin Close, Birchwood Science Park, Warrington, Cheshire WA3 7PB, UK. *Tropix,* 47 Wiggins Avenue, Bedford, Massachusetts 01730, USA.

United States Biochemical, PO Box 22400, Cleveland, OH 44122, USA.

Vector

Vector Laboratories Ltd, 3 Accent Park, Bakewell Road, Orton Southgate, Peterborough PE2 6XS, UK.

Vector Laboratories, 30 Ingold Road, Burlingame, CA 94010, USA. (http://www.vectorlabs.com/)

List of suppliers

Wellcome Reagents, Langley Court, Beckenham, Kent BR3 3BS, UK.
Whatman
Whatman International Ltd, Whatman House, St Leonard's Road, Maidstone, Kent ME16 0LS, UK.
Whatman Inc, 9 Bridewell Place, PO Box 1197, Clifton, NJ 07014, USA. (http://www.whatman.co.uk/)

Index

aciclovir 247
adenoviruses 72, 209, 287
   recombinant 288, 296
animals 1
antibody conjugation 171
antiserum production 222
antiviral testing 250, 256, 261

baculovirus expression 97, 189
β-galactosidase assay 133, 140, 236, 279, 281, 283

cell culture 2, 51, 216, 285
cell extracts 84, 150, 165, 172, 231, 236
chloramphenicol acetyl transferase (CAT) assay 131, 237
chromatography
   affinity 106
   ion exchange 193
   size exclusion 195
cidofovir 250
complementation 50, 68
cytomegalovirus (CMV) 254

DNA
   extraction 6, 8, 274, 299
   replication 83
   unwinding assay 89
DNA polymerase assay (Adenovirus) 88
DNAse I footprinting 101

eggs 1
ELISA assay 222
elongation assay 86

far-western blot 107
focus formation assay 218
foscarnet 249

ganciclovir 254
gel retardation assay 91, 148, 164, 167, 179
genome mapping 16
glutathione S-transferase: see GST
glycerol gradients 109
growth media 2
GST fusions 105, 239

hepatitis B virus (HBV) 260
herpesviruses 74, 247, 250, 267
   plaque assay 251
   recombinant 135, 273

immunofluorescence 232, 286
immunoprecipitation 182, 224
initiation assay 86
in situ hybridization (ISH) 124
in vitro translation 166

lamivudine 260
luciferase assay 159

marker rescue 10, 21
methylation protection 102
mobility shift assay: see gel retardation
mutagenesis
   chemical 56
   nucleoside analogues 59
   oligonucleotide insertion 143
   PCR 146
   site-directed 60, 144
mutation 44, 47, 141
   characterization 77

northern blotting 118, 161
nucleotide sequencing 22, 29
   alignment 42
   analysis 39
   assembly 31
   data collection 30
   homology searches 41
   libraries 23

parvoviruses 71
penciclovir 249
plaque reduction assay 251
plasmid miniprep 27
polyethylene glycol (PEG) precipitation 28
polymerase chain reaction (PCR) 62, 146, 259, 262
polyomaviruses 70, 211, 215
poxviruses 74, 177, 187
primer extension 163
primer walking 23

Index

protein
    expression 97, 187
    purification 92, 169, 178
proteolysis 108
pulsed field gel electrophoresis (PFGE) 19

reporter genes 128, 275
restriction enzyme digestion 9
reverse genetics 49
RNA isolation
    poly A+ 117
    total 115, 160
RT-PCR 122

scintillation proximity assay 199
sequencing: see nucleotide sequencing
slot blot 133
solid phase binding assay 197
Southern blot 17
south-western blot 153
surface plasmon resonance 184, 201
SV40 virus: see polyomaviruses

tissue culture: see cell culture
transcript mapping 11
transfection
    calcium phosphate 65, 129, 276
    DEAE dextran 66
    lipofection 67, 158

UV crosslinking 103

vaccinia virus: see poxviruses
virus
    purification 6, 294
    stocks 4, 52, 57, 250, 293
    titration 5, 53

western blot 234

yeast two hybrid method 240